Fundamentals of Mathematics

Fundamentals of Mathematics

Edited by
Bennett Perez

Larsen & Keller
www.larsen-keller.com

Fundamentals of Mathematics
Edited by Bennett Perez
ISBN: 978-1-63549-178-4 (Hardback)

☰ Larsen & Keller

Published by Larsen and Keller Education,
5 Penn Plaza,
19th Floor,
New York, NY 10001, USA

Cataloging-in-Publication Data

Fundamentals of mathematics / edited by Bennett Perez.
    p. cm.
Includes bibliographical references and index.
ISBN 978-1-63549-178-4
1. Mathematics. I. Perez, Bennett.
QA36 .F86 2017
510--dc23

Table of Contents

Permissions

Index

Preface

This book elucidates the concepts and innovative models around prospective developments with respect to mathematics. It talks in detail about the fundamental concepts and theories of this field. Mathematics refers to the study of space, quantity, structure and change. It encompasses methods like measurement, abstraction, counting, calculation, logic, etc. This text picks up individual branches of mathematics and explains their need and contribution in the context of the growth of the subject. The topics included in it on mathematics are of utmost significance and are bound to provide incredible insights to readers. Coherent flow of topics, student-friendly language and extensive use of concepts make this book an invaluable source of knowledge. Those in search of information to further their knowledge will be greatly assisted by this textbook.

A foreword of all Chapters of the book is provided below:

Chapter 1 - Mathematics is a science that has existed since Antiquity. It has developed on its own in many regions and can be seen as a science of abstraction. Mathematical inquiry relies on quantity and change; hence mathematical activity was highly praised by philosophers. The chapter on mathematics offers an insightful focus, keeping in mind the complex subject matter; **Chapter 2 -** The mathematical objects listed in this chapter are key to mathematical practice. They are conceptions of the abstract possibility that mathematicians seek to unlock and understand. Some of the concepts that are mentioned in this chapter are number, calculus and mathematical space. The chapter strategically encompasses and incorporates the major components and key concepts of mathematics, providing a complete understanding; **Chapter 3 -** The various branches of mathematics offer interpretations to questions that were of great concern to scholars. Algebra, geometry, trigonometry, all of which have been listed in the chapter, describes philosophical paradigms and their applications. This chapter is a compilation of the various branches of mathematics that form an integral part of the broader subject matter; **Chapter 4 -** Mathematical proof is patterned according to the stages that a mathematical equation or postulate follows while being examined. There are many components to mathematical proof such as calculation, mathematical proposition and measurement. The aspects elucidated in this chapter are of vital importance, and provide a better understanding of mathematics; **Chapter 5 -** Mathematics is applicable to all instances of life. Many a time, instances demand mathematical understanding and analysis. Theories listed in this chapter were developed with such analysis such as probability, graph theory, number theory, statistical models etc. The major categories of mathematics are dealt with great details in the chapter; **Chapter 6 -** These topics depend on mathematical logic and have become disciplines in their own right. Widely applicable, these branches of mathematics contribute greatly to mathematical knowledge. They also help to better understand and learn mathematics. This chapter provides a plethora of allied topics for better comprehension of mathematics; **Chapter 7 -** Mathematics is widely applicable. The disciplines listed here borrow from mathematical understanding in as much as theorems developed from pure mathematics are applied in these topics. Some of the topics listed in this chapter are mathematical physics and mathematical economics. The subjects discussed in the chapter are of great importance to broaden the existing knowledge on this field.

I would like to thank the entire editorial team who made sincere efforts for this book and my family who supported me in my efforts of working on this book. I take this opportunity to thank all those who have been a guiding force throughout my life.

Editor

Introduction to Mathematics

Mathematics is a science that has existed since Antiquity. It has developed on its own in many regions and can be seen as a science of abstraction. Mathematical inquiry relies on quantity and change; hence mathematical activity was highly praised by philosophers. The chapter on mathematics offers an insightful focus, keeping in mind the complex subject matter.

Mathematics is the study of topics such as quantity (numbers),structure,space, and change. There is a range of views among mathematicians and philosophers as to the exact scope and definition of mathematics.

Euclid (holding calipers), Greek mathematician, 3rd century BC, as imagined by Raphael
in this detail from *The School of Athens*.

Mathematicians seek out patterns and use them to formulate new conjectures. Mathematicians resolve the truth or falsity of conjectures by mathematical proof. When mathematical structures are good models of real phenomena, then mathematical reasoning can provide insight or predictions about nature. Through the use of abstraction and logic, mathematics developed from counting, calculation, measurement, and the systematic study of the shapes and motions of physical objects. Practical mathematics has been a human activity for as far back as written records exist. The research required to solve mathematical problems can take years or even centuries of sustained inquiry.

Rigorous arguments first appeared in Greek mathematics, most notably in Euclid's *Elements*. Since the pioneering work of Giuseppe Peano (1858–1932), David Hilbert (1862–1943), and others on axiomatic systems in the late 19th century, it has become customary to view mathematical research as establishing truth by rigorousdeduction from appropriately chosen axioms and definitions. Mathematics developed at a relatively slow pace until the Renaissance, when mathematical innovations interacting with new scientific discoveries led to a rapid increase in the rate of mathematical discovery that has continued to the present day.

Galileo Galilei (1564–1642) said, "The universe cannot be read until we have learned the language and become familiar with the characters in which it is written. It is written in mathematical language, and the letters are triangles, circles and other geometrical figures, without which means it is humanly impossible to comprehend a single word. Without these, one is wandering about in a dark labyrinth."Carl Friedrich Gauss (1777–1855) referred to mathematics as "the Queen of the Sciences".Benjamin Peirce (1809–1880) called mathematics "the science that draws necessary conclusions". David Hilbert said of mathematics: "We are not speaking here of arbitrariness in any sense. Mathematics is not like a game whose tasks are determined by arbitrarily stipulated rules. Rather, it is a conceptual system possessing internal necessity that can only be so and by no means otherwise."Albert Einstein (1879–1955) stated that "as far as the laws of mathematics refer to reality, they are not certain; and as far as they are certain, they do not refer to reality."

Mathematics is essential in many fields, including natural science, engineering, medicine, finance and the social sciences. Applied mathematics has led to entirely new mathematical disciplines, such as statistics and game theory. Mathematicians also engage in pure mathematics, or mathematics for its own sake, without having any application in mind. There is no clear line separating pure and applied mathematics, and practical applications for what began as pure mathematics are often discovered.

History

The history of mathematics can be seen as an ever-increasing series of abstractions. The first abstraction, which is shared by many animals, was probably that of numbers: the realization that a collection of two apples and a collection of two oranges (for example) have something in common, namely quantity of their members.

Greek mathematician Pythagoras (c. 570 – c. 495 BC), commonly credited with discovering the Pythagorean theorem

As evidenced by tallies found on bone, in addition to recognizing how to count physical objects, prehistoric peoples may have also recognized how to count abstract quantities, like time – days, seasons, years.

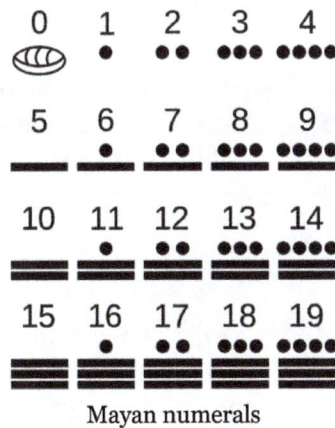

0	1	2	3	4
5	6	7	8	9
10	11	12	13	14
15	16	17	18	19

Mayan numerals

Evidence for more complex mathematics does not appear until around 3000 BC, when the Babylonians and Egyptians began using arithmetic, algebra and geometry for taxation and other financial calculations, for building and construction, and for astronomy. The earliest uses of mathematics were in trading, land measurement, painting and weaving patterns and the recording of time.

In Babylonian mathematicselementary arithmetic (addition, subtraction, multiplication and division) first appears in the archaeological record. Numeracy pre-dated writing and numeral systems have been many and diverse, with the first known written numerals created by Egyptians in Middle Kingdom texts such as the Rhind Mathematical Papyrus.

Between 600 and 300 BC the Ancient Greeks began a systematic study of mathematics in its own right with Greek mathematics.

Persian mathematician Al-Khwarizmi (c. 780 - c. 850), the inventor of the Algebra.

During the Golden Age of Islam, especially during the 9th and 10th centuries, mathematics saw many important innovations building on Greek mathematics: most of them include the contributions from Persian mathematicians such as Al-Khwarismi, Omar Khayyam and Sharaf al-Dīn al-Ṭūsī.

Mathematics has since been greatly extended, and there has been a fruitful interaction between mathematics and science, to the benefit of both. Mathematical discoveries continue to be made today. According to Mikhail B. Sevryuk, in the January 2006 issue of the *Bulletin of the American Mathematical Society*, "The number of papers and books included in the *Mathematical Reviews* database since 1940 (the first year of operation of MR) is now more than 1.9 million, and more than 75 thousand items are added to the database each year. The overwhelming majority of works in this ocean contain new mathematical theorems and their proofs."

Etymology

In Latin, and in English until around 1700, the term *mathematics* more commonly meant "astrology" (or sometimes "astronomy") rather than "mathematics"; the meaning gradually changed to its present one from about 1500 to 1800. This has resulted in several mistranslations: a particularly notorious one is Saint Augustine's warning that Christians should beware of *mathematici* meaning astrologers, which is sometimes mistranslated as a condemnation of mathematicians.

The apparent plural form in English, like the French plural form *les mathématiques,* used by Aristotle (384–322 BC), and meaning roughly "all things mathematical"; although it is plausible that English borrowed only the adjective *mathematic(al)* and formed the noun *mathematics* anew, after the pattern of physics and metaphysics, which were inherited from the Greek. In English, the noun *mathematics* takes singular verb forms. It is often shortened to *maths* or, in English-speaking North America, *math.*

Definitions of Mathematics

Leonardo Fibonacci, the Italian mathematician who established the Hindu–Arabic numeral system to the Western World

Aristotle defined mathematics as "the science of quantity", and this definition prevailed until the 18th century. Starting in the 19th century, when the study of mathematics increased in rigor and began to address abstract topics such as group theory and projective geometry, which have no clear-cut relation to quantity and measurement, mathematicians and philosophers began to propose a variety of new definitions. Some of these definitions emphasize the deductive character of much of mathematics, some emphasize its abstractness, some emphasize certain topics within mathematics. Today, no consensus on the definition of mathematics prevails, even among professionals. There is not even consensus on whether mathematics is an art or a science. A great many professional mathematicians take no interest in a definition of mathematics, or consider it undefinable. Some just say, "Mathematics is what mathematicians do."

Three leading types of definition of mathematics are called logicist, intuitionist, and formalist, each reflecting a different philosophical school of thought. All have severe problems, none has widespread acceptance, and no reconciliation seems possible.

An early definition of mathematics in terms of logic was Benjamin Peirce's "the science that draws necessary conclusions" (1870). In the *Principia Mathematica*, Bertrand Russell and Alfred North Whitehead advanced the philosophical program known as logicism, and attempted to prove that all mathematical concepts, statements, and principles can be defined and proved entirely in terms of symbolic logic. A logicist definition of mathematics is Russell's "All Mathematics is Symbolic Logic" (1903).

Intuitionist definitions, developing from the philosophy of mathematician L.E.J. Brouwer, identify mathematics with certain mental phenomena. An example of an intuitionist definition is "Mathematics is the mental activity which consists in carrying out constructs one after the other." A peculiarity of intuitionism is that it rejects some mathematical ideas considered valid according to other definitions. In particular, while other philosophies of mathematics allow objects that can be proved to exist even though they cannot be constructed, intuitionism allows only mathematical objects that one can actually construct.

Formalist definitions identify mathematics with its symbols and the rules for operating on them. Haskell Curry defined mathematics simply as "the science of formal systems". A formal system is a set of symbols, or *tokens*, and some *rules* telling how the tokens may be combined into *formulas*. In formal systems, the word *axiom* has a special meaning, different from the ordinary meaning of "a self-evident truth". In formal systems, an axiom is a combination of tokens that is included in a given formal system without needing to be derived using the rules of the system.

Mathematics as Science

Gauss referred to mathematics as "the Queen of the Sciences". In the original Latin *Regina Scientiarum*, as well as in German*Königin der Wissenschaften*, the word corresponding to *science* means a "field of knowledge", and this was the original meaning of "science" in English, also; mathematics is in this sense a field of knowledge. The specialization restricting the meaning of "science" to *natural science* follows the rise of Baconian science, which contrasted "natural science" to scholasticism, the Aristotelean method of inquiring from first principles. The role of empirical experimentation and observation is negligible in mathematics, compared to natural sciences such as biology, chemistry, or physics. Albert Einstein stated that "as far as the laws of mathematics refer to reality, they are not certain; and as far as they are certain, they do not refer to reality." More recently, Marcus du Sautoy has called mathematics "the Queen of Science ... the main driving force behind scientific discovery".

Carl Friedrich Gauss, known as the prince of mathematicians

Many philosophers believe that mathematics is not experimentally falsifiable, and thus not a science according to the definition of Karl Popper. However, in the 1930s Gödel's incompleteness theorems convinced many mathematicians that mathematics cannot be reduced to logic alone, and Karl Popper concluded that "most mathematical theories are, like those of physics and biology, hypothetico-deductive: pure mathematics therefore turns out to be much closer to the natural sciences whose hypotheses are conjectures, than it seemed even recently." Other thinkers, notably Imre Lakatos, have applied a version of falsificationism to mathematics itself.

An alternative view is that certain scientific fields (such as theoretical physics) are mathematics with axioms that are intended to correspond to reality. The theoretical physicist J.M. Ziman proposed that science is *public knowledge*, and thus includes mathematics. Mathematics shares much in common with many fields in the physical sciences, notably the exploration of the logical consequences of assumptions. Intuition and experimentation also play a role in the formulation of conjectures in both mathematics and the (other) sciences. Experimental mathematics continues to grow in importance within mathematics, and computation and simulation are playing an increasing role in both the sciences and mathematics.

The opinions of mathematicians on this matter are varied. Many mathematicians feel that to call their area a science is to downplay the importance of its aesthetic side, and its history in the traditional seven liberal arts; others feel that to ignore its connection to the sciences is to turn a blind eye to the fact that the interface between mathematics and its applications in science and engineering has driven much development in mathematics. One way this difference of viewpoint plays out is in the philosophical debate as to whether mathematics is *created* (as in art) or *discovered* (as in science). It is common to see universities divided into sections that include a division of *Science and Mathematics*, indicating that the fields are seen as being allied but that they do not coincide. In practice, mathematicians are typically grouped with scientists at the gross level but separated at finer levels. This is one of many issues considered in the philosophy of mathematics.

Inspiration, Pure and Applied Mathematics, and Aesthetics

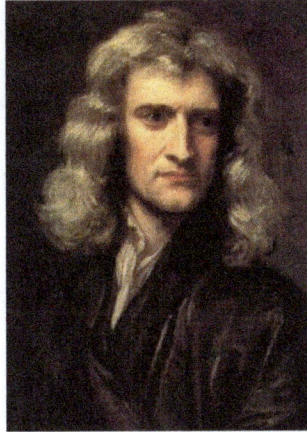

Isaac Newton (left) and Gottfried Wilhelm Leibniz (right), developers of infinitesimal calculus

Mathematics arises from many different kinds of problems. At first these were found in commerce, land measurement, architecture and later astronomy; today, all sciences suggest problems studied by mathematicians, and many problems arise within mathematics itself. For example, the physicistRichard Feynman invented the path integral formulation of quantum mechanics using a combination of mathematical reasoning and physical insight, and today's string theory, a still-developing scientific theory which attempts to unify the four fundamental forces of nature, continues to inspire new mathematics.

Some mathematics is relevant only in the area that inspired it, and is applied to solve further problems in that area. But often mathematics inspired by one area proves useful in many areas, and joins the general stock of mathematical concepts. A distinction is often made between pure mathematics and applied mathematics. However pure mathematics topics often turn out to have applications, e.g. number theory in cryptography. This remarkable fact, that even the "purest" mathematics often turns out to have practical applications, is what Eugene Wigner has called "the unreasonable effectiveness of mathematics". As in most areas of study, the explosion of knowledge in the scientific age has led to specialization: there are now hundreds of specialized areas in mathematics and the latest Mathematics Subject Classification runs to 46 pages. Several areas of applied mathematics have merged with related traditions outside of mathematics and become disciplines in their own right, including statistics, operations research, and computer science.

For those who are mathematically inclined, there is often a definite aesthetic aspect to much of mathematics. Many mathematicians talk about the *elegance* of mathematics, its intrinsic aesthetics and inner beauty. Simplicity and generality are valued. There is beauty in a simple and elegant proof, such as Euclid's proof that there are infinitely many prime numbers, and in an elegant numerical method that speeds calculation, such as the fast Fourier transform. G.H. Hardy in *A Mathematician's Apology* expressed the belief that these aesthetic considerations are, in themselves, sufficient to justify the study of pure mathematics. He identified criteria such as significance, unexpectedness, inevitability, and economy as factors that contribute to a mathematical aesthetic. Mathematicians often strive to find proofs that are particularly elegant, proofs from "The Book" of God according to Paul Erdős. The popularity of recreational mathematics is another sign of the pleasure many find in solving mathematical questions.

Notation, Language, and Rigor

Leonhard Euler, who created and popularized much of the mathematical notation used today

Most of the mathematical notation in use today was not invented until the 16th century. Before that, mathematics was written out in words, limiting mathematical discovery.Euler (1707–1783) was responsible for many of the notations in use today. Modern notation makes mathematics much easier for the professional, but beginners often find it daunting. It is compressed: a few symbols contain a great deal of information. Like musical notation, modern mathematical notation has a strict syntax and encodes information that would be difficult to write in any other way.

Mathematical language can be difficult to understand for beginners. Common words such as *or* and *only* have more precise meanings than in everyday speech. Moreover, words such as *open* and *field* have specialized mathematical meanings. Technical terms such as *homeomorphism* and *integrable* have precise meanings in mathematics. Additionally, shorthand phrases such as *iff* for "if and only if" belong to mathematical jargon. There is a reason for special notation and technical vocabulary: mathematics requires more precision than everyday speech. Mathematicians refer to this precision of language and logic as "rigor".

Mathematical proof is fundamentally a matter of rigor. Mathematicians want their theorems to follow from axioms by means of systematic reasoning. This is to avoid mistaken "theorems", based on fallible intuitions, of which many instances have occurred in the history of the subject. The level of rigor expected in mathematics has varied over time: the Greeks expected detailed arguments, but at the time of Isaac Newton the methods employed were less rigorous. Problems inherent in the definitions used by Newton would lead to a resurgence of careful analysis and formal proof in the 19th century. Misunderstanding the rigor is a cause for some of the common misconceptions of mathematics. Today, mathematicians continue to argue among themselves about computer-assisted proofs. Since large computations are hard to verify, such proofs may not be sufficiently rigorous.

Axioms in traditional thought were "self-evident truths", but that conception is problematic. At a formal level, an axiom is just a string of symbols, which has an intrinsic meaning only in the context of all derivable formulas of an axiomatic system. It was the goal of Hilbert's program to put all of mathematics on a firm axiomatic basis, but according to Gödel's incompleteness theorem every (sufficiently powerful) axiomatic system has undecidable formulas; and so a final axiomatization of mathematics is impossible. Nonetheless mathematics is often imagined to be (as far as its formal content) nothing but set theory in some axiomatization, in the sense that every mathematical statement or proof could be cast into formulas within set theory.

Fields of Mathematics

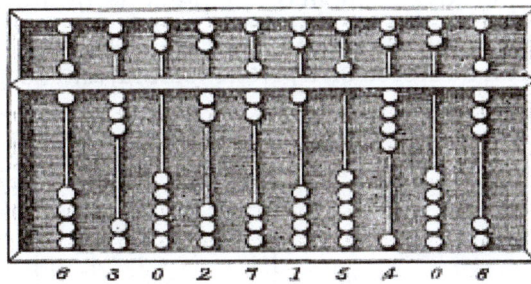

An abacus, a simple calculating tool used since ancient times

Mathematics can, broadly speaking, be subdivided into the study of quantity, structure, space, and change (i.e. arithmetic, algebra, geometry, and analysis). In addition to these main concerns, there are also subdivisions dedicated to exploring links from the heart of mathematics to other fields: to logic, to set theory (foundations), to the empirical mathematics of the various sciences (applied mathematics), and more recently to the rigorous study of uncertainty. While some areas might seem unrelated, the Langlands program has found connections between areas previously thought unconnected, such as Galois groups, Riemann surfaces and number theory.

Foundations and Philosophy

In order to clarify the foundations of mathematics, the fields of mathematical logic and set theory were developed. Mathematical logic includes the mathematical study of logic and the applications of formal logic to other areas of mathematics; set theory is the branch of mathematics that studies sets or collections of objects. Category theory, which deals in an abstract way with mathematical structures and relationships between them, is still in development. The phrase "crisis of foundations" describes the search for a rigorous foundation for mathematics that took place from approximately 1900 to 1930. Some disagreement about the foundations of mathematics continues to the present day. The crisis of foundations was stimulated by a number of controversies at the time, including the controversy over Cantor's set theory and the Brouwer–Hilbert controversy.

Mathematical logic is concerned with setting mathematics within a rigorous axiomatic framework, and studying the implications of such a framework. As such, it is home to Gödel's incompleteness theorems which (informally) imply that any effective formal system that contains basic arithmetic, if *sound* (meaning that all theorems that can be proved are true), is necessarily *incomplete* (meaning that there are true theorems which cannot be proved *in that system*). Whatever finite collection of number-theoretical axioms is taken as a foundation, Gödel showed how to construct a formal statement that is a true number-theoretical fact, but which does not follow from those axioms. Therefore, no formal system is a complete axiomatization of full number theory. Modern logic is divided into recursion theory, model theory, and proof theory, and is closely linked to theoretical computer science, as well as to category theory. In the context of recursion theory, the impossibility of a full axiomatization of number theory can also be formally demonstrated as a consequence of the MRDP theorem.

Theoretical computer science includes computability theory, computational complexity theory, and information theory. Computability theory examines the limitations of various theoretical models of the computer, including the most well-known model – the Turing machine. Complexity theory is the study of tractability by computer; some problems, although theoretically solvable by computer, are so expensive in terms of time or space that solving them is likely to remain practically unfeasible, even with the rapid advancement of computer hardware. A famous problem is the "P = NP?" problem, one of the Millennium Prize Problems. Finally, information theory is concerned with the amount of data that can be stored on a given medium, and hence deals with concepts such as compression and entropy.

$p \Rightarrow q$		$\begin{array}{ccc} X & \xrightarrow{f} & Y \\ & \searrow{g \circ f} & \downarrow{g} \\ & & Z \end{array}$	
Mathematical logic	*Set theory*	*Category theory*	*Theory of computation*

Pure Mathematics

Quantity

The study of quantity starts with numbers, first the familiar natural numbers and integers ("whole numbers") and arithmetical operations on them, which are characterized in arithmetic. The deeper properties of integers are studied in number theory, from which come such popular results as Fermat's Last Theorem. The twin prime conjecture and Goldbach's conjecture are two unsolved problems in number theory.

As the number system is further developed, the integers are recognized as a subset of the rational numbers ("fractions"). These, in turn, are contained within the real numbers, which are used to represent continuous quantities. Real numbers are generalized to complex numbers. These are the first steps of a hierarchy of numbers that goes on to include quaternions and octonions. Consideration of the natural numbers also leads to the transfinite numbers, which formalize the concept of "infinity". According to the fundamental theorem of algebra all solutions of equations in one unknown with complex coefficients are complex numbers, regardless of degree. Another area of study is the size of sets, which is described with the cardinal numbers. These include the aleph numbers, which allow meaningful comparison of the size of infinitely large sets.

$1, 2, 3, \dots$	$\dots, -2, -1, 0, 1, 2 \dots$	$-2, \dfrac{2}{3}, 1.21$	$-e, \sqrt{2}, 3, \pi$	$2, i, -2+3i, 2e^{i\frac{4\pi}{3}}$
Natural numbers	Integers	Rational numbers	Real numbers	Complex numbers

Structure

Many mathematical objects, such as sets of numbers and functions, exhibit internal structure as a consequence of operations or relations that are defined on the set. Mathematics then studies properties of those sets that can be expressed in terms of that structure; for instance number theory studies properties of the set of integers that can be expressed in terms of arithmetic operations. Moreover, it frequently happens that different such structured sets (or structures) exhibit similar properties, which makes it possible, by a further step of abstraction, to state axioms for a class of structures, and then study at once the whole class of structures satisfying these axioms. Thus one can study groups, rings, fields and other abstract systems; together such studies (for structures defined by algebraic operations) constitute the domain of abstract algebra.

By its great generality, abstract algebra can often be applied to seemingly unrelated problems; for instance a number of ancient problems concerning compass and straightedge constructions were finally solved using Galois theory, which involves field theory and group theory. Another example of an algebraic theory is linear algebra, which is the general study of vector spaces, whose elements called vectors have both quantity and direction, and can be used to model (relations between) points in space. This is one example of the phenomenon that the originally unrelated areas of geometry and algebra have very strong interactions in modern mathematics. Combinatorics studies ways of enumerating the number of objects that fit a given structure.

(1,2,3) (1,3,2) (2,1,3) (2,3,1) (3,1,2) (3,2,1)					
Combinatorics	Number theory	Group theory	Graph theory	Order theory	Algebra

Space

The study of space originates with geometry – in particular, Euclidean geometry, which combines space and numbers, and encompasses the well-known Pythagorean theorem. Trigonometry is the branch of mathematics that deals with relationships between the sides and the angles of triangles and with the trigonometric functions. The modern study of space generalizes these ideas to include higher-dimensional geometry, non-Euclidean geometries (which play a central role in general relativity) and topology. Quantity and space both play a role in analytic geometry, differential geometry, and algebraic geometry. Convex and discrete geometry were developed to solve problems in number theory and functional analysis but now are pursued with an eye on applications in optimization and computer science. Within differential geometry are the concepts of fiber bundles and calculus on manifolds, in particular, vector and tensor calculus. Within algebraic geometry is the description of geometric objects as solution sets of polynomial equations, combining the concepts of quantity and space, and also the study of topological groups, which combine structure and space. Lie groups are used to study space, structure, and change. Topology in all its many ramifications may have been the greatest growth area in 20th-century mathematics; it includes point-set topology, set-theoretic topology, algebraic topology and differential topology. In particular, instances of modern-day topology are metrizability theory, axiomatic set theory, homotopy theory, and Morse theory. Topology also includes the now solved Poincaré conjecture, and the still unsolved areas of the Hodge conjecture. Other results in geometry and topology, including the four color theorem and Kepler conjecture, have been proved only with the help of computers.

| Geometry | Trigonometry | Differential geometry | Topology | Fractal geometry | Measure theory |

Change

Understanding and describing change is a common theme in the natural sciences, and calculus was developed as a powerful tool to investigate it. Functions arise here, as a central concept describing a changing quantity. The rigorous study of real numbers and functions of a real variable is known as real analysis, with complex analysis the equivalent field for the complex numbers. Functional analysis focuses attention on (typically infinite-dimensional) spaces of functions. One of many applications of functional analysis is quantum mechanics. Many problems lead naturally to relationships between a quantity and its rate of change, and these are studied as differential equations. Many phenomena in nature can be described by dynamical systems; chaos theory makes precise the ways in which many of these systems exhibit unpredictable yet still deterministic behavior.

Calculus	Vector calculus	Differential equations	Dynamical systems	Chaos theory	Complex analysis

Applied Mathematics

Applied mathematics concerns itself with mathematical methods that are typically used in science, engineering, business, and industry. Thus, "applied mathematics" is a mathematical science with specialized knowledge. The term *applied mathematics* also describes the professional specialty in which mathematicians work on practical problems; as a profession focused on practical problems, *applied mathematics* focuses on the "formulation, study, and use of mathematical models" in science, engineering, and other areas of mathematical practice.

In the past, practical applications have motivated the development of mathematical theories, which then became the subject of study in pure mathematics, where mathematics is developed primarily for its own sake. Thus, the activity of applied mathematics is vitally connected with research in pure mathematics.

Statistics and Other Decision Sciences

Applied mathematics has significant overlap with the discipline of statistics, whose theory is formulated mathematically, especially with probability theory. Statisticians (working as part of a research project) "create data that makes sense" with random sampling and with randomized experiments; the design of a statistical sample or experiment specifies the analysis of the data (before the data be available). When reconsidering data from experiments and samples or when analyzing data from observational studies, statisticians "make sense of the data" using the art of modelling and the theory of inference – with model selection and estimation; the estimated models and consequential predictions should be tested on new data.

Statistical theory studies decision problems such as minimizing the risk (expected loss) of a statistical action, such as using a procedure in, for example, parameter estimation, hypothesis testing, and selecting the best. In these traditional areas of mathematical statistics, a statistical-decision problem is formulated by minimizing an objective function, like expected loss or cost, under specific constraints: For example, designing a survey often involves minimizing the cost of estimating a population mean with a given level of confidence. Because of its use of optimization, the mathematical theory of statistics shares concerns with other decision sciences, such as operations research, control theory, and mathematical economics.

Computational Mathematics

Computational mathematics proposes and studies methods for solving mathematical problems that are typically too large for human numerical capacity. Numerical analysis studies methods for problems in analysis using functional analysis and approximation theory; numerical analysis includes the study of approximation and discretization broadly with special concern for rounding errors. Numerical analysis and, more broadly, scientific computing also study non-analytic topics of mathematical science, especially algorithmicmatrix and graph theory. Other areas of computational mathematics include computer algebra and symbolic computation.

Mathematical physics	Fluid dynamics	Numerical analysis	Optimization	Probability theory	Statistics	Cryptography
Mathematical finance	Game theory	Mathematical biology	Mathematical chemistry	Mathematical economics	Control theory	

Mathematical Awards

Arguably the most prestigious award in mathematics is the Fields Medal, established in 1936 and awarded every four years (except around World War II) to as many as four individuals. The Fields Medal is often considered a mathematical equivalent to the Nobel Prize.

The Wolf Prize in Mathematics, instituted in 1978, recognizes lifetime achievement, and another major international award, the Abel Prize, was introduced in 2003. The Chern Medal was introduced in 2010 to recognize lifetime achievement. These accolades are awarded in recognition of a particular body of work, which may be innovational, or provide a solution to an outstanding problem in an established field.

A famous list of 23 open problems, called "Hilbert's problems", was compiled in 1900 by German mathematician David Hilbert. This list achieved great celebrity among mathematicians, and at least nine of the problems have now been solved. A new list of seven important problems, titled the "Millennium Prize Problems", was published in 2000. A solution to each of these problems carries a $1 million reward, and only one (the Riemann hypothesis) is duplicated in Hilbert's problems.

References

- Courant, Richard and H. Robbins, What Is Mathematics? : An Elementary Approach to Ideas and Methods, Oxford University Press, USA; 2 edition (July 18, 1996). ISBN 0-19-510519-2.
- Eves, Howard, An Introduction to the History of Mathematics, Sixth Edition, Saunders, 1990, ISBN 0-03-029558-0.

- Kline, Morris, Mathematical Thought from Ancient to Modern Times, Oxford University Press, USA; Paperback edition (March 1, 1990). ISBN 0-19-506135-7.

- Oxford English Dictionary, second edition, ed. John Simpson and Edmund Weiner, Clarendon Press, 1989, ISBN 0-19-861186-2.

- Pappas, Theoni, The Joy Of Mathematics, Wide World Publishing; Revised edition (June 1989). ISBN 0-933174-65-9.

- Peterson, Ivars, Mathematical Tourist, New and Updated Snapshots of Modern Mathematics, Owl Books, 2001, ISBN 0-8050-7159-8.

- Popper, Karl R. (1995). "On knowledge". In Search of a Better World: Lectures and Essays from Thirty Years. Routledge. ISBN 0-415-13548-6.

- Waltershausen, Wolfgang Sartorius von (1965) [first published 1856]. Gauss zum Gedächtniss. Sändig Reprint Verlag H. R. Wohlwend. ASIN B0000BN5SQ. ISBN 3-253-01702-8.

2

Key Components of Mathematics

The mathematical objects listed in this chapter are key to mathematical practice. They are conceptions of the abstract possibility that mathematicians seek to unlock and understand. Some of the concepts that are mentioned in this chapter are number, calculus and mathematical space. The chapter strategically encompasses and incorporates the major components and key concepts of mathematics, providing a complete understanding.

Number

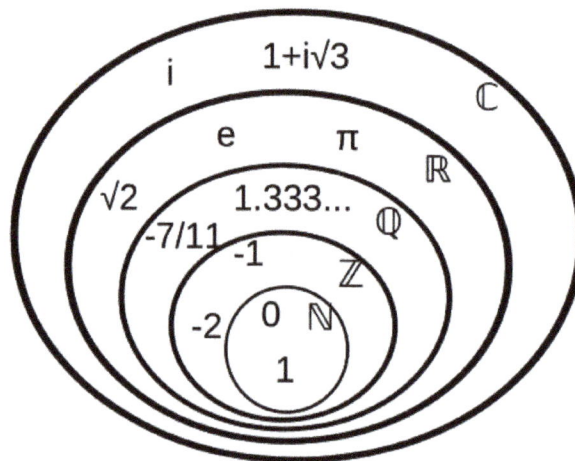

Subsets of the complex numbers.

A number is a mathematical object used to count, measure, and label. The original examples are the natural numbers 1, 2, 3, and so forth. A notational symbol that represents a number is called a numeral. In addition to their use in counting and measuring, numerals are often used for labels (as with telephone numbers), for ordering (as with serial numbers), and for codes (as with ISBNs). In common usage, *number* may refer to a symbol, a word, or a mathematical abstraction.

In mathematics, the notion of number has been extended over the centuries to include 0, negative numbers, rational numbers such as $\frac{1}{2}$ and $-\frac{2}{3}$, real numbers such as $\sqrt{2}$ and π,, complex numbers, which extend the real numbers by including $\sqrt{-1}$, and sometimes additional objects. Calculations with numbers are done with arithmetical operations, the most familiar being addition, subtraction, multiplication, division, and exponentiation. Their study or usage is called arithmetic. The same term may also refer to number theory, the study of the properties of the natural numbers.

Besides their practical uses, numbers have cultural significance throughout the world. For example, in Western society the number 13 is regarded as unlucky, and "a million" may signify "a lot." Though it is now regarded as pseudoscience, numerology, the belief in a mystical significance of numbers permeated ancient and medieval thought. Numerology heavily influenced the development of Greek mathematics, stimulating the investigation of many problems in number theory which are still of interest today.

During the 19th century, mathematicians began to develop many different abstractions which share certain properties of numbers and may be seen as extending the concept. Among the first were the hypercomplex numbers, which consist of various extensions or modifications of the complex number system. Today, number systems are considered important special examples of much more general categories such as rings and fields, and the application of the term "number" is a matter of convention, without fundamental significance.

Numerals

Numbers should be distinguished from numerals, the symbols used to represent numbers. Boyer showed that Egyptians created the first ciphered numeral system. Greeks followed by mapping their counting numbers onto Ionian and Doric alphabets. The number five can be represented by digit "5" or by the Roman numeral "V". Notations used to represent numbers are discussed in the article numeral systems. An important development in the history of numerals was the development of a positional system, like modern decimals, which have many advantages, such as representing large numbers with only a few symbols. The Roman numerals require extra symbols for larger numbers.

Main Classification

Different types of numbers have many different uses. Numbers can be classified into sets, called number systems, such as the natural numbers and the real numbers. The same number can be written in many different ways. For different methods of expressing numbers with symbols, such as the Roman numerals.

Main number systems		
	Natural	0, 1, 2, 3, 4, ... **or** 1, 2, 3, 4, ... $\mathbb{N}_0$ or $\mathbb{N}_1$ are sometimes used.
	Integer	..., −5, −4, −3, −2, −1, 0, 1, 2, 3, 4, 5, ...
	Rational	a/b where a and b are integers and b is not 0
	Real	The limit of a convergent sequence of rational numbers
	Complex	$a + bi$ where a and b are real numbers and i is the square root of −1

Natural Numbers

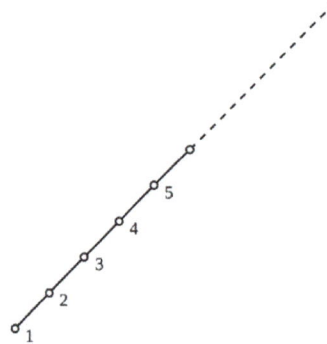

The natural numbers, starting with 1

The most familiar numbers are the natural numbers (sometimes called whole numbers or counting numbers): 1, 2, 3, and so on. Traditionally, the sequence of natural numbers started with 1 (0 was not even considered a number for the Ancient Greeks.) However, in the 19th century, set theorists and other mathematicians started including 0 (cardinality of the empty set, i.e. 0 elements, where 0 is thus the smallest cardinal number) in the set of natural numbers. Today, different mathematicians use the term to describe both sets, including 0 or not. The mathematical symbol for the set of all natural numbers is N, also written $\mathbb{N}$, and sometimes $\mathbb{N}_0$ or $\mathbb{N}_1$ when it is necessary to indicate whether the set should start with 0 or 1, respectively.

In the base 10 numeral system, in almost universal use today for mathematical operations, the symbols for natural numbers are written using ten digits: 0, 1, 2, 3, 4, 5, 6, 7, 8, and 9. The radix or base is the number of unique numerical digits, including zero, that a numeral system uses to represent numbers (for the decimal system, the radix is 10). In this base 10 system, the rightmost digit of a natural number has a place value of 1, and every other digit has a place value ten times that of the place value of the digit to its right.

In set theory, which is capable of acting as an axiomatic foundation for modern mathematics, natural numbers can be represented by classes of equivalent sets. For instance, the number 3 can be represented as the class of all sets that have exactly three elements. Alternatively, in Peano Arithmetic, the number 3 is represented as ssso, where s is the "successor" function (i.e., 3 is the third successor of 0). Many different representations are possible; all that is needed to formally represent 3 is to inscribe a certain symbol or pattern of symbols three times.

Integers

The negative of a positive integer is defined as a number that produces 0 when it is added to the corresponding positive integer. Negative numbers are usually written with a negative sign (a minus sign). As an example, the negative of 7 is written −7, and 7 + (−7) = 0. When the set of negative numbers is combined with the set of natural numbers (including 0), the result is defined as the set of integers, Z also written $\mathbb{Z}$. Here the letter Z comes from German*Zahl*, meaning "number". The set of integers forms a ring with the operations addition and multiplication.

The natural numbers form a subset of the integers. As there is no common standard for the inclusion or not of zero in the natural numbers, the natural numbers without zero are commonly referred to as positive integers, and the natural numbers with zero are referred to as non-negative integers.

Rational Numbers

A rational number is a number that can be expressed as a fraction with an integer numerator and a positive integer denominator. Negative denominators are allowed, but are commonly avoided, as every rational number is equal to a fraction with positive denominator. Fractions are written as two integers, the numerator and the denominator, with a dividing bar between them. The fraction m/n represents m parts of a whole divided into n equal parts. Two different fractions may correspond to the same rational number; for example 1/2 and 2/4 are equal, that is:

$$\frac{1}{2} = \frac{2}{4}.$$

If the absolute value of m is greater than n (supposed to be positive), then the absolute value of the fraction is greater than 1. Fractions can be greater than, less than, or equal to 1 and can also be positive, negative, or 0. The set of all rational numbers includes the integers, since every integer can be written as a fraction with denominator 1. For example −7 can be written −7/1. The symbol for the rational numbers is Q (for *quotient*), also written $\mathbb{Q}$.

Real Numbers

The real numbers include all the measuring numbers. The symbol for the real numbers is R, also written as $\mathbb{R}$. Real numbers are usually represented by using decimal numerals, in which a decimal point is placed to the right of the digit with place value 1. Each digit to the right of the decimal point has a place value one-tenth of the place value of the digit to its left. For example, 123.456 represents 123456/1000, or, in words, one hundred, two tens, three ones, four tenths, five hundredths, and six thousandths. A finite decimal representation allows us to represent exactly only the integers and those rational numbers whose denominators have only prime factors which are factors of ten. Thus one half is 0.5, one fifth is 0.2, one tenth is 0.1, and one fiftieth is 0.02. To represent the rest of the real numbers requires an infinite sequence of digits after the decimal point. Since it is impossible to write infinitely many digits, real numbers are commonly represented by rounding or truncating this sequence, or by establishing a pattern, such as 0.333..., with an ellipsis to indicate that the pattern continues. Thus 123.456 is an approximation of any real number between 1234555/10000 and 1234565/10000 (rounding) or any real number between 123456/1000 and 123457/1000 (truncation). Negative real numbers are written with a preceding minus sign: -123.456.

Every rational number is also a real number. It is not the case, however, that every real number is rational. A real number, which is not rational, is called irrational. A decimal represents a rational number if and only if has a finite number of digits or eventually repeats for ever, after any initial finite string digits. For example, 1/2 = 0.5 and 1/3 = 0.333... (forever repeating 3s, otherwise written 0.3). On the other hand, the real number π, the ratio of the circumference of any circle to its diameter, is

$$\pi = 3.14159265358979\ldots$$

Since the decimal neither ends nor eventually repeats forever it cannot be written as a fraction, and is an example of an irrational number. Other irrational numbers include

$$\sqrt{2} = 1.41421356237\ldots$$

(the square root of 2, that is, the positive number whose square is 2).

Just as the same fraction can be written in more than one way, the same decimal may have more than one representation. 1.0 and 0.999... are two different decimal numerals representing the natural number 1. There are infinitely many other ways of representing the number 1, for example 1.00, 1.000, and so on.

Every real number is either rational or irrational. Every real number corresponds to a point on the number line. The real numbers also have an important but highly technical property called the least upper bound property.

When a real number represents a measurement, there is always a margin of error. This is often indicated by rounding or truncating a decimal, so that digits that suggest a greater accuracy than the measurement itself are removed. The remaining digits are called significant digits. For example, measurements with a ruler can seldom be made without a margin of error of at least 0.001 meters. If the sides of a rectangle are measured as 1.23 meters and 4.56 meters, then multiplication gives an area for the rectangle of 5.6088 square meters. Since only the first two digits after the decimal place are significant, this is usually rounded to 5.61.

In abstract algebra, it can be shown that any completeordered field is isomorphic to the real numbers. The real numbers are not, however, an algebraically closed field, because they do not include the square root of minus one.

Complex Numbers

Moving to a greater level of abstraction, the real numbers can be extended to the complex numbers. This set of numbers arose historically from trying to find closed formulas for the roots of cubic and quartic polynomials. This led to expressions involving the square roots of negative numbers, and eventually to the definition of a new number: a square root of -1, denoted by i, a symbol assigned by Leonhard Euler, and called the imaginary unit. The complex numbers consist of all numbers of the form

$$a + bi$$

where a and b are real numbers. Because of this, complex numbers correspond to points on the complex plane, a vector space of two real dimensions. In the expression $a + bi$, the real number a is called the real part and b is called the imaginary part. If the real part of a complex number is 0, then the number is called an imaginary number or is referred to as *purely imaginary*; if the imaginary part is 0, then the number is a real number. Thus the real numbers are a subset of the complex numbers. If the real and imaginary parts of a complex number are both integers, then the number is called a Gaussian integer. The symbol for the complex numbers is C or $\mathbb{C}$.

In abstract algebra, the complex numbers are an example of an algebraically closed field, meaning that every polynomial with complex coefficients can be factored into linear factors. Like the real number system, the complex number system is a field and is complete, but unlike the real numbers, it is not ordered. That is, there is no meaning in saying that i is greater than 1, nor is there any meaning in saying that i is less than 1. In technical terms, the complex numbers lack the trichotomy property.

Each of the number systems mentioned above is a proper subset of the next number system. Symbolically, $\mathbb{N} \subset \mathbb{Z} \subset \mathbb{Q} \subset \mathbb{R} \subset \mathbb{C}$.

Subclasses of The Integers

Even and Odd Numbers

An even number is an integer that is "evenly divisible" by two, that is divisible by two without remainder; an odd number is an integer that is not even. (The old-fashioned term "evenly divisible" is now almost always shortened to "divisible".) Equivalently, another way of defining an odd number is that it is an integer of the form $n = 2k + 1$, where k is an integer, and an even number has the form $n = 2k$ where k is an integer.

Prime Numbers

A prime number is an integer greater than 1 that is not the product of two smaller positive integers. The first few prime numbers are 2, 3, 5, 7, and 11. The prime numbers have been widely studied for more than 2000 years and have led to many questions, only some of which have been answered. The study of these questions is called number theory. An example of a question that is still unanswered is whether every even number is the sum of two primes. This is called Goldbach's conjecture.

A question that has been answered is whether every integer greater than one is a product of primes in only one way, except for a rearrangement of the primes. This is called fundamental theorem of arithmetic. A proof appears in Euclid's Elements.

Other Classes of Integers

Many subsets of the natural numbers have been the subject of specific studies and have been named, often after the first mathematician that has studied them. Example of such sets of integers are Fibonacci numbers and perfect numbers. For more examples.

Subclasses of The Complex Numbers

Algebraic, Irrational and Transcendental Numbers

Algebraic numbers are those that are a solution to a polynomial equation with integer coefficients. Real numbers that are not rational numbers are called irrational numbers. Complex numbers which are not algebraic are called transcendental numbers. The algebraic numbers that are solutions of a monic polynomial equation with integer coefficients are called algebraic integers.

Computable Numbers

A computable number, also known as *recursive number*, is a real number such that there exists an algorithm which, given a positive number n as input, produces the first n digits of the computable number's decimal representation. Equivalent definitions can be given using μ-recursive functions, Turing machines or λ-calculus. The computable numbers are stable for all usual arithmetic operations, including the computation of the roots of a polynomial, and thus form a real closed field that contains the real algebraic numbers.

The computable numbers may be viewed as the real numbers that may be exactly represented in a computer: a computable number is exactly represented by its first digits and a program for computing further digits. However, the computable numbers are rarely used in practice. One reason is that there is no algorithm for testing the equality of two computable numbers. More precisely, there cannot exist any algorithm which takes any computable number as an input, and decides in every case if this number is equal to zero or not.

The set of computable numbers has the same cardinality as the natural numbers. Therefore, almost all real numbers are non-computable. However, it is very difficult to produce explicitly a real number that is not computable.

Extensions of The Concept

p-adic Numbers

The *p*-adic numbers may have infinitely long expansions to the left of the decimal point, in the same way that real numbers may have infinitely long expansions to the right. The number system that results depends on what base is used for the digits: any base is possible, but a prime number-base provides the best mathematical properties. The set of the *p*-adic numbers contains the rational numbers, but is not contained in the complex numbers.

The elements of an algebraic function field over a finite field and algebraic numbers have many similar properties. Therefore, they are often regarded as numbers by number theorists. The *p*-adic numbers play an important role in this analogy.

Hypercomplex Numbers

Some number systems that are not included in the complex numbers may be constructed from the real numbers in a way that generalize the construction of the complex numbers. They are sometimes called hypercomplex numbers. They include the quaternionsH, introduced by Sir William Rowan Hamilton, in which multiplication is not commutative, and the octonions, in which multiplication is not associative.

Transfinite Numbers

For dealing with infinite sets, the natural numbers have been generalized to the ordinal numbers and to the cardinal numbers. The former gives the ordering of the set, while the latter gives its size. For finite sets, both ordinal and cardinal numbers are identified with the natural numbers. In the infinite case, many ordinal numbers correspond to the same cardinal number.

Nonstandard Numbers

Superreal and surreal numbers extend the real numbers by adding infinitesimally small numbers and infinitely large numbers, but still form fields.

A relation number is defined as the class of relations consisting of all those relations that are similar to one member of the class.

History

First use of Numbers

Bones and other artifacts have been discovered with marks cut into them that many believe are tally marks. These tally marks may have been used for counting elapsed time, such as numbers of days, lunar cycles or keeping records of quantities, such as of animals.

A tallying system has no concept of place value (as in modern decimal notation), which limits its representation of large numbers. Nonetheless tallying systems are considered the first kind of abstract numeral system.

The first known system with place value was the Mesopotamian base 60 system (ca. 3400 BC) and the earliest known base 10 system dates to 3100 BC in Egypt.

Zero

The number 605 in Khmer numerals, from an inscription from 683 AD. An early use of zero as a decimal figure.

The use of 0 as a number should be distinguished from its use as a placeholder numeral in place-value systems. Many ancient texts used 0. Babylonian (Modern Iraq) and Egyptian texts used it. Egyptians used the word *nfr* to denote zero balance in double entry accounting entries. Indian texts used a Sanskrit word *Shunye* or *shunya* to refer to the concept of *void*. In mathematics texts this word often refers to the number zero.

Records show that the Ancient Greeks seemed unsure about the status of 0 as a number: they asked themselves "how can 'nothing' be something?" leading to interesting philosophical and, by the Medieval period, religious arguments about the nature and existence of 0 and the vacuum. The paradoxes of Zeno of Elea depend in large part on the uncertain interpretation of 0. (The ancient Greeks even questioned whether 1 was a number.)

The late Olmec people of south-central Mexico began to use a true zero (a shell glyph) in the New World possibly by the 4th century BC but certainly by 40 BC, which became an integral part of Maya numerals and the Maya calendar. Mayan arithmetic used base 4 and base 5 written as base 20. Sanchez in 1961 reported a base 4, base 5 "finger" abacus.

By 130 AD, Ptolemy, influenced by Hipparchus and the Babylonians, was using a symbol for 0 (a small circle with a long overbar) within a sexagesimal numeral system otherwise using alphabetic Greek numerals. Because it was used alone, not as just a placeholder, this Hellenistic zero was the first *documented* use of a true zero in the Old World. In later Byzantine manuscripts of his *Syntaxis Mathematica* (*Almagest*), the Hellenistic zero had morphed into the Greek letter omicron (otherwise meaning 70).

Another true zero was used in tables alongside Roman numerals by 525 (first known use by Dionysius Exiguus), but as a word, *nulla* meaning *nothing*, not as a symbol. When division produced 0 as a remainder, *nihil*, also meaning *nothing*, was used. These medieval zeros were used by all future medieval computists (calculators of Easter). An isolated use of their initial, N, was used in a table of Roman numerals by Bede or a colleague about 725, a true zero symbol.

An early documented use of the zero by Brahmagupta (in the *Brāhmasphuṭasiddhānta*) dates to 628. He treated 0 as a number and discussed operations involving it, including division. By this time (the 7th century) the concept had clearly reached Cambodia as Khmer numerals, and documentation shows the idea later spreading to China and the Islamic world.

Negative Numbers

The abstract concept of negative numbers was recognized as early as 100 BC – 50 BC in China. *The Nine Chapters on the Mathematical Art* contains methods for finding the areas of figures; red rods were used to denote positive coefficients, black for negative. The first reference in a Western work was in the 3rd century AD in Greece. Diophantus referred to the equation equivalent to $4x + 20 = 0$ (the solution is negative) in *Arithmetica*, saying that the equation gave an absurd result.

During the 600s, negative numbers were in use in India to represent debts. Diophantus' previous reference was discussed more explicitly by Indian mathematician Brahmagupta, in *Brāhmasphuṭasiddhānta* 628, who used negative numbers to produce the general form quadratic formula that remains in use today. However, in the 12th century in India, Bhaskara gives negative roots for quadratic equations but says the negative value "is in this case not to be taken, for it is inadequate; people do not approve of negative roots."

European mathematicians, for the most part, resisted the concept of negative numbers until the 17th century, although Fibonacci allowed negative solutions in financial problems where they could be interpreted as debts (chapter 13 of *Liber Abaci*, 1202) and later as losses (in *Flos*). At the same time, the Chinese were indicating negative numbers by drawing a diagonal stroke through the right-most non-zero digit of the corresponding positive number's numeral. The first use of negative numbers in a European work was by Nicolas Chuquet during the 15th century. He used them as exponents, but referred to them as "absurd numbers".

As recently as the 18th century, it was common practice to ignore any negative results returned by equations on the assumption that they were meaningless, just as René Descartes did with negative solutions in a Cartesian coordinate system.

Rational Numbers

It is likely that the concept of fractional numbers dates to prehistoric times. The Ancient Egyptians used their Egyptian fraction notation for rational numbers in mathematical texts such as the Rhind Mathematical Papyrus and the Kahun Papyrus. Classical Greek and Indian mathematicians made studies of the theory of rational numbers, as part of the general study of number theory. The best known of these is Euclid's *Elements*, dating to roughly 300 BC. Of the Indian texts, the most relevant is the Sthananga Sutra, which also covers number theory as part of a general study of mathematics.

The concept of decimal fractions is closely linked with decimal place-value notation; the two seem to have developed in tandem. For example, it is common for the Jain math sutra to include calculations of decimal-fraction approximations to pi or the square root of 2. Similarly, Babylonian math texts had always used sexagesimal (base 60) fractions with great frequency.

Irrational Numbers

The earliest known use of irrational numbers was in the IndianSulba Sutras composed between 800 and 500 BC. The first existence proofs of irrational numbers is usually attributed to Pythagoras, more specifically to the PythagoreanHippasus of Metapontum, who produced a (most likely geometrical) proof of the irrationality of the square root of 2. The story goes that Hippasus discovered irrational numbers when trying to represent the square root of 2 as a fraction. However Pythagoras believed in the absoluteness of numbers, and could not accept the existence of irrational numbers. He could not disprove their existence through logic, but he could not accept irrational numbers, so he sentenced Hippasus to death by drowning.

The 16th century brought final European acceptance of negative integral and fractional numbers. By the 17th century, mathematicians generally used decimal fractions with modern notation. It was not, however, until the 19th century that mathematicians separated irrationals into algebraic and transcendental parts, and once more undertook scientific study of irrationals. It had remained almost dormant since Euclid. In 1872, the publication of the theories of Karl Weierstrass (by his pupil Kossak), Heine (*Crelle*, 74), Georg Cantor (Annalen, 5), and Richard Dedekind was brought about. In 1869, Méray had taken the same point of departure as Heine, but the theory is generally referred to the year 1872. Weierstrass's method was completely set forth by Salvatore Pincherle (1880), and Dedekind's has received additional prominence through the author's later work (1888) and endorsement by Paul Tannery (1894). Weierstrass, Cantor, and Heine base their theories on infinite series, while Dedekind founds his on the idea of a cut (Schnitt) in the system of real numbers, separating all rational numbers into two groups having certain characteristic properties. The subject has received later contributions at the hands of Weierstrass, Kronecker (Crelle, 101), and Méray.

The search for roots of quintic and higher degree equations was an important development, the Abel–Ruffini theorem (Ruffini 1799, Abel 1824) showed that they could not be solved by radicals (formulas involving only arithmetical operations and roots). Hence it was necessary to consider the wider set of algebraic numbers (all solutions to polynomial equations). Galois (1832) linked polynomial equations to group theory giving rise to the field of Galois theory.

Continued fractions, closely related to irrational numbers (and due to Cataldi, 1613), received attention at the hands of Euler, and at the opening of the 19th century were brought into prominence through the writings of Joseph Louis Lagrange. Other noteworthy contributions have been made by Druckenmüller (1837), Kunze (1857), Lemke (1870), and Günther (1872). Ramus (1855) first connected the subject with determinants, resulting, with the subsequent contributions of Heine, Möbius, and Günther, in the theory of Kettenbruchdeterminanten.

Transcendental Numbers and Reals

The existence of transcendental numbers was first established by Liouville (1844, 1851). Hermite proved in 1873 that e is transcendental and Lindemann proved in 1882 that π is transcendental. Finally, Cantor showed that the set of all real numbers is uncountably infinite but the set of all algebraic numbers is countably infinite, so there is an uncountably infinite number of transcendental numbers.

Infinity and Infinitesimals

The earliest known conception of mathematical infinity appears in the Yajur Veda, an ancient Indian script, which at one point states, "If you remove a part from infinity or add a part to infinity, still what remains is infinity." Infinity was a popular topic of philosophical study among the Jain mathematicians c. 400 BC. They distinguished between five types of infinity: infinite in one and two directions, infinite in area, infinite everywhere, and infinite perpetually.

Aristotle defined the traditional Western notion of mathematical infinity. He distinguished between actual infinity and potential infinity—the general consensus being that only the latter had true value. Galileo Galilei's *Two New Sciences* discussed the idea of one-to-one correspondences between infinite sets. But the next major advance in the theory was made by Georg Cantor; in 1895 he published a book about his new set theory, introducing, among other things, transfinite numbers and formulating the continuum hypothesis.

In the 1960s, Abraham Robinson showed how infinitely large and infinitesimal numbers can be rigorously defined and used to develop the field of nonstandard analysis. The system of hyperreal numbers represents a rigorous method of treating the ideas about infinite and infinitesimal numbers that had been used casually by mathematicians, scientists, and engineers ever since the invention of infinitesimal calculus by Newton and Leibniz.

A modern geometrical version of infinity is given by projective geometry, which introduces "ideal points at infinity", one for each spatial direction. Each family of parallel lines in a given direction is postulated to converge to the corresponding ideal point. This is closely related to the idea of vanishing points in perspective drawing.

Complex Numbers

The earliest fleeting reference to square roots of negative numbers occurred in the work of the mathematician and inventor Heron of Alexandria in the 1st century AD, when he considered the volume of an impossible frustum of a pyramid. They became more prominent when in the 16th century closed formulas for the roots of third and fourth degree polynomials were discovered by Italian mathematicians such as Niccolò Fontana Tartaglia and Gerolamo Cardano. It was soon realized that these formulas, even if one was only interested in real solutions, sometimes required the manipulation of square roots of negative numbers.

This was doubly unsettling since they did not even consider negative numbers to be on firm ground at the time. When René Descartes coined the term "imaginary" for these quantities in 1637, he intended it as derogatory. A further source of confusion was that the equation

$$\left(\sqrt{-1}\right)^2 = \sqrt{-1}\sqrt{-1} = -1$$

seemed capriciously inconsistent with the algebraic identity

$$\sqrt{a}\sqrt{b} = \sqrt{ab},$$

which is valid for positive real numbers a and b, and was also used in complex number calculations with one of a, b positive and the other negative. The incorrect use of this identity, and the related identity

$$\frac{1}{\sqrt{a}} = \sqrt{\frac{1}{a}}$$

in the case when both a and b are negative even bedeviled Euler. This difficulty eventually led him to the convention of using the special symbol i in place of $\sqrt{-1}$ to guard against this mistake.

The 18th century saw the work of Abraham de Moivre and Leonhard Euler. De Moivre's formula (1730) states:

$$(\cos\theta + i\sin\theta)^n = \cos n\theta + i\sin n\theta$$

and to Euler (1748) Euler's formula of complex analysis:

$$\cos\theta + i\sin\theta = e^{i\theta}.$$

The existence of complex numbers was not completely accepted until Caspar Wessel described the geometrical interpretation in 1799. Carl Friedrich Gauss rediscovered and popularized it several years later, and as a result the theory of complex numbers received a notable expansion. The idea of the graphic representation of complex numbers had appeared, however, as early as 1685, in Wallis's *De Algebra tractatus*.

Also in 1799, Gauss provided the first generally accepted proof of the fundamental theorem of algebra, showing that every polynomial over the complex numbers has a full set of solutions in that realm. The general acceptance of the theory of complex numbers is due to the labors of Augustin Louis Cauchy and Niels Henrik Abel, and especially the latter, who was the first to boldly use complex numbers with a success that is well known.

Gauss studied complex numbers of the form $a + bi$, where a and b are integral, or rational (and i is one of the two roots of $x^2 + 1 = 0$). His student, Gotthold Eisenstein, studied the type $a + b\omega$, where ω is a complex root of $x^3 - 1 = 0$. Other such classes (called cyclotomic fields) of complex numbers derive from the roots of unity $x^k - 1 = 0$ for higher values of k. This generalization is largely due to Ernst Kummer, who also invented ideal numbers, which were expressed as geometrical entities by Felix Klein in 1893.

In 1850 Victor Alexandre Puiseux took the key step of distinguishing between poles and branch points, and introduced the concept of essential singular points. This eventually led to the concept of the extended complex plane.

Prime Numbers

Prime numbers have been studied throughout recorded history. Euclid devoted one book of the *Elements* to the theory of primes; in it he proved the infinitude of the primes and the fundamental theorem of arithmetic, and presented the Euclidean algorithm for finding the greatest common divisor of two numbers.

In 240 BC, Eratosthenes used the Sieve of Eratosthenes to quickly isolate prime numbers. But most further development of the theory of primes in Europe dates to the Renaissance and later eras.

In 1796, Adrien-Marie Legendre conjectured the prime number theorem, describing the asymptotic distribution of primes. Other results concerning the distribution of the primes include Euler's proof that the sum of the reciprocals of the primes diverges, and the Goldbach conjecture, which claims that any sufficiently large even number is the sum of two primes. Yet another conjecture related to the distribution of prime numbers is the Riemann hypothesis, formulated by Bernhard Riemann in 1859. The prime number theorem was finally proved by Jacques Hadamard and Charles de la Vallée-Poussin in 1896. Goldbach and Riemann's conjectures remain unproven and unrefuted.

Mathematical Structure

In mathematics, a structure on a set is an additional mathematical object that, in some manner, attaches (or relates) to that set to endow it with some additional meaning or significance.

A partial list of possible structures are measures, algebraic structures (groups, fields, etc.), topologies, metric structures (geometries), orders, events, equivalence relations, differential structures, and categories.

Sometimes, a set is endowed with more than one structure simultaneously; this enables mathematicians to study it more richly. For example, an ordering imposes a rigid form, shape, or topology on the set. As another example, if a set has both a topology and is a group, and these two structures are related in a certain way, the set becomes a topological group.

Mappings between sets which preserve structures (so that structures in the source or domain are mapped to equivalent structures in the destination or codomain) are of special interest in many fields of mathematics. Examples are homomorphisms, which preserve algebraic structures; homeomorphisms, which preserve topological structures; and diffeomorphisms, which preserve differential structures.

History

In 1939, the French group with the pseudonym Nicolas Bourbaki saw structures as the root of mathematics. They first mentioned them in their "Fascicule" of *Theory of Sets* and expanded it into Chapter IV of the 1957 edition. They identified three *mother structures*: algebraic, topological, and order.

Example: The Real Numbers

The set of real numbers has several standard structures:

- an order: each number is either less or more than any other number.

- algebraic structure: there are operations of multiplication and addition that make it into a field.

- a measure: intervals along the real line have a specific length, which can be extended to the Lebesgue measure on many of its subsets.

- a metric: there is a notion of distance between points.

- a geometry: it is equipped with a metric and is flat.

- a topology: there is a notion of open sets.

There are interfaces among these:

- Its order and, independently, its metric structure induce its topology.

- Its order and algebraic structure make it into an ordered field.

- Its algebraic structure and topology make it into a Lie group, a type of topological group.

Space (Mathematics)

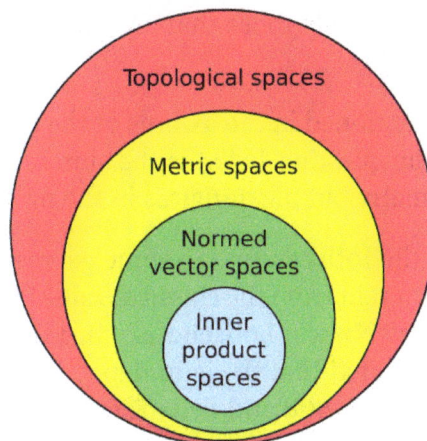

A hierarchy of mathematical spaces: The inner product induces a norm. The norm induces a metric. The metric induces a topology.

In mathematics, a space is a set (sometimes called a universe) with some added structure.

Mathematical spaces often form a hierarchy, i.e., one space may inherit all the characteristics of a parent space. For instance, all inner product spaces are also normed vector spaces, because the inner product *induces* a norm on the inner product space such that:

$$\|x\| = \sqrt{\langle x, x \rangle},$$

where the norm is indicated by enclosing in double vertical lines, and the inner product is indicated enclosing in by angle brackets.

Modern mathematics treats "space" quite differently compared to classical mathematics.

History

Before The Golden Age of Geometry

In the ancient mathematics, "space" was a geometric abstraction of the three-dimensional space observed in the everyday life. The axiomatic method had been the main research tool since Euclid (about 300 BC). The method of coordinates (analytic geometry) was adopted by René Descartes in 1637. At that time, geometric theorems were treated as an absolute objective truth knowable through intuition and reason, similar to objects of natural science; and axioms were treated as obvious implications of definitions.

Two equivalence relations between geometric figures were used: congruence and similarity. Translations, rotations and reflections transform a figure into congruent figures; homotheties — into similar figures. For example, all circles are mutually similar, but ellipses are not similar to circles. A third equivalence relation, introduced by projective geometry (Gaspard Monge, 1795), corresponds to projective transformations. Not only ellipses but also parabolas and hyperbolas turn into circles under appropriate projective transformations; they all are projectively equivalent figures.

The relation between the two geometries, Euclidean and projective, shows that mathematical objects are not given to us *with their structure*. Rather, each mathematical theory describes its objects by *some* of their properties, precisely those that are put as axioms at the foundations of the theory.

Distances and angles are never mentioned in the axioms of the projective geometry and therefore cannot appear in its theorems. The question "what is the sum of the three angles of a triangle" is meaningful in the Euclidean geometry but meaningless in the projective geometry.

A different situation appeared in the 19th century: in some geometries the sum of the three angles of a triangle is well-defined but different from the classical value (180 degrees). The non-Euclidean hyperbolic geometry, introduced by Nikolai Lobachevsky in 1829 and János Bolyai in 1832 (and Carl Gauss in 1816, unpublished) stated that the sum depends on the triangle and is always less than 180 degrees. Eugenio Beltrami in 1868 and Felix Klein in 1871 obtained Euclidean "models" of the non-Euclidean hyperbolic geometry, and thereby completely justified this theory.

This discovery forced the abandonment of the pretensions to the absolute truth of Euclidean geometry. It showed that axioms are not "obvious", nor "implications of definitions". Rather, they are hypotheses. To what extent do they correspond to an experimental reality? This important physical problem no longer has anything to do with mathematics. Even if a "geometry" does not correspond to an experimental reality, its theorems remain no less "mathematical truths".

A Euclidean model of a non-Euclidean geometry is a clever choice of some objects existing in Euclidean space and some relations between these objects that satisfy all axioms (therefore, all theorems) of the non-Euclidean geometry. These Euclidean objects and relations "play" the non-Euclidean geometry like contemporary actors playing an ancient performance. Relations between the actors only mimic relations between the characters in the play. Likewise, the chosen relations between the chosen objects of the Euclidean model only mimic the non-Euclidean relations. It shows that relations between objects are essential in mathematics, while the nature of the objects is not.

The Golden Age and Afterwards: Dramatic Change

According to Nicolas Bourbaki, the period between 1795 ("Geometrie descriptive" of Monge) and 1872 (the "Erlangen programme" of Klein) can be called the golden age of geometry. Analytic geometry made a great progress and succeeded in replacing theorems of classical geometry with computations via invariants of transformation groups. Since that time new theorems of classical geometry are of more interest to amateurs rather than to professional mathematicians.

However, it does not mean that the heritage of the classical geometry was lost. According to Bourbaki, "passed over in its role as an autonomous and living science, classical geometry is thus transfigured into a universal language of contemporary mathematics".

According to the famous inaugural lecture given by Bernhard Riemann in 1854, every mathematical object parametrized by n real numbers may be treated as a point of the -dimensional space of all such objects. Nowadays mathematicians follow this idea routinely and find it extremely suggestive to use the terminology of classical geometry nearly everywhere.

In order to fully appreciate the generality of this approach one should note that mathematics is "a pure theory of forms, which has as its purpose, not the combination of quantities, or of their images, the numbers, but objects of thought" (Hermann Hankel, 1867). This is a controversial characterization of the purpose of mathematics, which is not necessarily committed to the existence of "objects of thought."

Functions are important mathematical objects. Usually they form infinite-dimensional spaces, as noted already by Riemann and elaborated in the 20th century by functional analysis.

An object parametrized by n complex numbers may be treated as a point of a complex n-dimensional space. However, the same object is also parametrized by $2n$ real numbers (if c is a complex number, then c=a+bi, where a and b are real), thus, a point of a real $2n$-dimensional space. The complex dimension differs from the real dimension. This is only the tip of the iceberg. The "algebraic" concept of dimension applies to vector spaces. The "topological" concept of dimension applies to topological spaces. There is also Hausdorff dimension for metric spaces; this one can be non-integer (especially for fractals). Some kinds of spaces (for instance, measure spaces) admit no concept of dimension at all.

The original space investigated by Euclid is now called "the three-dimensional Euclidean space". Its axiomatization, started by Euclid 23 centuries ago, was finalized in the 20th century by David

Hilbert, with alternate treatments by Alfred Tarski and George Birkhoff among others. This approach describes the space via undefined primitives (such as "point", "between", "congruent") constrained by a number of axioms. Such a definition "from scratch" is now not often used, since it does not reveal the relation of this space to other spaces. The modern approach defines the three-dimensional Euclidean space more algebraically, via vector spaces and quadratic forms, namely, as an affine space whose difference space is a three-dimensional inner product space.

Also a three-dimensional projective space is now defined non-classically, as the space of all one-dimensional subspaces (that is, straight lines through the origin) of a four-dimensional vector space.

A space consists now of selected mathematical objects (for instance, functions on another space, or subspaces of another space, or just elements of a set) treated as points, and selected relationships between these points. It shows that spaces are just mathematical structures. One may expect that the structures called "spaces" are more geometric than others, but this is not always true. For example, a differentiable manifold (called also smooth manifold) is much more geometric than a measurable space, but no one calls it "differentiable space" (nor "smooth space").

Taxonomy of Spaces

Three Taxonomic Ranks

Spaces are classified on three levels. Given that each mathematical theory describes its objects by *some* of their properties, the first question to ask is: which properties?

For example, the upper-level classification distinguishes between Euclidean and projective spaces, since the distance between two points is defined in Euclidean spaces but undefined in projective spaces. These are spaces of different types.

Another example. The question "what is the sum of the three angles of a triangle" makes sense in a Euclidean space but not in a projective space; these are spaces of different types. In a non-Euclidean space the question makes sense but is answered differently, which is not an upper-level distinction.

Also, the distinction between a Euclidean plane and a Euclidean 3-dimensional space is not an upper-level distinction; the question "what is the dimension" makes sense in both cases.

In terms of Bourbaki the upper-level classification is related to "typical characterization" (or "typification"). However, it is not the same (since two equivalent structures may differ in typification).

On the second level of classification one takes into account answers to especially important questions (among the questions that make sense according to the first level). For example, this level distinguishes between Euclidean and non-Euclidean spaces; between finite-dimensional and infinite-dimensional spaces; between compact and non-compact spaces, etc.

In terms of Bourbaki the second-level classification is the classification by "species". Unlike biological taxonomy, a space may belong to several species.

On the third level of classification, roughly speaking, one takes into account answers to *all possible* questions (that make sense according to the first level). For example, this level distinguishes between spaces of different dimension, but does not distinguish between a plane of a three-dimensional Euclidean space, treated as a two-dimensional Euclidean space, and the set of all pairs of real numbers, also treated as a two-dimensional Euclidean space. Likewise it does not distinguish between different Euclidean models of the same non-Euclidean space.

More formally, the third level classifies spaces up to isomorphism. An isomorphism between two spaces is defined as a one-to-one correspondence between the points of the first space and the points of the second space, that preserves all relations between the points, stipulated by the given "typification". Mutually isomorphic spaces are thought of as copies of a single space. If one of them belongs to a given species then they all do.

The notion of isomorphism sheds light on the upper-level classification. Given a one-to-one correspondence between two spaces of the same type, one may ask whether it is an isomorphism or not. This question makes no sense for two spaces of different type.

Isomorphisms to itself are called automorphisms. Automorphisms of a Euclidean space are motions and reflections. Euclidean space is homogeneous in the sense that every point can be transformed into every other point by some automorphism.

Two Relations Between Spaces, and A Property of Spaces

Topological notions (continuity, convergence, open sets, closed sets etc.) are defined naturally in every Euclidean space. In other words, every Euclidean space is also a topological space. Every isomorphism between two Euclidean spaces is also an isomorphism between the corresponding topological spaces (called "homeomorphism"), but the converse is wrong: a homeomorphism may distort distances. In terms of Bourbaki, "topological space" is an underlying structure of the "Euclidean space" structure. Similar ideas occur in category theory: the category of Euclidean spaces is a concrete category over the category of topological spaces; the forgetful (or "stripping") functor maps the former category to the latter category.

A three-dimensional Euclidean space is a special case of a Euclidean space. In terms of Bourbaki, the species of three-dimensional Euclidean space is richer than the species of Euclidean space. Likewise, the species of compact topological space is richer than the species of topological space.

Euclidean axioms leave no freedom, they determine uniquely all geometric properties of the space. More exactly: all three-dimensional Euclidean spaces are mutually isomorphic. In this sense we have "the" three-dimensional Euclidean space. In terms of Bourbaki, the corresponding theory is univalent. In contrast, topological spaces are generally non-isomorphic, their theory is multivalent. A similar idea occurs in mathematical logic: a theory is called categorical if all its models of the same cardinality are mutually isomorphic. According to Bourbaki, the study of multivalent theories is the most striking feature which distinguishes modern mathematics from classical mathematics.

Types of Spaces

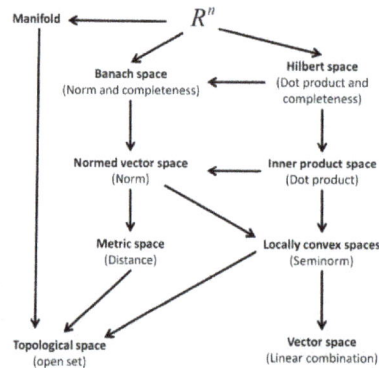

Overview of types of abstract spaces. An arrow from space A to space B implies that space A is also a kind of space B. That means, for instance, that a normed vector space is also a metric space.

Linear and Topological Spaces

Two basic spaces are linear spaces (also called vector spaces) and topological spaces.

Linear spaces are of algebraic nature; there are real linear spaces (over the field of real numbers), complex linear spaces (over the field of complex numbers), and more generally, linear spaces over any field. Every complex linear space is also a real linear space (the latter *underlies* the former), since each real number is also a complex number. Linear operations, given in a linear space by definition, lead to such notions as straight lines (and planes, and other linear subspaces); parallel lines; ellipses (and ellipsoids). However, orthogonal (perpendicular) lines cannot be defined, and circles cannot be singled out among ellipses. The dimension of a linear space is defined as the maximal number of linearly independent vectors or, equivalently, as the minimal number of vectors that span the space; it may be finite or infinite. Two linear spaces over the same field are isomorphic if and only if they are of the same dimension.

Topological spaces are of analytic nature. Open sets, given in a topological space by definition, lead to such notions as continuous functions, paths, maps; convergent sequences, limits; interior, boundary, exterior. However, uniform continuity, bounded sets, Cauchy sequences, differentiable functions (paths, maps) remain undefined. Isomorphisms between topological spaces are traditionally called homeomorphisms; these are one-to-one correspondences continuous in both directions. The open interval $(0,1)$ is homeomorphic to the whole real line $(-\infty, \infty)$ but not homeomorphic to the closed interval $[0,1]$, nor to a circle. The surface of a cube is homeomorphic to a sphere (the surface of a ball) but not homeomorphic to a torus. Euclidean spaces of different dimensions are not homeomorphic, which seems evident, but is not easy to prove. Dimension of a topological space is difficult to define; "inductive dimension" and "Lebesgue covering dimension" are used. Every subset of a topological space is itself a topological space (in contrast, only *linear* subsets of a linear space are linear spaces). Arbitrary topological spaces, investigated by general topology (called also point-set topology) are too diverse for a complete classification (up to homeomorphism). They are inhomogeneous (in general). Compact topological spaces are an important class of topological spaces ("species" of this "type"). Every continuous function is bounded on such space. The closed interval $[0,1]$ and the extended real line $[-\infty, \infty]$ are compact; the open interval

$(0,1)$ and the line $(-\infty,\infty)$ are not. Geometric topology investigates manifolds (another "species" of this "type"); these are topological spaces locally homeomorphic to Euclidean spaces. Low-dimensional manifolds are completely classified (up to homeomorphism).

The two structures discussed above (linear and topological) are both underlying structures of the "linear topological space" structure. That is, a linear topological space is both a linear (real or complex) space and a (homogeneous, in fact) topological space. However, an arbitrary combination of these two structures is generally not a linear topological space; the two structures must conform, namely, the linear operations must be continuous.

Every finite-dimensional (real or complex) linear space is a linear topological space in the sense that it carries one and only one topology that makes it a linear topological space. The two structures, "finite-dimensional (real or complex) linear space" and "finite-dimensional linear topological space", are thus equivalent, that is, mutually underlying. Accordingly, every invertible linear transformation of a finite-dimensional linear topological space is a homeomorphism. In the infinite dimension, however, different topologies conform to a given linear structure, and invertible linear transformations are generally not homeomorphisms.

Affine and Projective Spaces

It is convenient to introduce affine and projective spaces by means of linear spaces, as follows. An n-dimensional linear subspace of an $(n+1)$-dimensional linear space, being itself an n-dimensional linear space, is not homogeneous; it contains a special point, the origin. Shifting it by a vector external to it, one obtains an n-dimensional affine space. It is homogeneous. In the words of John Baez, "an affine space is a vector space that's forgotten its origin". A straight line in the affine space is, by definition, its intersection with a two-dimensional linear subspace (plane through the origin) of the $(n+1)$-dimensional linear space. Every linear space is also an affine space.

Every point of the affine space is its intersection with a one-dimensional linear subspace (line through the origin) of the $(n+1)$-dimensional linear space. However, some one-dimensional subspaces are parallel to the affine space; in some sense, they intersect it at infinity. The set of all one-dimensional linear subspaces of an $(n+1)$-dimensional linear space is, by definition, an n-dimensional projective space. Choosing an n-dimensional affine space as before one observes that the affine space is embedded as a proper subset into the projective space. However, the projective space itself is homogeneous. A straight line in the projective space, by definition, corresponds to a two-dimensional linear subspace of the $(n+1)$-dimensional linear space.

Defined this way, affine and projective spaces are of algebraic nature; they can be real, complex, and more generally, over any field.

Every real (or complex) affine or projective space is also a topological space. An affine space is a non-compact manifold; a projective space is a compact manifold.

Metric and Uniform Spaces

Distances between points are defined in a metric space. Every metric space is also a topological space. Bounded sets and Cauchy sequences are defined in a metric space (but not just in a topological space). Isomorphisms between metric spaces are called isometries. A metric space is called complete if all Cauchy sequences converge. Every incomplete space is isometrically embedded into its completion. Every compact metric space is complete; the real line is non-compact but complete; the open interval $(0,1)$ is incomplete.

A topological space is called metrizable, if it underlies a metric space. All manifolds are metrizable.

Every Euclidean space is also a complete metric space. Moreover, all geometric notions immanent to a Euclidean space can be characterized in terms of its metric. For example, the straight segment connecting two given points A and C consists of all points B such that the distance between A and C is equal to the sum of two distances, between A and B and between B and C.

Uniform spaces do not introduce distances, but still allow one to use uniform continuity, Cauchy sequences, completeness and completion. Every uniform space is also a topological space. Every *linear* topological space (metrizable or not) is also a uniform space. More generally, every commutative topological group is also a uniform space. A non-commutative topological group, however, carries two uniform structures, one left-invariant, the other right-invariant. Linear topological spaces are complete in finite dimension but generally incomplete in infinite dimension.

Normed, Banach, Inner Product, and Hilbert Spaces

Vectors in a Euclidean space are a linear space, but each vector x has also a length, in other words, norm, $\| x \|$. A (real or complex) linear space endowed with a norm is a normed space. Every normed space is both a linear topological space and a metric space. A Banach space is a complete normed space. Many spaces of sequences or functions are infinite-dimensional Banach spaces.

The set of all vectors of norm less than one is called the unit ball of a normed space. It is a convex, centrally symmetric set, generally not an ellipsoid; for example, it may be a polygon (on the plane). The parallelogram law (called also parallelogram identity) $\| x - y \|^2 + \| x + y \|^2 = 2 \| x \|^2 + 2 \| y \|^2$ generally fails in normed spaces, but holds for vectors in Euclidean spaces, which follows from the fact that the squared Euclidean norm of a vector is its inner product to itself.

An inner product space is a (real or complex) linear space endowed with a bilinear (or sesquilinear) form satisfying some conditions and called inner product. Every inner product space is also a normed space. A normed space underlies an inner product space if and only if it satisfies the parallelogram law, or equivalently, if its unit ball is an ellipsoid. Angles between vectors are defined in inner product spaces. A Hilbert space is defined as a complete inner product space. (Some authors insist that it must be complex, others admit also real Hilbert spaces.) Many spaces of sequences or functions are infinite-dimensional Hilbert spaces. Hilbert spaces are very important for quantum theory.

All n-dimensional real inner product spaces are mutually isomorphic. One may say that the n-dimensional Euclidean space is the n-dimensional real inner product space that's forgotten its origin.

Smooth and Riemannian Manifolds (Spaces)

Smooth manifolds are not called "spaces", but could be. Smooth (differentiable) functions, paths, maps, given in a smooth manifold by definition, lead to tangent spaces. Every smooth manifold is a (topological) manifold. Smooth surfaces in a finite-dimensional linear space (like the surface of an ellipsoid, not a polytope) are smooth manifolds. Every smooth manifold can be embedded into a finite-dimensional linear space. A smooth path in a smooth manifold has (at every point) the tangent vector, belonging to the tangent space (attached to this point). Tangent spaces to an n-dimensional smooth manifold are n-dimensional linear spaces. A smooth function has (at every point) the differential, – a linear functional on the tangent space. Real (or complex) finite-dimensional linear, affine and projective spaces are also smooth manifolds.

A Riemannian manifold, or Riemann space, is a smooth manifold whose tangent spaces are endowed with inner product (satisfying some conditions). Euclidean spaces are also Riemann spaces. Smooth surfaces in Euclidean spaces are Riemann spaces. A hyperbolic non-Euclidean space is also a Riemann space. A curve in a Riemann space has the length. A Riemann space is both a smooth manifold and a metric space; the length of the shortest curve is the distance. The angle between two curves intersecting at a point is the angle between their tangent lines.

Waiving positivity of inner product on tangent spaces one gets pseudo-Riemann (especially, Lorentzian) spaces very important for general relativity.

Measurable, Measure, and Probability Spaces

Waiving distances and angles while retaining volumes (of geometric bodies) one moves toward measure theory. Besides the volume, a measure generalizes area, length, mass (or charge) distribution, and also probability distribution, according to Andrey Kolmogorov's approach to probability theory.

A "geometric body" of classical mathematics is much more regular than just a set of points. The boundary of the body is of zero volume. Thus, the volume of the body is the volume of its interior, and the interior can be exhausted by an infinite sequence of cubes. In contrast, the boundary of an arbitrary set of points can be of non-zero volume (an example: the set of all rational points inside a given cube). Measure theory succeeded in extending the notion of volume (or another measure) to a vast class of sets, so-called measurable sets. Indeed, non-measurable sets almost never occur in applications, but anyway, the theory must restrict itself to measurable sets (and functions).

Measurable sets, given in a measurable space by definition, lead to measurable functions and maps. In order to turn a topological space into a measurable space one endows it with a σ-algebra. The σ-algebra of Borel sets is most popular, but not the only choice (Baire sets, universally measurable sets etc. are used sometimes). Alternatively, a σ-algebra can be generated by a given collection of sets (or functions) irrespective of any topology. Quite often, different topologies lead to the same σ-algebra (for example, the norm topology and the weak topology on a separable Hilbert space). Every subset of a measurable space is itself a measurable space.

Standard measurable spaces (called also standard Borel spaces) are especially useful. Every Borel set (in particular, every closed set and every open set) in a Euclidean space (and more generally, in a complete separable metric space) is a standard measurable space. All uncountable standard measurable spaces are mutually isomorphic.

A measure space is a measurable space endowed with a measure. A Euclidean space with Lebesgue measure is a measure space. Integration theory defines integrability and integrals of measurable functions on a measure space.

Sets of measure 0, called null sets, are negligible. Accordingly, a mod 0 isomorphism is defined as isomorphism between subsets of full measure (that is, with negligible complement).

A probability space is a measure space such that the measure of the whole space is equal to 1. The product of any family (finite or not) of probability spaces is a probability space. In contrast, for measure spaces in general, only the product of finitely many spaces is defined. Accordingly, there are many infinite-dimensional probability measures (especially, Gaussian measures), but no infinite-dimensional Lebesgue measure.

Standard probability spaces are especially useful. Every probability measure on a standard measurable space leads to a standard probability space. The product of a sequence (finite or not) of standard probability spaces is a standard probability space. All non-atomic standard probability spaces are mutually isomorphic mod 0; one of them is the interval $(0,1)$ with Lebesgue measure.

These spaces are less geometric. In particular, the idea of dimension, applicable (in one form or another) to all other spaces, does not apply to measurable, measure and probability spaces.

A topological space becomes also a measurable space when endowed with the Borel σ-algebra. However, the topology is not uniquely determined by its Borel σ-algebra; and not every σ-algebra is the Borel σ-algebra of some topology.

Calculus

Calculus is the mathematical study of change, in the same way that geometry is the study of shape and algebra is the study of operations and their application to solving equations. It has two major branches, differential calculus (concerning rates of change and slopes of curves), and integral calculus (concerning accumulation of quantities and the areas under and between curves); these two branches are related to each other by the fundamental theorem of calculus. Both branches make use of the fundamental notions of convergence of infinite sequences and infinite series to a well-defined limit. Generally, modern calculus is considered to have been developed in the 17th century by Isaac Newton and Gottfried Leibniz. Today, calculus has widespread uses in science, engineering and economics and can solve many problems that elementary algebra alone cannot.

Calculus is a part of modern mathematics education. A course in calculus is a gateway to other, more advanced courses in mathematics devoted to the study of functions and limits, broadly called mathematical analysis. Calculus has historically been called "the calculus of infinitesimals", or "infinitesimal calculus". *Calculus* (plural *calculi*) is also used for naming some methods of calculation or theories of computation, such as propositional calculus, calculus of variations, lambda calculus, and process calculus.

History

Modern calculus was developed in 17th-century Europe by Isaac Newton and Gottfried Wilhelm Leibniz, but elements of it have appeared in ancient India, Greece, China, medieval Europe, and the Middle East.

Ancient

The ancient period introduced some of the ideas that led to integral calculus, but does not seem to have developed these ideas in a rigorous and systematic way. Calculations of volume and area, one goal of integral calculus, can be found in the EgyptianMoscow papyrus (c. 1820 BC), but the formulas are simple instructions, with no indication as to method, and some of them lack major components. From the age of Greek mathematics, Eudoxus (c. 408–355 BC) used the method of exhaustion, which foreshadows the concept of the limit, to calculate areas and volumes, while Archimedes (c. 287–212 BC) developed this idea further, inventing heuristics which resemble the methods of integral calculus. The method of exhaustion was later reinvented in China by Liu Hui in the 3rd century AD in order to find the area of a circle. In the 5th century AD, Zu Chongzhi established a method that would later be called Cavalieri's principle to find the volume of a sphere.

Medieval

Indian mathematicians gave a semi-rigorous method of differentiation of some trigonometric functions. In the Middle East, Alhazen derived a formula for the sum of fourth powers. He used the results to carry out what would now be called an integration, where the formulae for the sums of integral squares and fourth powers allowed him to calculate the volume of a paraboloid. In the 14th century, Indian mathematician Madhava of Sangamagrama and the Kerala school of astronomy and mathematics stated components of calculus such as the Taylor series and infinite series approximations. However, they were not able to "combine many differing ideas under the two unifying themes of the derivative and the integral, show the connection between the two, and turn calculus into the great problem-solving tool we have today".

Modern

> "The calculus was the first achievement of modern mathematics and it is difficult to overestimate its importance. I think it defines more unequivocally than anything else the inception of modern mathematics, and the system of mathematical analysis, which is its logical development, still constitutes the greatest technical advance in exact thinking." — John von Neumann

In Europe, the foundational work was a treatise due to Bonaventura Cavalieri, who argued that volumes and areas should be computed as the sums of the volumes and areas of infinitesimally thin cross-sections. The ideas were similar to Archimedes' in The Method, but this treatise is believed to have been lost in the 13th century, and was only rediscovered in the early 20th century, and so would have been unknown to Cavalieri. Cavalieri's work was not well respected since his methods could lead to erroneous results, and the infinitesimal quantities he introduced were disreputable at first.

The formal study of calculus brought together Cavalieri's infinitesimals with the calculus of finite differences developed in Europe at around the same time. Pierre de Fermat, claiming that he borrowed from Diophantus, introduced the concept of adequality, which represented equality up to an infinitesimal error term. The combination was achieved by John Wallis, Isaac Barrow, and James Gregory, the latter two proving the second fundamental theorem of calculus around 1670.

The product rule and chain rule, the notion of higher derivatives, Taylor series, and analytical functions were introduced by Isaac Newton in an idiosyncratic notation which he used to solve problems of mathematical physics. In his works, Newton rephrased his ideas to suit the mathematical idiom of the time, replacing calculations with infinitesimals by equivalent geometrical arguments which were considered beyond reproach. He used the methods of calculus to solve the problem of planetary motion, the shape of the surface of a rotating fluid, the oblateness of the earth, the motion of a weight sliding on a cycloid, and many other problems discussed in his *Principia Mathematica* (1687). In other work, he developed series expansions for functions, including fractional and irrational powers, and it was clear that he understood the principles of the Taylor series. He did not publish all these discoveries, and at this time infinitesimal methods were still considered disreputable.

These ideas were arranged into a true calculus of infinitesimals by Gottfried Wilhelm Leibniz, who was originally accused of plagiarism by Newton. He is now regarded as an independent inventor of and contributor to calculus. His contribution was to provide a clear set of rules for working with infinitesimal quantities, allowing the computation of second and higher derivatives, and providing the product rule and chain rule, in their differential and integral forms. Unlike Newton, Leibniz paid a lot of attention to the formalism, often spending days determining appropriate symbols for concepts.

Leibniz and Newton are usually both credited with the invention of calculus. Newton was the first to apply calculus to general physics and Leibniz developed much of the notation used in calculus today. The basic insights that both Newton and Leibniz provided were the laws of differentiation and integration, second and higher derivatives, and the notion of an approximating polynomial series. By Newton's time, the fundamental theorem of calculus was known.

When Newton and Leibniz first published their results, there was great controversy over which mathematician (and therefore which country) deserved credit. Newton derived his results first (later to be published in his *Method of Fluxions*), but Leibniz published his *Nova Methodus pro Maximis et Minimis* first. Newton claimed Leibniz stole ideas from his unpublished notes, which Newton had shared with a few members of the Royal Society. This controversy divided English-speaking mathematicians from continental European mathematicians for many years, to the detriment of English mathematics. A careful examination of the papers of Leibniz and Newton shows that they arrived at their results independently, with Leibniz starting first with integration and Newton with differentiation. Today, both Newton and Leibniz are given credit for developing calculus independently. It is Leibniz, however, who gave the new discipline its name. Newton called his calculus "the science of fluxions".

Since the time of Leibniz and Newton, many mathematicians have contributed to the continuing development of calculus. One of the first and most complete works on both infinitesimal and integral calculus was written in 1748 by Maria Gaetana Agnesi.

Maria Gaetana Agnesi

Foundations

In calculus, *foundations* refers to the rigorous development of the subject from axioms and definitions. In early calculus the use of infinitesimal quantities was thought unrigorous, and was fiercely criticized by a number of authors, most notably Michel Rolle and Bishop Berkeley. Berkeley famously described infinitesimals as the ghosts of departed quantities in his book *The Analyst* in 1734. Working out a rigorous foundation for calculus occupied mathematicians for much of the century following Newton and Leibniz, and is still to some extent an active area of research today.

Several mathematicians, including Maclaurin, tried to prove the soundness of using infinitesimals, but it would not be until 150 years later when, due to the work of Cauchy and Weierstrass, a way was finally found to avoid mere "notions" of infinitely small quantities. The foundations of differential and integral calculus had been laid. In Cauchy's Cours d'Analyse, we find a broad range of foundational approaches, including a definition of continuity in terms of infinitesimals, and a (somewhat imprecise) prototype of an (ε, δ)-definition of limit in the definition of differentiation. In his work Weierstrass formalized the concept of limit and eliminated infinitesimals. Following the work of Weierstrass, it eventually became common to base calculus on limits instead of infinitesimal quantities, though the subject is still occasionally called "infinitesimal calculus". Bernhard Riemann used these ideas to give a precise definition of the integral. It was also during this period that the ideas of calculus were generalized to Euclidean space and the complex plane.

In modern mathematics, the foundations of calculus are included in the field of real analysis, which contains full definitions and proofs of the theorems of calculus. The reach of calculus has also been greatly extended. Henri Lebesgue invented measure theory and used it to define integrals of all but the most pathological functions. Laurent Schwartz introduced distributions, which can be used to take the derivative of any function whatsoever.

Limits are not the only rigorous approach to the foundation of calculus. Another way is to use Abraham Robinson's non-standard analysis. Robinson's approach, developed in the 1960s, uses technical machinery from mathematical logic to augment the real number system with infinitesimal and infinite numbers, as in the original Newton-Leibniz conception. The resulting numbers are called hyperreal numbers, and they can be used to give a Leibniz-like development of the usual rules of calculus.

Significance

While many of the ideas of calculus had been developed earlier in Greece, China, India, Iraq, Persia, and Japan, the use of calculus began in Europe, during the 17th century, when Isaac Newton and Gottfried Wilhelm Leibniz built on the work of earlier mathematicians to introduce its basic principles. The development of calculus was built on earlier concepts of instantaneous motion and area underneath curves.

Applications of differential calculus include computations involving velocity and acceleration, the slope of a curve, and optimization. Applications of integral calculus include computations involving area, volume, arc length, center of mass, work, and pressure. More advanced applications include power series and Fourier series.

Calculus is also used to gain a more precise understanding of the nature of space, time, and motion. For centuries, mathematicians and philosophers wrestled with paradoxes involving division by zero or sums of infinitely many numbers. These questions arise in the study of motion and area. The ancient Greek philosopher Zeno of Elea gave several famous examples of such paradoxes. Calculus provides tools, especially the limit and the infinite series, which resolve the paradoxes.

Principles

Limits and Infinitesimals

Calculus is usually developed by working with very small quantities. Historically, the first method of doing so was by infinitesimals. These are objects which can be treated like real numbers but which are, in some sense, "infinitely small". For example, an infinitesimal number could be greater than 0, but less than any number in the sequence 1, 1/2, 1/3, ... and thus less than any positive real number. From this point of view, calculus is a collection of techniques for manipulating infinitesimals. The symbols dx and dy were taken to be infinitesimal, and the derivative dy/dx was simply their ratio.

The infinitesimal approach fell out of favor in the 19th century because it was difficult to make the notion of an infinitesimal precise. However, the concept was revived in the 20th century with the introduction of non-standard analysis and smooth infinitesimal analysis, which provided solid foundations for the manipulation of infinitesimals.

In the 19th century, infinitesimals were replaced by the epsilon, delta approach to limits. Limits describe the value of a function at a certain input in terms of its values at a nearby input. They capture small-scale behavior in the context of the real number system. In this treatment, calculus is a collection of techniques for manipulating certain limits. Infinitesimals get replaced by very small numbers, and the infinitely small behavior of the function is found by taking the limiting behavior for smaller and smaller numbers. Limits were the first way to provide rigorous foundations for calculus, and for this reason they are the standard approach.

Differential Calculus

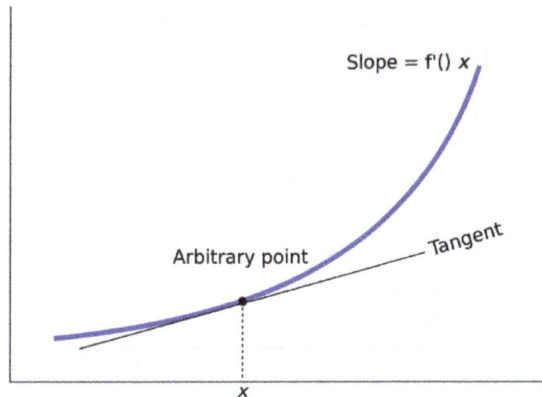

Tangent line at $(x, f(x))$. The derivative $f'(x)$ of a curve at a point is the slope (rise over run) of the line tangent to that curve at that point.

Differential calculus is the study of the definition, properties, and applications of the derivative of a function. The process of finding the derivative is called *differentiation*. Given a function and a point in the domain, the derivative at that point is a way of encoding the small-scale behavior of the function near that point. By finding the derivative of a function at every point in its domain, it is possible to produce a new function, called the *derivative function* or just the *derivative* of the original function. In mathematical jargon, the derivative is a linear operator which inputs a function and outputs a second function. This is more abstract than many of the processes studied in elementary algebra, where functions usually input a number and output another number. For example, if the doubling function is given the input three, then it outputs six, and if the squaring function is given the input three, then it outputs nine. The derivative, however, can take the squaring function as an input. This means that the derivative takes all the information of the squaring function—such as that two is sent to four, three is sent to nine, four is sent to sixteen, and so on—and uses this information to produce another function. (The function it produces turns out to be the doubling function.)

The most common symbol for a derivative is an apostrophe-like mark called prime. Thus, the derivative of the function of f is f', pronounced "f prime." For instance, if $f(x) = x^2$ is the squaring function, then $f'(x) = 2x$ is its derivative, the doubling function.

If the input of the function represents time, then the derivative represents change with respect to time. For example, if f is a function that takes a time as input and gives the position of a ball at that time as output, then the derivative of f is how the position is changing in time, that is, it is the velocity of the ball.

If a function is linear (that is, if the graph of the function is a straight line), then the function can be written as $y = mx + b$, where x is the independent variable, y is the dependent variable, b is the y-intercept, and:

$$m = \frac{\text{rise}}{\text{run}} = \frac{\text{change in } y}{\text{change in } x} = \frac{\Delta y}{\Delta x}.$$

This gives an exact value for the slope of a straight line. If the graph of the function is not a straight line, however, then the change in y divided by the change in x varies. Derivatives give an exact meaning to the notion of change in output with respect to change in input. To be concrete, let f be a function, and fix a point a in the domain of f. $(a, f(a))$ is a point on the graph of the function. If h is a number close to zero, then $a + h$ is a number close to a. Therefore, $(a + h, f(a + h))$ is close to $(a, f(a))$. The slope between these two points is

$$m = \frac{f(a+h)-f(a)}{(a+h)-a} = \frac{f(a+h)-f(a)}{h}.$$

This expression is called a *difference quotient*. A line through two points on a curve is called a *secant line*, so m is the slope of the secant line between $(a, f(a))$ and $(a + h, f(a + h))$. The secant line is only an approximation to the behavior of the function at the point a because it does not account for what happens between a and $a + h$. It is not possible to discover the behavior at a by setting h to zero because this would require dividing by zero, which is undefined. The derivative is defined by taking the limit as h tends to zero, meaning that it considers the behavior of f for all small values of h and extracts a consistent value for the case when h equals zero:

$$\lim_{h \to 0} \frac{f(a+h)-f(a)}{h}.$$

Geometrically, the derivative is the slope of the tangent line to the graph of f at a. The tangent line is a limit of secant lines just as the derivative is a limit of difference quotients. For this reason, the derivative is sometimes called the slope of the function f.

Here is a particular example, the derivative of the squaring function at the input 3. Let $f(x) = x^2$ be the squaring function.

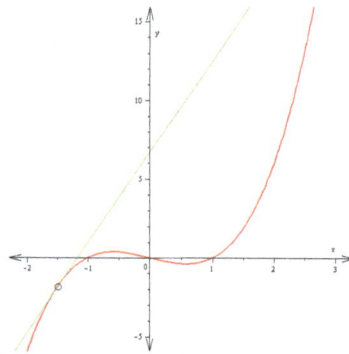

The derivative $f'(x)$ of a curve at a point is the slope of the line tangent to that curve at that point. This slope is determined by considering the limiting value of the slopes of secant lines. Here the function involved (drawn in red) is $f(x) = x^3 - x$. The tangent line (in green) which passes through the point $(-3/2, -15/8)$ has a slope of 23/4. Note that the vertical and horizontal scales in this image are different.

$$
\begin{aligned}
f'(3) &= \lim_{h \to 0} \frac{(3+h)^2 - 3^2}{h} \\
&= \lim_{h \to 0} \frac{9 + 6h + h^2 - 9}{h} \\
&= \lim_{h \to 0} \frac{6h + h^2}{h} \\
&= \lim_{h \to 0} (6 + h) \\
&= 6.
\end{aligned}
$$

The slope of the tangent line to the squaring function at the point (3, 9) is 6, that is to say, it is going up six times as fast as it is going to the right. The limit process just described can be performed for any point in the domain of the squaring function. This defines the *derivative function* of the squaring function, or just the *derivative* of the squaring function for short. A similar computation to the one above shows that the derivative of the squaring function is the doubling function.

Leibniz Notation

A common notation, introduced by Leibniz, for the derivative in the example above is

$$y = x^2$$
$$\frac{dy}{dx} = 2x.$$

In an approach based on limits, the symbol dy/dx is to be interpreted not as the quotient of two numbers but as a shorthand for the limit computed above. Leibniz, however, did intend it to represent the quotient of two infinitesimally small numbers, dy being the infinitesimally small change in y caused by an infinitesimally small change dx applied to x. We can also think of d/dx as a differentiation operator, which takes a function as an input and gives another function, the derivative, as the output. For example:

$$\frac{d}{dx}(x^2) = 2x.$$

In this usage, the dx in the denominator is read as "with respect to x". Even when calculus is developed using limits rather than infinitesimals, it is common to manipulate symbols like dx and dy as if they were real numbers; although it is possible to avoid such manipulations, they are sometimes notationally convenient in expressing operations such as the total derivative.

Integral Calculus

Integral calculus is the study of the definitions, properties, and applications of two related concepts, the *indefinite integral* and the *definite integral*. The process of finding the value of an integral is called *integration*. In technical language, integral calculus studies two related linear operators.

The *indefinite integral* is the *antiderivative*, the inverse operation to the derivative. *F* is an indefinite integral of *f* when *f* is a derivative of *F*. (This use of lower- and upper-case letters for a function and its indefinite integral is common in calculus.)

The *definite integral* inputs a function and outputs a number, which gives the algebraic sum of areas between the graph of the input and the x-axis. The technical definition of the definite integral involves the limit of a sum of areas of rectangles, called a Riemann sum.

A motivating example is the distances traveled in a given time.

Distance = Speed · Time

If the speed is constant, only multiplication is needed, but if the speed changes, a more powerful method of finding the distance is necessary. One such method is to approximate the distance traveled by breaking up the time into many short intervals of time, then multiplying the time elapsed in each interval by one of the speeds in that interval, and then taking the sum (a Riemann sum) of the approximate distance traveled in each interval. The basic idea is that if only a short time elapses, then the speed will stay more or less the same. However, a Riemann sum only gives an approximation of the distance traveled. We must take the limit of all such Riemann sums to find the exact distance traveled.

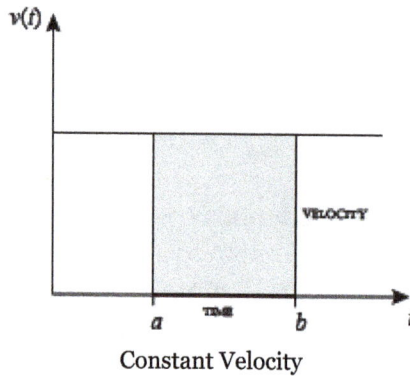

Constant Velocity

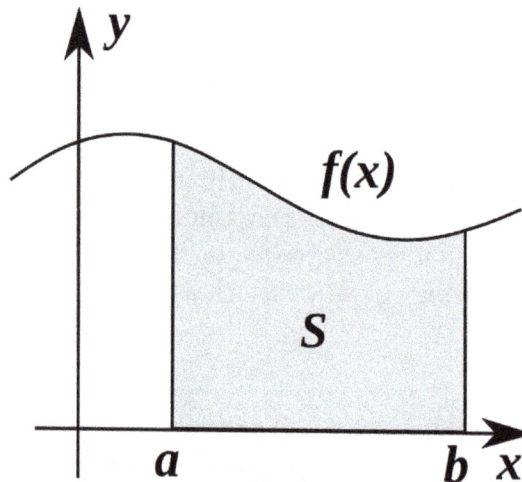

Integration can be thought of as measuring the area under a curve, defined by $f(x)$, between two points (here a and b).

When velocity is constant, the total distance traveled over the given time interval can be computed by multiplying velocity and time. For example, travelling a steady 50 mph for 3 hours results in a total distance of 150 miles. In the diagram on the left, when constant velocity and time are graphed, these two values form a rectangle with height equal to the velocity and width equal to the time elapsed. Therefore, the product of velocity and time also calculates the rectangular area under the (constant) velocity curve. This connection between the area under a curve and distance traveled can be extended to *any* irregularly shaped region exhibiting a fluctuating velocity over a given time period. If $f(x)$ in the diagram on the right represents speed as it varies over time, the distance traveled (between the times represented by a and b) is the area of the shaded region s.

To approximate that area, an intuitive method would be to divide up the distance between a and

b into a number of equal segments, the length of each segment represented by the symbol Δx. For each small segment, we can choose one value of the function $f(x)$. Call that value h. Then the area of the rectangle with base Δx and height h gives the distance (time Δx multiplied by speed h) traveled in that segment. Associated with each segment is the average value of the function above it, $f(x) = h$. The sum of all such rectangles gives an approximation of the area between the axis and the curve, which is an approximation of the total distance traveled. A smaller value for Δx will give more rectangles and in most cases a better approximation, but for an exact answer we need to take a limit as Δx approaches zero.

The symbol of integration is $\int$, an elongated S (the S stands for "sum"). The definite integral is written as:

$$\int_a^b f(x)dx.$$

and is read "the integral from a to b of f-of-x with respect to x." The Leibniz notation dx is intended to suggest dividing the area under the curve into an infinite number of rectangles, so that their width Δx becomes the infinitesimally small dx. In a formulation of the calculus based on limits, the notation

$$\int_a^b \cdots dx$$

is to be understood as an operator that takes a function as an input and gives a number, the area, as an output. The terminating differential, dx, is not a number, and is not being multiplied by $f(x)$, although, serving as a reminder of the Δx limit definition, it can be treated as such in symbolic manipulations of the integral. Formally, the differential indicates the variable over which the function is integrated and serves as a closing bracket for the integration operator.

The indefinite integral, or antiderivative, is written:

$$\int f(x)dx.$$

Functions differing by only a constant have the same derivative, and it can be shown that the antiderivative of a given function is actually a family of functions differing only by a constant. Since the derivative of the function $y = x^2 + C$, where C is any constant, is $y' = 2x$, the antiderivative of the latter given by:

$$\int 2x\,dx = x^2 + C.$$

The unspecified constant C present in the indefinite integral or antiderivative is known as the constant of integration.

Fundamental Theorem

The fundamental theorem of calculus states that differentiation and integration are inverse operations. More precisely, it relates the values of antiderivatives to definite integrals. Because it is usually easier to compute an antiderivative than to apply the definition of a definite integral, the fundamental theorem of calculus provides a practical way of computing definite integrals. It can also be interpreted as a precise statement of the fact that differentiation is the inverse of integration.

The fundamental theorem of calculus states: If a function f is continuous on the interval $[a, b]$ and if F is a function whose derivative is f on the interval (a, b), then

$$\int_a^b f(x)dx = F(b) - F(a).$$

Furthermore, for every x in the interval (a, b),

$$\frac{d}{dx}\int_a^x f(t)dt = f(x).$$

This realization, made by both Newton and Leibniz, who based their results on earlier work by Isaac Barrow, was key to the proliferation of analytic results after their work became known. The fundamental theorem provides an algebraic method of computing many definite integrals—without performing limit processes—by finding formulas for antiderivatives. It is also a prototype solution of a differential equation. Differential equations relate an unknown function to its derivatives, and are ubiquitous in the sciences.

Applications

The logarithmic spiral of the Nautilus shell is a classical image used to depict the growth and change related to calculus

Calculus is used in every branch of the physical sciences, actuarial science, computer science, statistics, engineering, economics, business, medicine, demography, and in other fields wherever a problem can be mathematically modeled and an optimal solution is desired. It allows one to go from (non-constant) rates of change to the total change or vice versa, and many times in studying a problem we know one and are trying to find the other.

Physics makes particular use of calculus; all concepts in classical mechanics and electromagnetism are related through calculus. The mass of an object of known density, the moment of inertia of objects, as well as the total energy of an object within a conservative field can be found by the use of calculus. An example of the use of calculus in mechanics is Newton's second law of motion: historically stated it expressly uses the term "rate of change" which refers to the derivative saying *The rate of change of momentum of a body is equal to the resultant force acting on the body and is in the same direction.* Commonly expressed today as Force = Mass × acceleration, it involves differential calculus because acceleration is the time derivative of velocity or second time derivative of trajectory or spatial position. Starting from knowing how an object is accelerating, we use calculus to derive its path.

Maxwell's theory of electromagnetism and Einstein's theory of general relativity are also expressed in the language of differential calculus. Chemistry also uses calculus in determining reaction rates and radioactive decay. In biology, population dynamics starts with reproduction and death rates to model population changes.

Calculus can be used in conjunction with other mathematical disciplines. For example, it can be used with linear algebra to find the "best fit" linear approximation for a set of points in a domain. Or it can be used in probability theory to determine the probability of a continuous random variable from an assumed density function. In analytic geometry, the study of graphs of functions, calculus is used to find high points and low points (maxima and minima), slope, concavity and inflection points.

Green's Theorem, which gives the relationship between a line integral around a simple closed curve C and a double integral over the plane region D bounded by C, is applied in an instrument known as a planimeter, which is used to calculate the area of a flat surface on a drawing. For example, it can be used to calculate the amount of area taken up by an irregularly shaped flower bed or swimming pool when designing the layout of a piece of property.

Discrete Green's Theorem, which gives the relationship between a double integral of a function around a simple closed rectangular curve C and a linear combination of the antiderivative's values at corner points along the edge of the curve, allows fast calculation of sums of values in rectangular domains. For example, it can be used to efficiently calculate sums of rectangular domains in images, in order to rapidly extract features and detect object; another algorithm that could be used is the summed area table.

In the realm of medicine, calculus can be used to find the optimal branching angle of a blood vessel so as to maximize flow. From the decay laws for a particular drug's elimination from the body, it is used to derive dosing laws. In nuclear medicine, it is used to build models of radiation transport in targeted tumor therapies.

In economics, calculus allows for the determination of maximal profit by providing a way to easily calculate both marginal cost and marginal revenue.

Calculus is also used to find approximate solutions to equations; in practice it is the standard way to solve differential equations and do root finding in most applications. Examples are methods such as Newton's method, fixed point iteration, and linear approximation. For instance, spacecraft use a variation of the Euler method to approximate curved courses within zero gravity environments.

Varieties

Over the years, many reformulations of calculus have been investigated for different purposes.

Non-standard Calculus

Imprecise calculations with infinitesimals were widely replaced with the rigorous (ε, δ)-definition of limit starting in the 1870s. Meanwhile, calculations with infinitesimals persisted and often led to correct results. This led Abraham Robinson to investigate if it were possible to develop a number system with infinitesimal quantities over which the theorems of calculus were still valid. In 1960, building upon the work of Edwin Hewitt and Jerzy Łoś, he succeeded in developing non-standard analysis. The theory of non-standard analysis is rich enough to be applied in many branches of mathematics. As such, books and articles dedicated solely to the traditional theorems of calculus often go by the title non-standard calculus.

Smooth Infinitesimal Analysis

This is another reformulation of the calculus in terms of infinitesimals. Based on the ideas of F. W. Lawvere and employing the methods of category theory, it views all functions as being continuous and incapable of being expressed in terms of discrete entities. One aspect of this formulation is that the law of excluded middle does not hold in this formulation.

Constructive Analysis

Constructive mathematics is a branch of mathematics that insists that proofs of the existence of a number, function, or other mathematical object should give a construction of the object. As such constructive mathematics also rejects the law of excluded middle. Reformulations of calculus in a constructive framework are generally part of the subject of constructive analysis.

References

- Steven Galovich, Introduction to Mathematical Structures, Harcourt Brace Javanovich, 23 January 1989, ISBN 0-15-543468-3.

- Gowers, Timothy; Barrow-Green, June; Leader, Imre, eds. (2008), The Princeton Companion to Mathematics, Princeton University Press, ISBN 978-0-691-11880-2 .

- Dun, Liu; Fan, Dainian; Cohen, Robert Sonné (1966). "A comparison of Archimdes' and Liu Hui's studies of circles". Chinese studies in the history and philosophy of science and technology. 130. Springer: 279. ISBN 0-7923-3463-9. ,Chapter , p. 279

- Zill, Dennis G.; Wright, Scott; Wright, Warren S. (2009). Calculus: Early Transcendentals (3 ed.). Jones & Bartlett Learning. p. xxvii. ISBN 0-7637-5995-3. Extract of page 27

- von Neumann, J., "The Mathematician", in Heywood, R. B., ed., The Works of the Mind, University of Chicago Press, 1947, pp. 180–196. Reprinted in Bródy, F., Vámos, T., eds., The Neumann Compedium, World Scientific Publishing Co. Pte. Ltd., 1995, ISBN 981-02-2201-7.

- André Weil: Number theory. An approach through history. From Hammurapi to Legendre. Birkhauser Boston, Inc., Boston, MA, 1984, ISBN 0-8176-4565-9, p. 28.

- Allaire, Patricia R. (2007). Foreword. A Biography of Maria Gaetana Agnesi, an Eighteenth-century Woman Mathematician. By Cupillari, Antonella (illustrated ed.). Edwin Mellen Press. p. iii. ISBN 978-0-7734-5226-8.

- Corry, Leo (September 1992). "Nicolas Bourbaki and the concept of mathematical structure". Synthese. 92 (3): 315–348. doi:10.1007/bf00414286. Retrieved 7 April 2016.

- Wells, Richard B. (2010). Biological signal processing and computational neuroscience (PDF). pp. 296–335. Retrieved 7 April 2016.

3

Branches of Mathematics

The various branches of mathematics offer interpretations to questions that were of great concern to scholars. Algebra, geometry, trigonometry, all of which have been listed in the chapter, describes philosophical paradigms and their applications. This chapter is a compilation of the various branches of mathematics that form an integral part of the broader subject matter.

Applied Mathematics

Applied mathematics is a branch of mathematics that deals with mathematical methods that find use in science, engineering, business, computer science, and industry. Thus, applied mathematics is a combination of mathematical science and specialized knowledge. The term "applied mathematics" also describes the professional specialty in which mathematicians work on practical problems by formulating and studying mathematical models. In the past, practical applications have motivated the development of mathematical theories, which then became the subject of study in pure mathematics where abstract concepts are studied for their own sake. The activity of applied mathematics is thus intimately connected with research in pure mathematics.

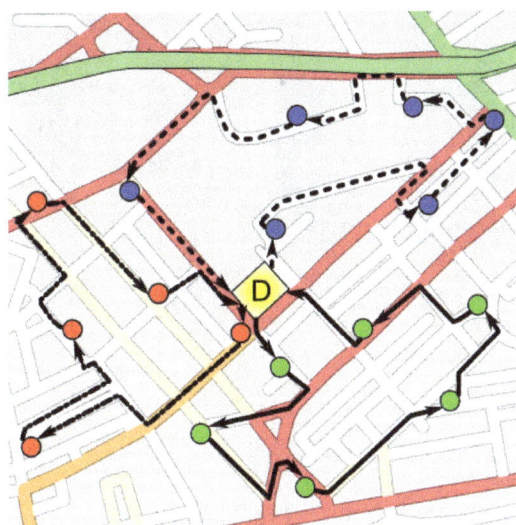

Efficient solutions to the vehicle routing problem require tools from combinatorial optimization and integer programming.

History

Historically, applied mathematics consisted principally of applied analysis, most notably differential equations; approximation theory (broadly construed, to include representations, asymptotic methods, variational methods, and numerical analysis); and applied probability. These areas of mathematics related directly to the development of Newtonian physics, and in fact, the distinction between mathematicians and physicists was not sharply drawn before the mid-19th century. This history left a pedagogical legacy in the United States: until the early 20th century, subjects such as classical mechanics were often taught in applied mathematics departments at American universities rather than in physics departments, and fluid mechanics may still be taught in applied mathematics departments.Quantitative finance is now taught in mathematics departments across universities and mathematical finance is considered a full branch of applied mathematics.Engineering and computer science departments have traditionally made use of applied mathematics.

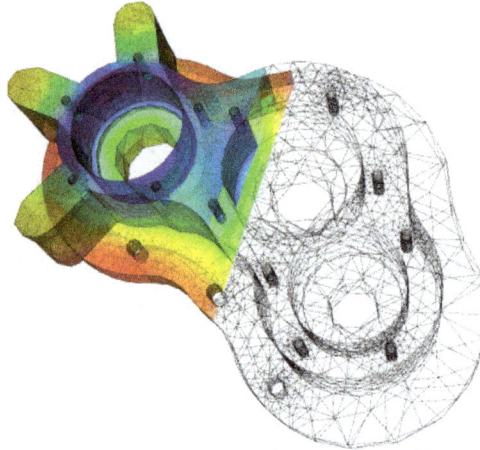

A numerical solution to the heat equation on a pump casing model using the finite element method.

Divisions

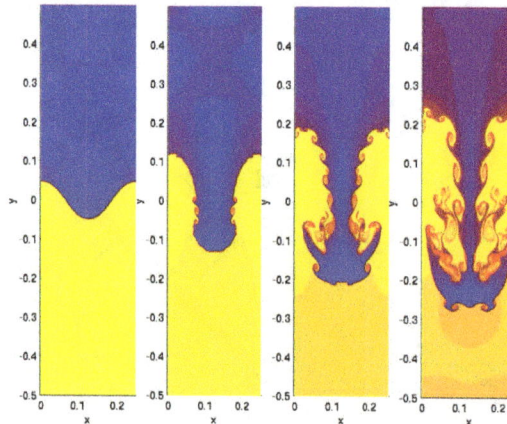

Fluid mechanics is often considered a branch of applied mathematics and mechanical engineering.

Today, the term "applied mathematics" is used in a broader sense. It includes the classical areas noted above as well as other areas that have become increasingly important in applications. Even fields such as number theory that are part of pure mathematics are now important in applications (such as cryptography), though they are not generally considered to be part of the field of applied mathematics *per se*. Sometimes, the term "applicable mathematics" is used to distinguish between the traditional applied mathematics that developed alongside physics and the many areas of mathematics that are applicable to real-world problems today.

There is no consensus as to what the various branches of applied mathematics are. Such categorizations are made difficult by the way mathematics and science change over time, and also by the way universities organize departments, courses, and degrees.

Many mathematicians distinguish between "applied mathematics," which is concerned with mathematical methods, and the "applications of mathematics" within science and engineering. A biologist using a population model and applying known mathematics would not be *doing* applied mathematics, but rather *using* it; however, mathematical biologists have posed problems that have stimulated the growth of pure mathematics. Mathematicians such as Poincaré and Arnold deny the existence of "applied mathematics" and claim that there are only "applications of mathematics." Similarly, non-mathematicians blend applied mathematics and applications of mathematics. The use and development of mathematics to solve industrial problems is also called "industrial mathematics".

The success of modern numerical mathematical methods and software has led to the emergence of computational mathematics, computational science, and computational engineering, which use high-performance computing for the simulation of phenomena and the solution of problems in the sciences and engineering. These are often considered interdisciplinary.

Utility

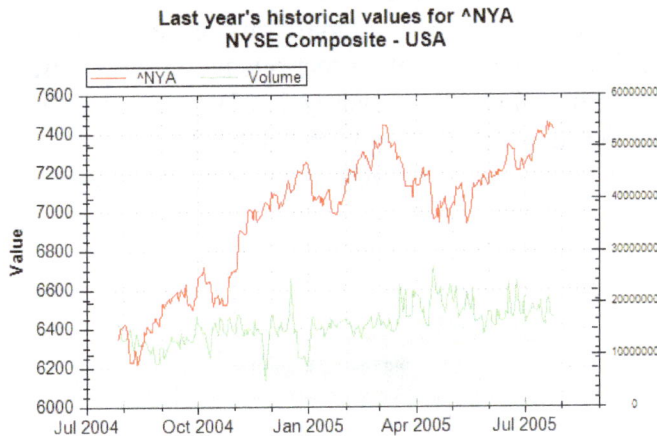

Mathematical finance is concerned with the modelling of financial markets.

Historically, mathematics was most important in the natural sciences and engineering. However, since World War II, fields outside of the physical sciences have spawned the creation of new areas of mathematics, such as game theory and social choice theory, which grew out of economic considerations.

The advent of the computer has enabled new applications: studying and using the new computer technology itself (computer science) to study problems arising in other areas of science (computational science) as well as the mathematics of computation (for example, theoretical computer science, computer algebra, numerical analysis). Statistics is probably the most widespread mathematical science used in the social sciences, but other areas of mathematics, most notably economics, are proving increasingly useful in these disciplines.

Status in Academic Departments

Academic institutions are not consistent in the way they group and label courses, programs, and degrees in applied mathematics. At some schools, there is a single mathematics department, whereas others have separate departments for Applied Mathematics and (Pure) Mathematics. It is very common for Statistics departments to be separated at schools with graduate programs, but many undergraduate-only institutions include statistics under the mathematics department.

Many applied mathematics programs (as opposed to departments) consist of primarily cross-listed courses and jointly appointed faculty in departments representing applications. Some Ph.D. programs in applied mathematics require little or no coursework outside of mathematics, while others require substantial coursework in a specific area of application. In some respects this difference reflects the distinction between "application of mathematics" and "applied mathematics".

Some universities in the UK host departments of *Applied Mathematics and Theoretical Physics*, but it is now much less common to have separate departments of pure and applied mathematics. A notable exception to this is the Department of Applied Mathematics and Theoretical Physics at the University of Cambridge, housing the Lucasian Professor of Mathematics whose past holders include Isaac Newton, Charles Babbage, James Lighthill, Paul Dirac and Stephen Hawking.

Schools with separate applied mathematics departments range from Brown University, which has a large Division of Applied Mathematics that offers degrees through the doctorate, to Santa Clara University, which offers only the M.S. in applied mathematics. Research universities dividing their mathematics department into pure and applied sections include MIT. Brigham Young University also has an Applied and Computational Emphasis (ACME), a program that allows student to graduate with a Mathematics degree, with an emphasis in Applied Maths. Students in this program also learn another skill (Computer Science, Engineering, Physics, Pure Math, etc.) to supplement their applied maths skills.

Associated Mathematical Sciences

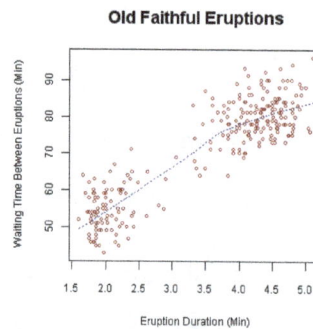

Applied mathematics has substantial overlap with statistics.

Applied mathematics is closely related to other mathematical sciences.

Scientific Computing

Scientific computing includes applied mathematics (especially numerical analysis), computing science (especially high-performance computing), and mathematical modelling in a scientific discipline.

Computer Science

Computer science relies on logic, algebra, and combinatorics.

Operations Research and Management Science

Operations research and management science are often taught in faculties of engineering, business, and public policy.

Statistics

Applied mathematics has substantial overlap with the discipline of statistics. Statistical theorists study and improve statistical procedures with mathematics, and statistical research often raises mathematical questions. Statistical theory relies on probability and decision theory, and makes extensive use of scientific computing, analysis, and optimization; for the design of experiments, statisticians use algebra and combinatorial design. Applied mathematicians and statisticians often work in a department of mathematical sciences (particularly at colleges and small universities).

Actuarial Science

Actuarial science applies probability, statistics, and economic theory to assess risk in insurance, finance and other industries and professions.

Mathematical Economics

Mathematical economics is the application mathematical methods to represent theories and analyze problems in economics. The applied methods usually refer to nontrivial mathematical techniques or approaches. Mathematical economics is based on statistics, probability, mathematical programming (as well as other computational methods), operations research, game theory, and some methods from mathematical analysis. In this regard, it resembles (but is distinct from) financial mathematics, another part of applied mathematics.

According to the Mathematics Subject Classification (MSC), mathematical economics falls into the Applied mathematics/other classification of category 91:

Game theory, economics, social and behavioral sciences

with MSC2010 classifications for 'Game theory' at codes 91Axx and for 'Mathematical economics' at codes 91Bxx.

Other Disciplines

The line between applied mathematics and specific areas of application is often blurred. Many universities teach mathematical and statistical courses outside of the respective departments, in departments and areas including business, engineering, physics, chemistry, psychology, biology, computer science, scientific computation, and mathematical physics.

Geometry

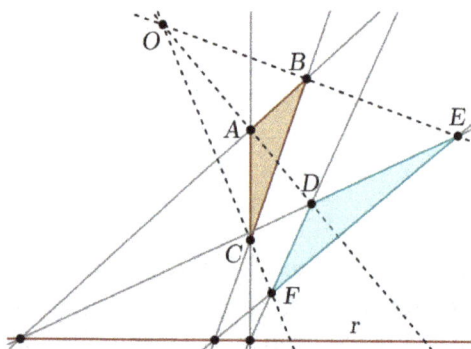

An illustration of Desargues' theorem, an important result in Euclidean and projective geometry

Geometry is a branch of mathematics concerned with questions of shape, size, relative position of figures, and the properties of space. A mathematician who works in the field of geometry is called a geometer. Geometry arose independently in a number of early cultures as a body of practical knowledge concerning lengths, areas, and volumes, with elements of formal mathematical science emerging in the West as early as Thales (6th century BC). By the 3rd century BC, geometry was put into an axiomatic form by Euclid, whose treatment—Euclidean geometry—set a standard for many centuries to follow.Archimedes developed ingenious techniques for calculating areas and volumes, in many ways anticipating modern integral calculus. The field of astronomy, especially as it relates to mapping the positions of stars and planets on the celestial sphere and describing the relationship between movements of celestial bodies, served as an important source of geometric problems during the next one and a half millennia. In the classical world, both geometry and astronomy were considered to be part of the Quadrivium, a subset of the seven liberal arts considered essential for a free citizen to master.

The introduction of coordinates by René Descartes and the concurrent developments of algebra marked a new stage for geometry, since geometric figures such as plane curves could now be represented analytically in the form of functions and equations. This played a key role in the emergence of infinitesimal calculus in the 17th century. Furthermore, the theory of perspective showed that there is more to geometry than just the metric properties of figures: perspective is the origin of projective geometry. The subject of geometry was further enriched by the study of the intrinsic structure of geometric objects that originated with Euler and Gauss and led to the creation of topology and differential geometry.

In Euclid's time, there was no clear distinction between physical and geometrical space. Since the 19th-century discovery of non-Euclidean geometry, the concept of space has undergone a radical transformation and raised the question of which geometrical space best fits physical space. With the rise of formal mathematics in the 20th century, 'space' (whether 'point', 'line', or 'plane') lost its intuitive contents, so today one has to distinguish between physical space, geometrical spaces (in which 'space', 'point' etc. still have their intuitive meanings) and abstract spaces. Contemporary geometry considers manifolds, spaces that are considerably more abstract than the familiar Euclidean space, which they only approximately resemble at small scales. These spaces may be endowed with additional structure which allow one to speak about length. Modern geometry has many ties to physics as is exemplified by the links between pseudo-Riemannian geometry and general relativity. One of the youngest physical theories, string theory, is also very geometric in flavour.

While the visual nature of geometry makes it initially more accessible than other mathematical areas such as algebra or number theory, geometric language is also used in contexts far removed from its traditional, Euclidean provenance (for example, in fractal geometry and algebraic geometry).

Overview

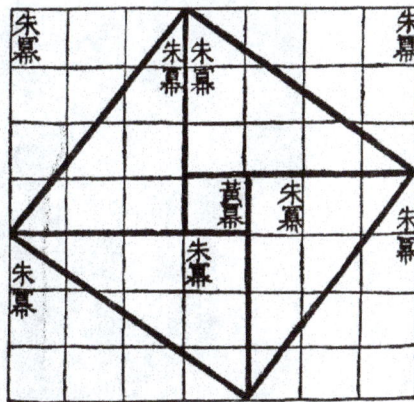

Visual checking of the Pythagorean theorem for the (3, 4, 5) triangle as in the Chou Pei Suan Ching 500–200 BC.

Because the recorded development of geometry spans more than two millennia, perceptions of what constitutes geometry have evolved throughout the ages:

Practical Geometry

Geometry originated as a practical science concerned with surveys, measurements, areas, and volumes. Among other highlights, notable accomplishments include formulas for lengths, areas and volumes, such as the Pythagorean theorem, circumference and area of a circle, area of a triangle, volume of a cylinder, sphere, and a pyramid. A method of computing certain inaccessible distances or heights based on similarity of geometric figures is attributed to Thales. The development of astronomy led to the emergence of trigonometry and spherical trigonometry, together with the attendant computational techniques.

Axiomatic Geometry

Euclid took a more abstract approach in his Elements, one of the most influential books ever written. Euclid introduced certain axioms, or postulates, expressing primary or self-evident properties of points, lines, and planes. He proceeded to rigorously deduce other properties by mathematical reasoning. The characteristic feature of Euclid's approach to geometry was its rigor, and it has come to be known as *axiomatic* or *synthetic* geometry. At the start of the 19th century, the discovery of non-Euclidean geometries by Nikolai Ivanovich Lobachevsky (1792–1856), János Bolyai (1802–1860) and Carl Friedrich Gauss (1777–1855) and others led to a revival of interest in this discipline, and in the 20th century, David Hilbert (1862–1943) employed axiomatic reasoning in an attempt to provide a modern foundation of geometry.

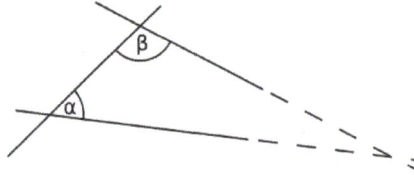

An illustration of Euclid's parallel postulate

Geometry lessons in the 20th century

Geometric Constructions

Classical geometers paid special attention to constructing geometric objects that had been described in some other way. Classically, the only instruments allowed in geometric constructions are the compass and straightedge. Also, every construction had to be complete in a finite number of steps. However, some problems turned out to be difficult or impossible to solve by these means alone, and ingenious constructions using parabolas and other curves, as well as mechanical devices, were found.

Numbers in Geometry

In ancient Greece the Pythagoreans considered the role of numbers in geometry. However, the discovery of incommensurable lengths, which contradicted their philosophical views, made them abandon abstract numbers in favor of concrete geometric quantities, such as length and area of figures. Numbers were reintroduced into geometry in the form of coordinates by Descartes, who realized that the study of geometric shapes can be facilitated by their algebraic representation, and for whom the Cartesian plane is named. Analytic geometry applies methods of algebra to geometric questions, typically by relating geometric curves to algebraic equations. These ideas played a key role in the development of calculus in the 17th century and led to the discovery of many new properties of plane curves. Modern algebraic geometry considers similar questions on a vastly more abstract level.

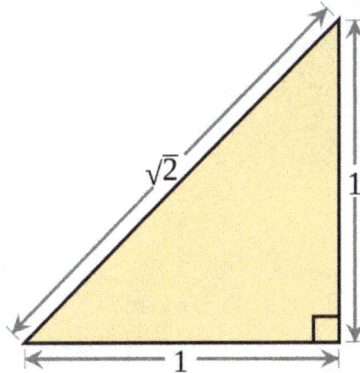

The Pythagoreans discovered that the sides of a triangle could have incommensurable lengths.

Geometry of Position

Even in ancient times, geometers considered questions of relative position or spatial relationship of geometric figures and shapes. Some examples are given by inscribed and circumscribed circles of polygons, lines intersecting and tangent to conic sections, the Pappus and Menelaus configurations of points and lines. In the Middle Ages, new and more complicated questions of this type were considered: What is the maximum number of spheres simultaneously touching a given sphere of the same radius (kissing number problem)? What is the densest packing of spheres of equal size in space (Kepler conjecture)? Most of these questions involved 'rigid' geometrical shapes, such as lines or spheres. Projective, convex, and discrete geometry are three sub-disciplines within present day geometry that deal with these types of questions.

Leonhard Euler, in studying problems like the Seven Bridges of Königsberg, considered the most fundamental properties of geometric figures based solely on shape, independent of their metric properties. Euler called this new branch of geometry *geometria situs* (geometry of place), but it is now known as topology. Topology grew out of geometry, but turned into a large independent discipline. It does not differentiate between objects that can be continuously deformed into each other. The objects may nevertheless retain some geometry, as in the case of hyperbolic knots.

Geometry Beyond Euclid

In the nearly two thousand years since Euclid, while the range of geometrical questions asked and answered inevitably expanded, the basic understanding of space remained essentially the same. Immanuel Kant argued that there is only one, *absolute*, geometry, which is known to be true *a priori* by an inner faculty of mind: Euclidean geometry was synthetic a priori. This dominant view was overturned by the revolutionary discovery of non-Euclidean geometry in the works of Bolyai, Lobachevsky, and Gauss (who never published his theory). They demonstrated that ordinary Euclidean space is only one possibility for development of geometry. A broad vision of the subject of geometry was then expressed by Riemann in his 1867 inauguration lecture *Über die Hypothesen, welche der Geometrie zu Grunde liegen* (*On the hypotheses on which geometry is based*), published only after his death. Riemann's new idea of space proved crucial in Einstein's general relativity theory, and Riemannian geometry, that considers very general spaces in which the notion of length is defined, is a mainstay of modern geometry.

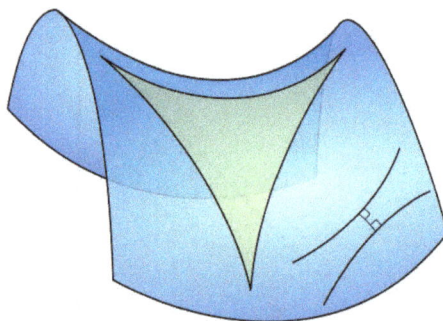

Differential geometry uses tools from calculus to study problems involving curvature.

Dimension

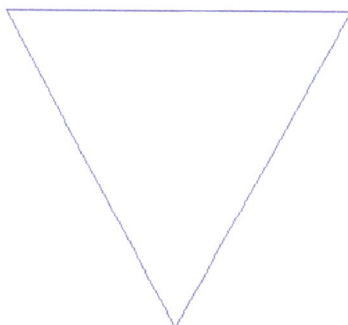

The Koch snowflake, with fractal dimension=log4/log3 and topological dimension=1

Where the traditional geometry allowed dimensions 1 (a line), 2 (a plane) and 3 (our ambient world conceived of as three-dimensional space), mathematicians have used higher dimensions for nearly two centuries. Dimension has gone through stages of being any natural number n, possibly infinite with the introduction of Hilbert space, and any positive real number in fractal geometry. Dimension theory is a technical area, initially within general topology, that discusses *definitions*; in common with most mathematical ideas, dimension is now defined rather than an intuition. Connected topological manifolds have a well-defined dimension; this is a theorem (invariance of domain) rather than anything *a priori*.

The issue of dimension still matters to geometry, in the absence of complete answers to classic questions. Dimensions 3 of space and 4 of space-time are special cases in geometric topology. Dimension 10 or 11 is a key number in string theory. Research may bring a satisfactory *geometric* reason for the significance of 10 and 11 dimensions.

Symmetry

The theme of symmetry in geometry is nearly as old as the science of geometry itself. Symmetric shapes such as the circle, regular polygons and platonic solids held deep significance for many ancient philosophers and were investigated in detail before the time of Euclid. Symmetric patterns occur in nature and were artistically rendered in a multitude of forms, including the graphics of M. C. Escher. Nonetheless, it was not until the second half of 19th century that the unifying role of symmetry in foundations of geometry was recognized. Felix Klein's Erlangen program proclaimed that, in a very precise sense, symmetry, expressed via the notion of a transformation group, determines what geometry *is*. Symmetry in classical Euclidean geometry is represented by congruences and rigid motions, whereas in projective geometry an analogous role is played by collineations, geometric transformations that take straight lines into straight lines. However it was in the new geometries of Bolyai and Lobachevsky, Riemann, Clifford and Klein, and Sophus Lie that Klein's idea to 'define a geometry via its symmetry group' proved most influential. Both discrete and continuous symmetries play prominent roles in geometry, the former in topology and geometric group theory, the latter in Lie theory and Riemannian geometry.

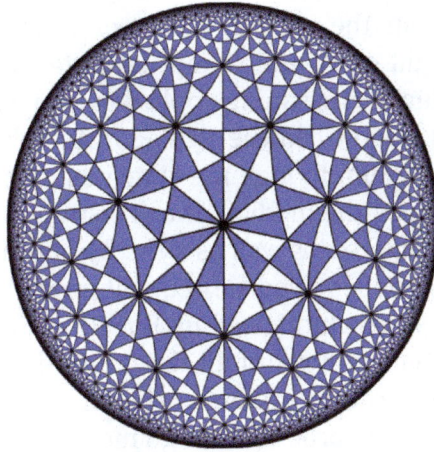

A tiling of the hyperbolic plane

A different type of symmetry is the principle of duality in projective geometry (Duality (projective geometry)) among other fields. This meta-phenomenon can roughly be described as follows: in any theorem, exchange *point* with *plane*, *join* with *meet*, *lies in* with *contains*, and you will get an equally true theorem. A similar and closely related form of duality exists between a vector space and its dual space.

History

A European and an Arab practicing geometry in the 15th century.

The earliest recorded beginnings of geometry can be traced to ancient Mesopotamia and Egypt in the 2nd millennium BC. Early geometry was a collection of empirically discovered principles concerning lengths, angles, areas, and volumes, which were developed to meet some practical need in surveying, construction, astronomy, and various crafts. The earliest known texts on geometry are the Egyptian*Rhind Papyrus* (2000–1800 BC) and *Moscow Papyrus* (c. 1890 BC), the Babylonian clay tablets such as Plimpton 322 (1900 BC). For example, the Moscow Papyrus gives a formula for calculating the volume of a truncated pyramid, or frustum. Later clay tablets (350–50 BC) demonstrate that Babylonian astronomers implemented trapezoid procedures for computing Jupiter's position and motion within time-velocity space. These geometric procedures anticipated the Oxford Calculators, including the mean speed theorem, by 14 centuries. South of Egypt the ancient Nubians established a system of geometry including early versions of sun clocks.

In the 7th century BC, the Greek mathematician Thales of Miletus used geometry to solve problems such as calculating the height of pyramids and the distance of ships from the shore. He is credited with the first use of deductive reasoning applied to geometry, by deriving four corollaries to Thales' Theorem. Pythagoras established the Pythagorean School, which is credited with the first proof of the Pythagorean theorem, though the statement of the theorem has a long history-Eudoxus (408–c. 355 BC) developed the method of exhaustion, which allowed the calculation of areas and volumes of curvilinear figures, as well as a theory of ratios that avoided the problem of incommensurable magnitudes, which enabled subsequent geometers to make significant advances. Around 300 BC, geometry was revolutionized by Euclid, whose *Elements*, widely considered the most successful and influential textbook of all time, introduced mathematical rigor through the axiomatic method and is the earliest example of the format still used in mathematics today, that of definition, axiom, theorem, and proof. Although most of the contents of the *Elements* were already known, Euclid arranged them into a single, coherent logical framework. The *Elements* was known to all educated people in the West until the middle of the 20th century and its contents are still taught in geometry classes today.Archimedes (c. 287–212 BC) of Syracuse used the method of exhaustion to calculate the area under the arc of a parabola with the summation of an infinite series, and gave remarkably accurate approximations of Pi. He also studied the spiral bearing his name and obtained formulas for the volumes of surfaces of revolution.

Woman teaching geometry. Illustration at the beginning of a medieval translation of Euclid's Elements, (c. 1310)

Indian mathematicians also made many important contributions in geometry. The *Satapatha Brahmana* (3rd century BC) contains rules for ritual geometric constructions that are similar to the *Sulba Sutras*. According to (Hayashi 2005, p. 363), the *Śulba Sūtras* contain "the earliest extant verbal expression of the Pythagorean Theorem in the world, although it had already been known to the Old Babylonians. They contain lists of Pythagorean triples, which are particular cases of Diophantine equations. In the Bakhshali manuscript, there is a handful of geometric problems (including problems about volumes of irregular solids). The Bakhshali manuscript also "employs a decimal place value system with a dot for zero." Aryabhata's *Aryabhatiya* (499) includes the computation of areas and volumes. Brahmagupta wrote his astronomical work *Brāhma Sphuṭa Siddhānta* in 628. Chapter 12, containing 66 Sanskrit verses, was divided into two sections: "basic operations" (including cube roots, fractions, ratio and proportion, and barter) and "practical mathematics" (including mixture, mathematical series, plane figures, stacking bricks, sawing of timber, and piling of grain). In the latter section, he stated his famous theorem on the diagonals of a cyclic quadrilateral. Chapter 12 also included a formula for the area of a cyclic quadrilateral (a generalization of Heron's formula), as well as a complete description of rational triangles (*i.e.* triangles with rational sides and rational areas).

In the Middle Ages, mathematics in medieval Islam contributed to the development of geometry, especially algebraic geometry. Al-Mahani (b. 853) conceived the idea of reducing geometrical problems such as duplicating the cube to problems in algebra. Thābit ibn Qurra (known as Thebit in Latin) (836–901) dealt with arithmetic operations applied to ratios of geometrical quantities, and contributed to the development of analytic geometry. Omar Khayyám (1048–1131) found geometric solutions to cubic equations. The theorems of Ibn al-Haytham (Alhazen), Omar Khayyam and Nasir al-Din al-Tusi on quadrilaterals, including the Lambert quadrilateral and Saccheri quadrilateral, were early results in hyperbolic geometry, and along with their alternative postulates, such as Playfair's axiom, these works had a considerable influence on the development of non-Euclidean geometry among later European geometers, including Witelo (c. 1230–c. 1314), Gersonides (1288–1344), Alfonso, John Wallis, and Giovanni Girolamo Saccheri.

In the early 17th century, there were two important developments in geometry. The first was the creation of analytic geometry, or geometry with coordinates and equations, by René Descartes (1596–1650) and Pierre de Fermat (1601–1665). This was a necessary precursor to the development of calculus and a precise quantitative science of physics. The second geometric development of this period was the systematic study of projective geometry by Girard Desargues (1591–1661). Projective geometry is a geometry without measurement or parallel lines, just the study of how points are related to each other.

Two developments in geometry in the 19th century changed the way it had been studied previously. These were the discovery of non-Euclidean geometries by Nikolai Ivanovich Lobachevsky, János Bolyai and Carl Friedrich Gauss and of the formulation of symmetry as the central consideration in the Erlangen Programme of Felix Klein (which generalized the Euclidean and non-Euclidean geometries). Two of the master geometers of the time were Bernhard Riemann (1826–1866), working primarily with tools from mathematical analysis, and introducing the Riemann surface, and Henri Poincaré, the founder of algebraic topology and the geometric theory of dynamical systems. As a consequence of these major changes in the conception of geometry, the concept of "space" became something rich and varied, and the natural background for theories as different as complex analysis and classical mechanics.

Contemporary Geometry

Euclidean Geometry

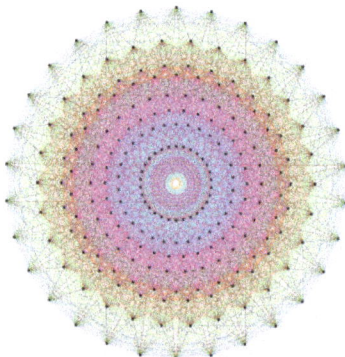

The 4$_{21}$polytope, orthogonally projected into the E$_8$Lie groupCoxeter plane

Euclidean geometry has become closely connected with computational geometry, computer graphics, convex geometry, incidence geometry, finite geometry, discrete geometry, and some areas of combinatorics. Attention was given to further work on Euclidean geometry and the Euclidean groups by crystallography and the work of H. S. M. Coxeter, and can be seen in theories of Coxeter groups and polytopes. Geometric group theory is an expanding area of the theory of more general discrete groups, drawing on geometric models and algebraic techniques.

Differential Geometry

Differential geometry has been of increasing importance to mathematical physics due to Einstein's general relativity postulation that the universe is curved. Contemporary differential geometry is *intrinsic*, meaning that the spaces it considers are smooth manifolds whose geometric structure is governed by a Riemannian metric, which determines how distances are measured near each point, and not *a priori* parts of some ambient flat Euclidean space.

Topology and Geometry

A thickening of the trefoil knot

The field of topology, which saw massive development in the 20th century, is in a technical sense a type of transformation geometry, in which transformations are homeomorphisms. This has often been expressed in the form of the dictum 'topology is rubber-sheet geometry'. Contemporary geometric topology and differential topology, and particular subfields such as Morse theory, would be counted by most mathematicians as part of geometry. Algebraic topology and general topology have gone their own ways.

Algebraic Geometry

Quintic Calabi–Yau threefold

The field of algebraic geometry is the modern incarnation of the Cartesian geometry of co-ordinates. From late 1950s through mid-1970s it had undergone major foundational development, largely due to work of Jean-Pierre Serre and Alexander Grothendieck. This led to the introduction of schemes and greater emphasis on topological methods, including various cohomology theories. One of seven Millennium Prize problems, the Hodge conjecture, is a question in algebraic geometry.

The study of low-dimensional algebraic varieties, algebraic curves, algebraic surfaces and algebraic varieties of dimension 3 ("algebraic threefolds"), has been far advanced. Gröbner basis theory and real algebraic geometry are among more applied subfields of modern algebraic geometry. Arithmetic geometry is an active field combining algebraic geometry and number theory. Other directions of research involve moduli spaces and complex geometry. Algebro-geometric methods are commonly applied in string and brane theory.

Trigonometry

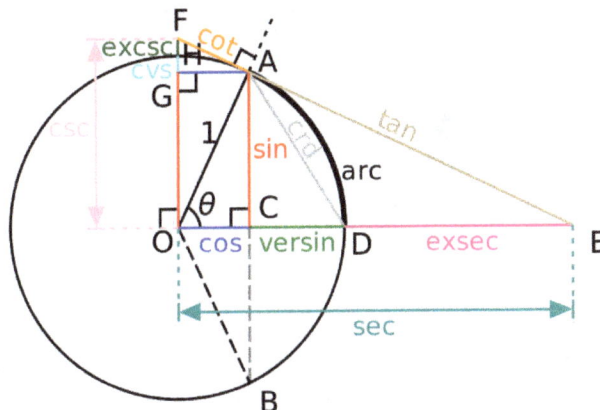

All of the trigonometric functions of an angle θ can be constructed geometrically in terms of a unit circle centered at O.

Trigonometry is a branch of mathematics that studies relationships involving lengths and angles of triangles. The field emerged in the Hellenistic world during the 3rd century BC from applications of geometry to astronomical studies.

The 3rd-century astronomers first noted that the lengths of the sides of a right-angle triangle and the angles between those sides have fixed relationships: that is, if at least the length of one side and the value of one angle is known, then all other angles and lengths can be determined algorithmically. These calculations soon came to be defined as the trigonometric functions and today are pervasive in both pure and applied mathematics: fundamental methods of analysis such as the Fourier transform, for example, or the wave equation, use trigonometric functions to understand cyclical phenomena across many applications in fields as diverse as physics, mechanical and electrical engineering, music and acoustics, astronomy, ecology, and biology. Trigonometry is also the foundation of surveying.

Trigonometry is most simply associated with planarright-angle triangles (each of which is a two-dimensional triangle with one angle equal to 90 degrees). The applicability to non-right-angle triangles exists, but, since any non-right-angle triangle (on a flat plane) can be bisected to create two right-angle triangles, most problems can be reduced to calculations on right-angle triangles. Thus the majority of applications relate to right-angle triangles. One exception to this is spherical trigonometry, the study of triangles on spheres, surfaces of constant positive curvature, in elliptic geometry (a fundamental part of astronomy and navigation). Trigonometry on surfaces of negative curvature is part of hyperbolic geometry.

Trigonometry basics are often taught in schools, either as a separate course or as a part of a precalculus course.

History

Hipparchus, credited with compiling the first trigonometric table, is known as "the father of trigonometry".

Sumerian astronomers studied angle measure, using a division of circles into 360 degrees. They, and later the Babylonians, studied the ratios of the sides of similar triangles and discovered some properties of these ratios but did not turn that into a systematic method for finding sides and angles of triangles. The ancient Nubians used a similar method.

In the 3rd century BC, Hellenistic mathematicians such as Euclid (from Alexandria, Egypt) and Archimedes (from Syracuse, Sicily) studied the properties of chords and inscribed angles in circles, and they proved theorems that are equivalent to modern trigonometric formulae, although they presented them geometrically rather than algebraically. In 140 BC, Hipparchus (from Iznik, Turkey) gave the first tables of chords, analogous to modern tables of sine values, and used them to solve problems in trigonometry and spherical trigonometry. In the 2nd century AD, the Greco-Egyptian astronomer Ptolemy (from Alexandria, Egypt) printed detailed trigonometric tables (Ptolemy's table of chords) in Book 1, chapter 11 of his Almagest. Ptolemy used chord length to define his trigonometric functions, a minor difference from the sine convention we use today. (The value we call sin(θ) can be found by looking up the chord length for twice the angle of interest (2θ) in Ptolemy's table, and then dividing that value by two.) Centuries passed before more detailed tables were produced, and Ptolemy's treatise remained in use for performing trigonometric calculations in astronomy throughout the next 1200 years in the medieval Byzantine, Islamic, and, later, Western European worlds.

The modern sine convention is first attested in the *Surya Siddhanta*, and its properties were further documented by the 5th century (AD) Indian mathematician and astronomer Aryabhata. These Greek and Indian works were translated and expanded by medieval Islamic mathematicians. By the 10th century, Islamic mathematicians were using all six trigonometric functions, had tabulated their values, and were applying them to problems in spherical geometry. At about the same time, Chinese mathematicians developed trigonometry independently, although it was not a major field of study for them. Knowledge of trigonometric functions and methods reached Western Europe via Latin translations of Ptolemy's Greek Almagest as well as the works of Persian and Arabic astronomers such as Al Battani and Nasir al-Din al-Tusi. One of the earliest works on trigonometry by a northern European mathematician is *De Triangulis* by the 15th century German mathematician Regiomontanus, who was encouraged to write, and provided with a copy of the Almagest, by the Byzantine Greek scholar cardinal Basilios Bessarion with whom he lived for several years. At the same time, another translation of the Almagest from Greek into Latin was completed by the Cretan George of Trebizond. Trigonometry was still so little known in 16th-century northern Europe that Nicolaus Copernicus devoted two chapters of *De revolutionibus orbium coelestium* to explain its basic concepts.

Driven by the demands of navigation and the growing need for accurate maps of large geographic areas, trigonometry grew into a major branch of mathematics.Bartholomaeus Pitiscus was the first to use the word, publishing his *Trigonometria* in 1595.Gemma Frisius described for the first time the method of triangulation still used today in surveying. It was Leonhard Euler who fully incorporated complex numbers into trigonometry. The works of the Scottish mathematicians James Gregory in the 17th century and Colin Maclaurin in the 18th century were influential in the development of trigonometric series. Also in the 18th century, Brook Taylor defined the general Taylor series.

Overview

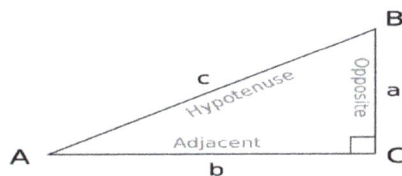

In this right triangle: $\sin A = a/c$; $\cos A = b/c$; $\tan A = a/b$.

If one angle of a triangle is 90 degrees and one of the other angles is known, the third is thereby fixed, because the three angles of any triangle add up to 180 degrees. The two acute angles therefore add up to 90 degrees: they are complementary angles. The shape of a triangle is completely determined, except for similarity, by the angles. Once the angles are known, the ratios of the sides are determined, regardless of the overall size of the triangle. If the length of one of the sides is known, the other two are determined. These ratios are given by the following trigonometric functions of the known angle A, where a, b and c refer to the lengths of the sides in the accompanying figure:

- Sine function (sin), defined as the ratio of the side opposite the angle to the hypotenuse.

$$\sin A = \frac{\text{opposite}}{\text{hypotenuse}} = \frac{a}{c}.$$

- Cosine function (cos), defined as the ratio of the adjacent leg to the hypotenuse.

$$\cos A = \frac{\text{adjacent}}{\text{hypotenuse}} = \frac{b}{c}.$$

- Tangent function (tan), defined as the ratio of the opposite leg to the adjacent leg.

$$\tan A = \frac{\text{opposite}}{\text{adjacent}} = \frac{a}{b} = \frac{a}{c} * \frac{c}{b} = \frac{a}{c} / \frac{b}{c} = \frac{\sin A}{\cos A}.$$

The hypotenuse is the side opposite to the 90 degree angle in a right triangle; it is the longest side of the triangle and one of the two sides adjacent to angle A. The adjacent leg is the other side that is adjacent to angle A. The opposite side is the side that is opposite to angle A. The terms perpendicular and base are sometimes used for the opposite and adjacent sides respectively. Many people find it easy to remember what sides of the right triangle are equal to sine, cosine, or tangent, by memorizing the word SOH-CAH-TOA.

The reciprocals of these functions are named the cosecant (csc or cosec), secant (sec), and cotangent (cot), respectively:

$$\csc A = \frac{1}{\sin A} = \frac{\text{hypotenuse}}{\text{opposite}} = \frac{c}{a},$$

$$\sec A = \frac{1}{\cos A} = \frac{\text{hypotenuse}}{\text{adjacent}} = \frac{c}{b},$$

$$\cot A = \frac{1}{\tan A} = \frac{\text{adjacent}}{\text{opposite}} = \frac{\cos A}{\sin A} = \frac{b}{a}.$$

The inverse functions are called the arcsine, arccosine, and arctangent, respectively. There are arithmetic relations between these functions, which are known as trigonometric identities. The cosine, cotangent, and cosecant are so named because they are respectively the sine, tangent, and secant of the complementary angle abbreviated to "co-".

With these functions, one can answer virtually all questions about arbitrary triangles by using the law of sines and the law of cosines. These laws can be used to compute the remaining angles and sides of any triangle as soon as two sides and their included angle or two angles and a side or three sides are known. These laws are useful in all branches of geometry, since every polygon may be described as a finite combination of triangles.

Extending The Definitions

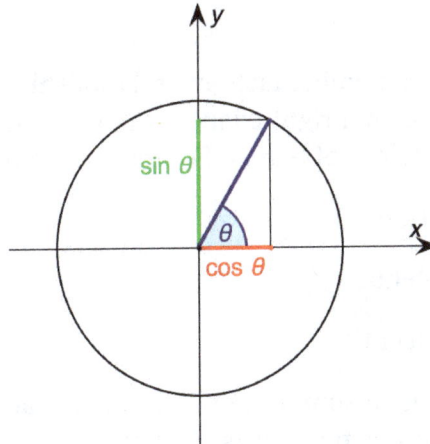

Fig. 1a – Sine and cosine of an angle θ defined using the unit circle.

The above definitions only apply to angles between 0 and 90 degrees (0 and $\pi/2$ radians). Using the unit circle, one can extend them to all positive and negative arguments. The trigonometric functions are periodic, with a period of 360 degrees or 2π radians. That means their values repeat at those intervals. The tangent and cotangent functions also have a shorter period, of 180 degrees or π radians.

The trigonometric functions can be defined in other ways besides the geometrical definitions above, using tools from calculus and infinite series. With these definitions the trigonometric functions can be defined for complex numbers. The complex exponential function is particularly useful.

$$e^{x+iy} = e^x(\cos y + i \sin y).$$

Euler's and De Moivre's formulas.

Graphing process of $y = \sin(x)$ using a unit circle.

Graphing process of $y = \csc(x)$, the reciprocal of sine, using a unit circle.

Graphing process of $y = \tan(x)$ using a unit circle.

Mnemonics

A common use of mnemonics is to remember facts and relationships in trigonometry. For example, the *sine*, *cosine*, and *tangent* ratios in a right triangle can be remembered by representing them and their corresponding sides as strings of letters. For instance, a mnemonic is SOH-CAH-TOA:

Sine = Opposite ÷ Hypotenuse

Cosine = Adjacent ÷ Hypotenuse

Tangent = Opposite ÷ Adjacent

One way to remember the letters is to sound them out phonetically (i.e., *SOH-CAH-TOA*, which is pronounced 'so-kə-toe-uh'. Another method is to expand the letters into a sentence, such as "Some Old Hippie Caught Another Hippie Trippin' On Acid".

Calculating Trigonometric Functions

Trigonometric functions were among the earliest uses for mathematical tables. Such tables were incorporated into mathematics textbooks and students were taught to look up values and how to interpolate between the values listed to get higher accuracy. Slide rules had special scales for trigonometric functions.

Today, scientific calculators have buttons for calculating the main trigonometric functions (sin, cos, tan, and sometimes cis and their inverses). Most allow a choice of angle measurement methods: degrees, radians, and sometimes gradians. Most computer programming languages provide function libraries that include the trigonometric functions. The floating point unit hardware incorporated into the microprocessor chips used in most personal computers has built-in instructions for calculating trigonometric functions.

Applications of Trigonometry

Sextants are used to measure the angle of the sun or stars with respect to the horizon. Using trigonometry and a marine chronometer, the position of the ship can be determined from such measurements.

There is an enormous number of uses of trigonometry and trigonometric functions. For instance, the technique of triangulation is used in astronomy to measure the distance to nearby stars, in geography to measure distances between landmarks, and in satellite navigation systems. The sine and cosine functions are fundamental to the theory of periodic functions, such as those that describe sound and light waves.

Fields that use trigonometry or trigonometric functions include astronomy (especially for locating apparent positions of celestial objects, in which spherical trigonometry is essential) and hence navigation (on the oceans, in aircraft, and in space), music theory, audio synthesis, acoustics, optics, electronics, biology, medical imaging (CAT scans and ultrasound), pharmacy, chemistry, number theory (and hence cryptology), seismology, meteorology, oceanography, many physical sciences, land surveying and geodesy, architecture, image compression, phonetics, economics, electrical engineering, mechanical engineering, civil engineering, computer graphics, cartography, crystallography and game development.

Pythagorean Identities

The following identities are related to the Pythagorean theorem and hold for any value:

$$\sin^2 A + \cos^2 A = 1$$

$$\tan^2 A + 1 = \sec^2 A$$

$$1 + \cot^2 A = \csc^2 A$$

Angle Transformation Formulae

$$\sin(A \pm B) = \sin A \cos B \pm \cos A \sin B$$

$$\cos(A \pm B) = \cos A \cos B \mp \sin A \sin B$$

$$\tan(A \pm B) = \frac{\tan A \pm \tan B}{1 \mp \tan A \tan B}$$

$$\cot(A \pm B) = \frac{\cot A \cot B \mp 1}{\cot B \pm \cot A}$$

Common Formulae

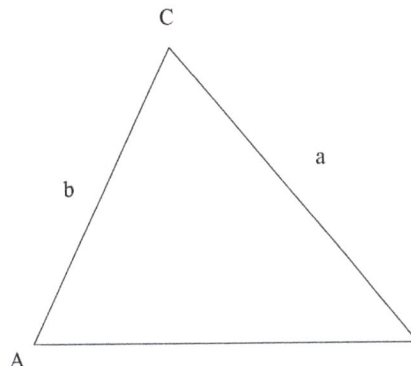

Triangle with sides a,b,c and respectively opposite angles A,B,C

Certain equations involving trigonometric functions are true for all angles and are known as *trigonometric identities*. Some identities equate an expression to a different expression involving the same angles. These are listed in List of trigonometric identities. Triangle identities that relate the sides and angles of a given triangle are listed below.

In the following identities, A, B and C are the angles of a triangle and a, b and c are the lengths of sides of the triangle opposite the respective angles.

Law of Sines

The law of sines (also known as the "sine rule") for an arbitrary triangle states:

$$\frac{a}{\sin A} = \frac{b}{\sin B} = \frac{c}{\sin C} = 2R = \frac{abc}{2},$$

where Δ is the area of the triangle and R is the radius of the circumscribed circle of the triangle:

$$R = \frac{abc}{\sqrt{(a+b+c)(a-b+c)(a+b-c)(b+c-a)}}.$$

Another law involving sines can be used to calculate the area of a triangle. Given two sides a and b and the angle between the sides C, the area of the triangle is given by half the product of the lengths of two sides and the sine of the angle between the two sides:

$$\text{Area} = \sin\frac{1}{2}ab \quad C$$

Law of Cosines

The law of cosines (known as the cosine formula, or the "cos rule") is an extension of the Pythagorean theorem to arbitrary triangles:

$$c^2 = a^2 + b^2 - 2ab\cos C,$$

or equivalently:

$$\cos C = \frac{a^2 + b^2 - c^2}{2ab}.$$

The law of cosines may be used to prove Heron's formula, which is another method that may be used to calculate the area of a triangle. This formula states that if a triangle has sides of lengths a, b, and c, and if the semiperimeter is

$$s = \frac{1}{2}(a+b+c),$$

then the area of the triangle is:

$$\text{Area} = \Delta = \sqrt{s(s-a)(s-b)(s-c)} = \frac{abc}{4R},$$

where R is the radius of the circumcircle of the triangle.

Law of Tangents

The law of tangents:

$$\frac{a-b}{a+b} = \frac{\tan\left[\frac{1}{2}(A-B)\right]}{\tan\left[\frac{1}{2}(A+B)\right]}$$

Euler's Formula

Euler's formula, which states that $e^{ix} = \cos x + i\sin x$, produces the following analytical identities for sine, cosine, and tangent in terms of e and the imaginary unit i:

$$\sin x = \frac{e^{ix} - e^{-ix}}{2i}, \qquad \cos x = \frac{e^{ix} + e^{-ix}}{2}, \qquad \tan x = \frac{i(e^{-ix} - e^{ix})}{e^{ix} + e^{-ix}}.$$

Algebra

$$x = \frac{-b \pm \sqrt{b^2 - 4ac}}{2a}$$

The quadratic formula expresses the solution of the degree two equation $ax^2 + bx + c = 0$ in terms of its coefficients a, b, c, where a is not zero.

Algebra (from Arabic"*al-jabr*" meaning "reunion of broken parts") is one of the broad parts of mathematics, together with number theory, geometry and analysis. In its most general form, algebra is the study of mathematical symbols and the rules for manipulating these symbols; it is a unifying thread of almost all of mathematics. As such, it includes everything from elementary equation solving to the study of abstractions such as groups, rings, and fields. The more basic parts of algebra are called elementary algebra, the more abstract parts are called abstract algebra or modern algebra. Elementary algebra is generally considered to be essential for any study of mathematics, science, or engineering, as well as such applications as medicine and economics. Abstract algebra is a major area in advanced mathematics, studied primarily by professional mathematicians. Much early work in algebra, as the Arabic origin of its name suggests, was done in the Middle East, by Persian mathematicians such as al-Khwārizmī (780–850) and Omar Khayyam (1048–1131).

Elementary algebra differs from arithmetic in the use of abstractions, such as using letters to stand for numbers that are either unknown or allowed to take on many values. For example, in $x + 2 = 5$ the letter x is unknown, but the law of inverses can be used to discover its value: $x = 3$. In $E = mc^2$, the letters E and m are variables, and the letter c is a constant, the speed of light in a vacuum. Algebra gives methods for solving equations and expressing formulas that are much easier (for those who know how to use them) than the older method of writing everything out in words.

The word *algebra* is also used in certain specialized ways. A special kind of mathematical object in abstract algebra is called an "algebra", and the word is used, for example, in the phrases linear algebra and algebraic topology.

A mathematician who does research in algebra is called an algebraist.

Etymology

The word *algebra* comes from the Arabic (*al-jabr* "restoration") from the title of the book *Ilm al-jabr wa'l-muḳābala* by al-Khwarizmi. The word entered the English language during the fifteenth century, from either Spanish, Italian, or Medieval Latin. It originally referred to the surgical procedure of setting broken or dislocated bones. The mathematical meaning was first recorded in the sixteenth century.

Different Meanings of "Algebra"

The word "algebra" has several related meanings in mathematics, as a single word or with qualifiers.

- As a single word without article, "algebra" names a broad part of mathematics.

- As a single word with article or in plural, "algebra" denotes a specific mathematical structure, whose precise definition depends on the author. Usually the structure has an addition, multiplication, and a scalar multiplication. When some au-thors use the term "algebra", they make a subset of the following additional assumptions: associative, commutative, unital, and/or finite-dimensional. In universal algebra, the word "algebra" refers to a generalization of the above concept, which allows for n-ary operations.

- With a qualifier, there is the same distinction:

  o Without article, it means a part of algebra, such as linear algebra, elementary algebra (the symbol-manipulation rules taught in elementary courses of mathematics as part of primary and secondary education), or abstract algebra (the study of the algebraic structures for themselves).

  o With an article, it means an instance of some abstract structure, like a Lie algebra, associative algebra, or vertex operator algebra.

  o Sometimes both meanings exist for the same qualifier, as in the sentence: *Commutative algebra is the study of commutative rings, which are commutative algebras over the integers.*

Algebra as a Branch of Mathematics

Algebra began with computations similar to those of arithmetic, with letters standing for numbers. This allowed proofs of properties that are true no matter which numbers are involved. For example, in the quadratic equation

$$ax^2 + bx + c = 0,$$

a, b, c can be any numbers whatsoever (except that a cannot be 0), and the quadratic formula can be used to quickly and easily find the value of the unknown quantity x.

As it developed, algebra was extended to other non-numerical objects, such as vectors, matrices, and polynomials. Then the structural properties of these non-numerical objects were abstracted to define algebraic structures such as groups, rings, and fields.

Before the 16th century, mathematics was divided into only two subfields, arithmetic and geometry. Even though some methods, which had been developed much earlier, may be considered nowadays as algebra, the emergence of algebra and, soon thereafter, of infinitesimal calculus as subfields of mathematics only dates from the 16th or 17th century. From the second half of 19th century on, many new fields of mathematics appeared, most of which made use of both arithmetic and geometry, and almost all of which used algebra.

Today, algebra has grown until it includes many branches of mathematics, as can be seen in the Mathematics Subject Classification where none of the first level areas (two digit entries) is called *algebra*. Today algebra includes section 08-General algebraic systems, 12-Field theory and polynomials, 13-Commutative algebra, 15-Linear and multilinear algebra; matrix theory, 16-Associative rings and algebras, 17-Nonassociative rings and algebras, 18-Category theory; homological algebra, 19-K-theory and 20-Group theory. Algebra is also used extensively in 11-Number theory and 14-Algebraic geometry.

History

Early history of Algebra

The roots of algebra can be traced to the ancient Babylonians, who developed an advanced arithmetical system with which they were able to do calculations in an algorithmic fashion. The Babylonians developed formulas to calculate solutions for problems typically solved today by using linear equations, quadratic equations, and indeterminate linear equations. By contrast, most Egyptians of this era, as well as Greek and Chinese mathematics in the 1st millennium BC, usually solved such equations by geometric methods, such as those described in the *Rhind Mathematical Papyrus*, Euclid's *Elements*, and *The Nine Chapters on the Mathematical Art*. The geometric work of the Greeks, typified in the *Elements*, provided the framework for generalizing formulae beyond the solution of particular problems into more general systems of stating and solving equations, although this would not be realized until mathematics developed in medieval Islam.

By the time of Plato, Greek mathematics had undergone a drastic change. The Greeks created a geometric algebra where terms were represented by sides of geometric objects, usually lines, that had letters associated with them.Diophantus (3rd century AD) was an AlexandrianGreek mathematician and the author of a series of books called *Arithmetica*. These texts deal with solving algebraic equations, and have led, in number theory to the modern notion of Diophantine equation·

Earlier traditions discussed above had a direct influence on the PersianMuḥammad ibn Mūsā al-Khwārizmī (c. 780–850). He later wrote *The Compendious Book on Calculation by Completion and Balancing*, which established algebra as a mathematical discipline that is independent of geometry and arithmetic.

The Hellenistic mathematicians Hero of Alexandria and Diophantus as well as Indian mathematicians such as Brahmagupta continued the traditions of Egypt and Babylon, though Diophantus' *Arithmetica* and Brahmagupta's *Brāhmasphuṭasiddhānta* are on a higher level. For example, the first complete arithmetic solution (including zero and negative solutions) to quadratic equations was described by Brahmagupta in his book *Brahmasphutasiddhanta*. Later, Persian and

Arabic mathematicians developed algebraic methods to a much higher degree of sophistication. Although Diophantus and the Babylonians used mostly special *ad hoc* methods to solve equations, Al-Khwarizmi's contribution was fundamental. He solved linear and quadratic equations without algebraic symbolism, negative numbers or zero, thus he had to distinguish several types of equations.

In the context where algebra is identified with the theory of equations, the Greek mathematician Diophantus has traditionally been known as the "father of algebra" but in more recent times there is much debate over whether al-Khwarizmi, who founded the discipline of *al-jabr*, deserves that title instead. Those who support Diophantus point to the fact that the algebra found in *Al-Jabr* is slightly more elementary than the algebra found in *Arithmetica* and that *Arithmetica* is syncopated while *Al-Jabr* is fully rhetorical. Those who support Al-Khwarizmi point to the fact that he introduced the methods of "reduction" and "balancing" (the transposition of subtracted terms to the other side of an equation, that is, the cancellation of like terms on opposite sides of the equation) which the term *al-jabr* originally referred to, and that he gave an exhaustive explanation of solving quadratic equations, supported by geometric proofs, while treating algebra as an independent discipline in its own right. His algebra was also no longer concerned "with a series of problems to be resolved, but an exposition which starts with primitive terms in which the combinations must give all possible prototypes for equations, which henceforward explicitly constitute the true object of study". He also studied an equation for its own sake and "in a generic manner, insofar as it does not simply emerge in the course of solving a problem, but is specifically called on to define an infinite class of problems".

Another Persian mathematician Omar Khayyam is credited with identifying the foundations of algebraic geometry and found the general geometric solution of the cubic equation. Yet another Persian mathematician, Sharaf al-Dīn al-Tūsī, found algebraic and numerical solutions to various cases of cubic equations. He also developed the concept of a function. The Indian mathematicians Mahavira and Bhaskara II, the Persian mathematician Al-Karaji, and the Chinese mathematician Zhu Shijie, solved various cases of cubic, quartic, quintic and higher-order polynomial equations using numerical methods. In the 13th century, the solution of a cubic equation by Fibonacci is representative of the beginning of a revival in European algebra. As the Islamic world was declining, the European world was ascending. And it is here that algebra was further developed.

History of Algebra

Italian mathematician Girolamo Cardano published the solutions to the cubic and quartic equations in his 1545 book *Ars magna*.

François Viète's work on new algebra at the close of the 16th century was an important step towards modern algebra. In 1637, René Descartes published *La Géométrie*, inventing analytic geometry and introducing modern algebraic notation. Another key event in the further development of algebra was the general algebraic solution of the cubic and quartic equations, developed in the mid-16th century. The idea of a determinant was developed by Japanese mathematicianSeki Kōwa in the 17th century, followed independently by Gottfried Leibniz ten years later, for the purpose of solving systems of simultaneous linear equations using matrices. Gabriel Cramer also did some work on matrices and determinants in the 18th century. Permutations were studied by Joseph-Louis Lagrange in his 1770 paper *Réflexions sur la résolution algébrique des équations* devoted to solutions of algebraic equations, in which he introduced Lagrange resolvents. Paolo Ruffini was the first person to develop the theory of permutation groups, and like his predecessors, also in the context of solving algebraic equations.

Abstract algebra was developed in the 19th century, deriving from the interest in solving equations, initially focusing on what is now called Galois theory, and on constructibility issues.George Peacock was the founder of axiomatic thinking in arithmetic and algebra. Augustus De Morgan discovered relation algebra in his *Syllabus of a Proposed System of Logic*. Josiah Willard Gibbs developed an algebra of vectors in three-dimensional space, and Arthur Cayley developed an algebra of matrices (this is a noncommutative algebra).

Areas of Mathematics With The Word Algebra in Their Name

Some areas of mathematics that fall under the classification abstract algebra have the word algebra in their name; linear algebra is one example. Others do not: group theory, ring theory, and field theory are examples. In this section, we list some areas of mathematics with the word "algebra" in the name.

- Elementary algebra, the part of algebra that is usually taught in elementary courses of mathematics.

- Abstract algebra, in which algebraic structures such as groups, rings and fields are axiomatically defined and investigated.

- Linear algebra, in which the specific properties of linear equations, vector spaces and matrices are studied.

- Commutative algebra, the study of commutative rings.

- Computer algebra, the implementation of algebraic methods as algorithms and computer programs.

- Homological algebra, the study of algebraic structures that are fundamental to study topological spaces.

- Universal algebra, in which properties common to all algebraic structures are studied.

- Algebraic number theory, in which the properties of numbers are studied from an algebraic point of view.

- Algebraic geometry, a branch of geometry, in its primitive form specifying curves and surfaces as solutions of polynomial equations.

- Algebraic combinatorics, in which algebraic methods are used to study combinatorial questions.

Many mathematical structures are called algebras:

- Algebra over a field or more generally algebra over a ring. Many classes of algebras over a field or over a ring have a specific name:
  - Associative algebra
  - Non-associative algebra
  - Lie algebra
  - Hopf algebra
  - C*-algebra
  - Symmetric algebra
  - Exterior algebra
  - Tensor algebra
- In measure theory,
  - Sigma-algebra
  - Algebra over a set
- In category theory
  - F-algebra and F-coalgebra
  - T-algebra
- In logic,
  - Relational algebra: a set of finitary relations that is closed under certain operators.
  - Boolean algebra, a structure abstracting the computation with the truth values*false* and *true*. The structures also have the same name.
  - Heyting algebra

Elementary Algebra

$$3x^2 - 2xy + c$$

Algebraic expression notation:
1 – power (exponent)

2 – coefficient
3 – term
4 – operator
5 – constant term
xyc – variables/constants

Elementary algebra is the most basic form of algebra. It is taught to students who are presumed to have no knowledge of mathematics beyond the basic principles of arithmetic. In arithmetic, only numbers and their arithmetical operations (such as +, −, ×, ÷) occur. In algebra, numbers are often represented by symbols called variables (such as a, n, x, y or z). This is useful because:

- It allows the general formulation of arithmetical laws (such as $a + b = b + a$ for all a and b), and thus is the first step to a systematic exploration of the properties of the real number system.

- It allows the reference to "unknown" numbers, the formulation of equations and the study of how to solve these. (For instance, "Find a number x such that $3x + 1 = 10$" or going a bit further "Find a number x such that $ax + b = c$". This step leads to the conclusion that it is not the nature of the specific numbers that allows us to solve it, but that of the operations involved.)

- It allows the formulation of functional relationships. (For instance, "If you sell x tickets, then your profit will be $3x - 10$ dollars, or $f(x) = 3x - 10$, where f is the function, and x is the number to which the function is applied".)

Polynomials

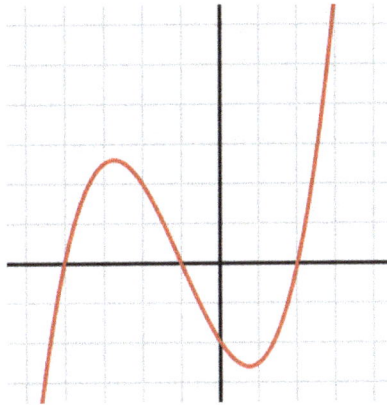

The graph of a polynomial function of degree 3.

A polynomial is an expression that is the sum of a finite number of non-zero terms, each term consisting of the product of a constant and a finite number of variables raised to whole number powers. For example, $x^2 + 2x - 3$ is a polynomial in the single variable x. A polynomial expression is an expression that may be rewritten as a polynomial, by using commutativity, associativity and distributivity of addition and multiplication. For example, $(x - 1)(x + 3)$ is a polynomial expression, that, properly speaking, is not a polynomial. A polynomial function is a function that is defined by a polynomial, or, equivalently, by a polynomial expression. The two preceding examples define the same polynomial function.

Two important and related problems in algebra are the factorization of polynomials, that is, expressing a given polynomial as a product of other polynomials that can not be factored any further, and the computation of polynomial greatest common divisors. The example polynomial above can be factored as $(x - 1)(x + 3)$. A related class of problems is finding algebraic expressions for the roots of a polynomial in a single variable.

Education

It has been suggested that elementary algebra should be taught to students as young as eleven years old, though in recent years it is more common for public lessons to begin at the eighth grade level ($\approx$ 13 y.o. $\pm$) in the United States.

Since 1997, Virginia Tech and some other universities have begun using a personalized model of teaching algebra that combines instant feedback from specialized computer software with one-on-one and small group tutoring, which has reduced costs and increased student achievement.

Abstract Algebra

Abstract algebra extends the familiar concepts found in elementary algebra and arithmetic of numbers to more general concepts. Here are listed fundamental concepts in abstract algebra.

Sets: Rather than just considering the different types of numbers, abstract algebra deals with the more general concept of *sets*: a collection of all objects (called elements) selected by property specific for the set. All collections of the familiar types of numbers are sets. Other examples of sets include the set of all two-by-two matrices, the set of all second-degree polynomials ($ax^2 + bx + c$), the set of all two dimensional vectors in the plane, and the various finite groups such as the cyclic groups, which are the groups of integers modulo n. Set theory is a branch of logic and not technically a branch of algebra.

Binary operations: The notion of addition (+) is abstracted to give a *binary operation*, $*$ say. The notion of binary operation is meaningless without the set on which the operation is defined. For two elements a and b in a set S, $a*b$ is another element in the set; this condition is called closure. Addition (+), subtraction (−), multiplication (×), and division (÷) can be binary operations when defined on different sets, as are addition and multiplication of matrices, vectors, and polynomials.

Identity elements: The numbers zero and one are abstracted to give the notion of an *identity element* for an operation. Zero is the identity element for addition and one is the identity element for multiplication. For a general binary operator $*$ the identity element e must satisfy $a*e = a$ and $e*a = a$, and is necessarily unique, if it exists. This holds for addition as $a + 0 = a$ and $0 + a = a$ and multiplication $a \times 1 = a$ and $1 \times a = a$. Not all sets and operator combinations have an identity element; for example, the set of positive natural numbers (1, 2, 3, ...) has no identity element for addition.

Inverse elements: The negative numbers give rise to the concept of *inverse elements*. For addition, the inverse of a is written $-a$, and for multiplication the inverse is written a^{-1}. A general two-sided inverse element a^{-1} satisfies the property that $a*a^{-1} = e$ and $a^{-1}*a = e$, where e is the identity element.

Associativity: Addition of integers has a property called associativity. That is, the grouping of the numbers to be added does not affect the sum. For example: $(2 + 3) + 4 = 2 + (3 + 4)$. In general, this becomes $(a*b)*c = a*(b*c)$. This property is shared by most binary operations, but not subtraction or division or octonion multiplication.

Commutativity: Addition and multiplication of real numbers are both commutative. That is, the order of the numbers does not affect the result. For example: $2 + 3 = 3 + 2$. In general, this becomes $a*b = b*a$. This property does not hold for all binary operations. For example, matrix multiplication and quaternion multiplication are both non-commutative.

Groups

Combining the above concepts gives one of the most important structures in mathematics: a group. A group is a combination of a set S and a single binary operation*, defined in any way you choose, but with the following properties:

- An identity element e exists, such that for every member a of S, $e*a$ and $a*e$ are both identical to a.

- Every element has an inverse: for every member a of S, there exists a member a^{-1} such that $a*a^{-1}$ and $a^{-1}*a$ are both identical to the identity element.

- The operation is associative: if a, b and c are members of S, then $(a*b)*c$ is identical to $a*(b*c)$.

If a group is also commutative—that is, for any two members a and b of S, $a*b$ is identical to $b*a$—then the group is said to be abelian.

For example, the set of integers under the operation of addition is a group. In this group, the identity element is 0 and the inverse of any element a is its negation, $-a$. The associativity requirement is met, because for any integers a, b and c, $(a + b) + c = a + (b + c)$

The nonzero rational numbers form a group under multiplication. Here, the identity element is 1, since $1 \times a = a \times 1 = a$ for any rational number a. The inverse of a is $1/a$, since $a \times 1/a = 1$.

The integers under the multiplication operation, however, do not form a group. This is because, in general, the multiplicative inverse of an integer is not an integer. For example, 4 is an integer, but its multiplicative inverse is ¼, which is not an integer.

The theory of groups is studied in group theory. A major result in this theory is the classification of finite simple groups, mostly published between about 1955 and 1983, which separates the finite simple groups into roughly 30 basic types.

Semigroups, quasigroups, and monoids are structures similar to groups, but more general. They comprise a set and a closed binary operation, but do not necessarily satisfy the other conditions. A semigroup has an *associative* binary operation, but might not have an identity element. A monoid is a semigroup which does have an identity but might not have an inverse for every element. A quasigroup satisfies a requirement that any element can be turned into any other by either a unique left-multiplication or right-multiplication; however the binary operation might not be associative.

All groups are monoids, and all monoids are semigroups.

Set	Natural numbers N		Integers Z		Rational numbers Q (also real R and complex C numbers)				Integers modulo 3: Z_3 = {0, 1, 2}	
									Examples	
Operation	+	× (w/o zero)	+	× (w/o zero)	+	−	× (w/o zero)	÷ (w/o zero)	+	× (w/o zero)
Closed	Yes	Yes	Yes	Yes	Yes	Yes	Yes	Yes	Yes	Yes
Identity	0	1	0	1	0	N/A	1	N/A	0	1
Inverse	N/A	N/A	−a	N/A	−a	N/A	$1/a$	N/A	0, 2, 1, respectively	N/A, 1, 2, respectively
Associative	Yes	Yes	Yes	Yes	Yes	No	Yes	No	Yes	Yes
Commutative	Yes	Yes	Yes	Yes	Yes	No	Yes	No	Yes	Yes
Structure	monoid	monoid	abelian group	monoid	abelian group	quasigroup	abelian group	quasigroup	abelian group	abelian group (Z_2)

Rings and Fields

Groups just have one binary operation. To fully explain the behaviour of the different types of numbers, structures with two operators need to be studied. The most important of these are rings, and fields.

A ring has two binary operations (+) and (×), with × distributive over +. Under the first operator (+) it forms an *abelian group*. Under the second operator (×) it is associative, but it does not need to have identity, or inverse, so division is not required. The additive (+) identity element is written as 0 and the additive inverse of a is written as $-a$.

Distributivity generalises the *distributive law* for numbers. For the integers $(a + b) \times c = a \times c + b \times c$ and $c \times (a + b) = c \times a + c \times b$, and × is said to be *distributive* over +.

The integers are an example of a ring. The integers have additional properties which make it an integral domain.

A field is a *ring* with the additional property that all the elements excluding 0 form an *abelian group* under ×. The multiplicative (×) identity is written as 1 and the multiplicative inverse of a is written as a^{-1}.

The rational numbers, the real numbers and the complex numbers are all examples of fields.

Algebraic Geometry

Algebraic geometry is a branch of mathematics, classically studying zeros of multivariate polynomials. Modern algebraic geometry is based on the use of abstract algebraic techniques, mainly from commutative algebra, for solving geometrical problems about these sets of zeros.

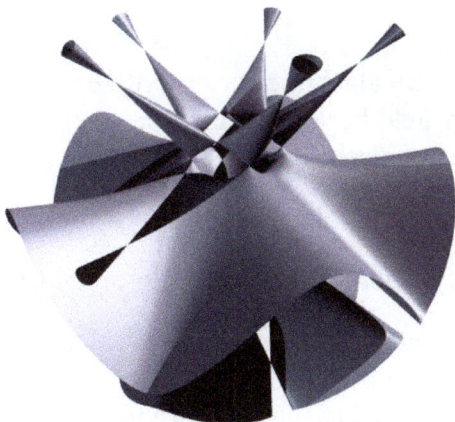

This Togliatti surface is an algebraic surface of degree five. The picture represents a portion of its real locus.

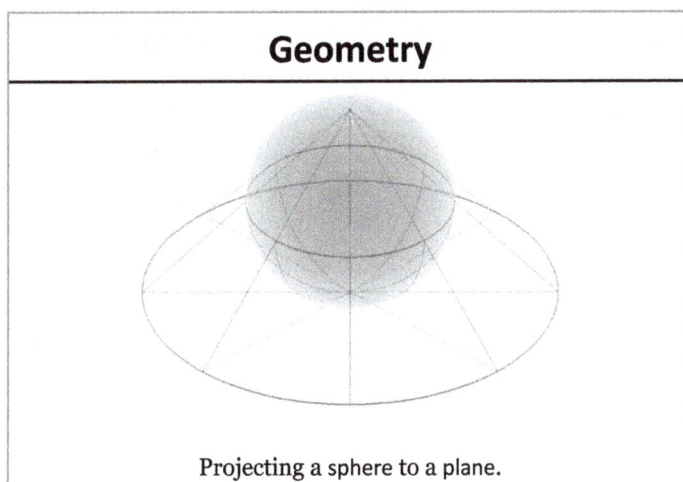

Geometry

Projecting a sphere to a plane.

The fundamental objects of study in algebraic geometry are algebraic varieties, which are geometric manifestations of solutions of systems of polynomial equations. Examples of the most studied classes of algebraic varieties are: plane algebraic curves, which include lines, circles, parabolas, ellipses, hyperbolas, cubic curves like elliptic curves and quartic curves like lemniscates, and Cassini ovals. A point of the plane belongs to an algebraic curve if its coordinates satisfy a given polynomial equation. Basic questions involve the study of the points of special interest like the singular points, the inflection points and the points at infinity. More advanced questions involve the topology of the curve and relations between the curves given by different equations.

Algebraic geometry occupies a central place in modern mathematics and has multiple conceptual connections with such diverse fields as complex analysis, topology and number theory. Initially a study of systems of polynomial equations in several variables, the subject of algebraic geometry starts where equation solving leaves off, and it becomes even more important to understand the intrinsic properties of the totality of solutions of a system of equations, than to find a specific solution; this leads into some of the deepest areas in all of mathematics, both conceptually and in terms of technique.

In the 20th century, algebraic geometry split into several subareas.

- The mainstream of algebraic geometry is devoted to the study of the complex points of the algebraic varieties and more generally to the points with coordinates in an algebraically closed field.

- The study of the points of an algebraic variety with coordinates in the field of the rational numbers or in a number field became arithmetic geometry (or more classically Diophantine geometry), a subfield of algebraic number theory.

- The study of the real points of an algebraic variety is the subject of real algebraic geometry.

- A large part of singularity theory is devoted to the singularities of algebraic varieties.

- With the rise of the computers, a computational algebraic geometry area has emerged, which lies at the intersection of algebraic geometry and computer algebra. It consists essentially in developing algorithms and software for studying and finding the properties of explicitly given algebraic varieties.

Much of the development of the mainstream of algebraic geometry in the 20th century occurred within an abstract algebraic framework, with increasing emphasis being placed on "intrinsic" properties of algebraic varieties not dependent on any particular way of embedding the variety in an ambient coordinate space; this parallels developments in topology, differential and complex geometry. One key achievement of this abstract algebraic geometry is Grothendieck's scheme theory which allows one to use sheaf theory to study algebraic varieties in a way which is very similar to its use in the study of differential and analytic manifolds. This is obtained by extending the notion of point: In classical algebraic geometry, a point of an affine variety may be identified, through Hilbert's Nullstellensatz, with a maximal ideal of the coordinate ring, while the points of the corresponding affine scheme are all prime ideals of this ring. This means that a point of such a scheme may be either a usual point or a subvariety. This approach also enables a unification of the language and the tools of classical algebraic geometry, mainly concerned with complex points, and of algebraic number theory. Wiles's proof of the longstanding conjecture called Fermat's last theorem is an example of the power of this approach.

Basic Notions

Zeros of Simultaneous Polynomials

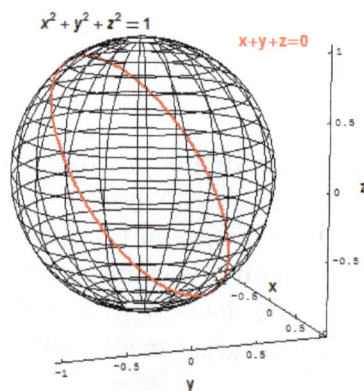

Sphere and slanted circle

In classical algebraic geometry, the main objects of interest are the vanishing sets of collections of polynomials, meaning the set of all points that simultaneously satisfy one or more polynomial equations. For instance, the two-dimensionalsphere in three-dimensional Euclidean spaceR³ could be defined as the set of all points (x,y,z) with

$$x^2 + y^2 + z^2 - 1 = 0.$$

A "slanted" circle in R³ can be defined as the set of all points (x,y,z) which satisfy the two polynomial equations

$$x^2 + y^2 + z^2 - 1 = 0,$$

$$x + y + z = 0.$$

Affine Varieties

First we start with a fieldk. In classical algebraic geometry, this field was always the complex numbers C, but many of the same results are true if we assume only that k is algebraically closed. We consider the affine space of dimension n over k, denoted $A^n(k)$ (or more simply A^n, when k is clear from the context). When one fixes a coordinates system, one may identify $A^n(k)$ with k^n. The purpose of not working with k^n is to emphasize that one "forgets" the vector space structure that k^n carries.

A function $f : A^n \to A^1$ is said to be *polynomial* (or *regular*) if it can be written as a polynomial, that is, if there is a polynomial p in $k[x_1,...,x_n]$ such that $f(M) = p(t_1,...,t_n)$ for every point M with coordinates $(t_1,...,t_n)$ in A^n. The property of a function to be polynomial (or regular) does not depend on the choice of a coordinate system in A^n.

When a coordinate system is chosen, the regular functions on the affine n-space may be identified with the ring of polynomial functions in n variables over k. Therefore, the set of the regular functions on A^n is a ring, which is denoted $k[A^n]$.

We say that a polynomial *vanishes* at a point if evaluating it at that point gives zero. Let S be a set of polynomials in $k[A^n]$. The *vanishing set of S* (or *vanishing locus* or *zero set*) is the set $V(S)$ of all points in A^n where every polynomial in S vanishes. In other words,

$$V(S) = \{(t_1,\ldots,t_n) \mid \forall p \in S, p(t_1,\ldots,t_n) = 0\}.$$

A subset of A^n which is $V(S)$, for some S, is called an *algebraic set*. The V stands for *variety* (a specific type of algebraic set to be defined below).

Given a subset U of A^n, can one recover the set of polynomials which generate it? If U is *any* subset of A^n, define $I(U)$ to be the set of all polynomials whose vanishing set contains U. The I stands for ideal: if two polynomials f and g both vanish on U, then $f+g$ vanishes on U, and if h is any polynomial, then hf vanishes on U, so $I(U)$ is always an ideal of the polynomial ring $k[A^n]$.

Two natural questions to ask are:

- Given a subset U of A^n, when is $U = V(I(U))$?

- Given a set S of polynomials, when is $S = I(V(S))$?

The answer to the first question is provided by introducing the Zariski topology, a topology on A^n whose closed sets are the algebraic sets, and which directly reflects the algebraic structure of $k[A^n]$. Then $U = V(I(U))$ if and only if U is an algebraic set or equivalently a Zariski-closed set. The answer to the second question is given by Hilbert's Nullstellensatz. In one of its forms, it says that $I(V(S))$ is the radical of the ideal generated by S. In more abstract language, there is a Galois connection, giving rise to two closure operators; they can be identified, and naturally play a basic role in the theory; the example is elaborated at Galois connection.

For various reasons we may not always want to work with the entire ideal corresponding to an algebraic set U. Hilbert's basis theorem implies that ideals in $k[A^n]$ are always finitely generated.

An algebraic set is called *irreducible* if it cannot be written as the union of two smaller algebraic sets. Any algebraic set is a finite union of irreducible algebraic sets and this decomposition is unique. Thus its elements are called the *irreducible components* of the algebraic set. An irreducible algebraic set is also called a *variety*. It turns out that an algebraic set is a variety if and only if it may be defined as the vanishing set of a prime ideal of the polynomial ring.

Some authors do not make a clear distinction between algebraic sets and varieties and use *irreducible variety* to make the distinction when needed.

Regular Functions

Just as continuous functions are the natural maps on topological spaces and smooth functions are the natural maps on differentiable manifolds, there is a natural class of functions on an algebraic set, called *regular functions* or *polynomial functions*. A regular function on an algebraic set V contained in A^n is the restriction to V of a regular function on A^n. For an algebraic set defined on the field of the complex numbers, the regular functions are smooth and even analytic.

It may seem unnaturally restrictive to require that a regular function always extend to the ambient space, but it is very similar to the situation in a normaltopological space, where the Tietze extension theorem guarantees that a continuous function on a closed subset always extends to the ambient topological space.

Just as with the regular functions on affine space, the regular functions on V form a ring, which we denote by $k[V]$. This ring is called the *coordinate ring of V*.

Since regular functions on V come from regular functions on A^n, there is a relationship between the coordinate rings. Specifically, if a regular function on V is the restriction of two functions f and g in $k[A^n]$, then $f - g$ is a polynomial function which is null on V and thus belongs to $I(V)$. Thus $k[V]$ may be identified with $k[A^n]/I(V)$.

Morphism of Affine Varieties

Using regular functions from an affine variety to A^1, we can define regular maps from one affine variety to another. First we will define a regular map from a variety into affine space: Let V be a variety contained in A^n. Choose m regular functions on V, and call them $f_1, ..., f_m$. We define a *reg-*

ular map f from V to A^m by letting $f = (f_1, ..., f_m)$. In other words, each f_i determines one coordinate of the range of f.

If V' is a variety contained in A^m, we say that f is a *regular map* from V to V' if the range of f is contained in V'.

The definition of the regular maps apply also to algebraic sets. The regular maps are also called *morphisms*, as they make the collection of all affine algebraic sets into a category, where the objects are the affine algebraic sets and the morphisms are the regular maps. The affine varieties is a subcategory of the category of the algebraic sets.

Given a regular map g from V to V' and a regular function f of $k[V']$, then $f \circ g \in k[V]$. The map $f \rightarrow f \circ g$ is a ring homomorphism from $k[V']$ to $k[V]$. Conversely, every ring homomorphism from $k[V']$ to $k[V]$ defines a regular map from V to V'. This defines an equivalence of categories between the category of algebraic sets and the opposite category of the finitely generated reduced k-algebras. This equivalence is one of the starting points of scheme theory.

Rational Function and Birational Equivalence

Contrarily to the preceding ones, this section concerns only varieties and not algebraic sets. On the other hand, the definitions extend naturally to projective varieties (next section), as an affine variety and its projective completion have the same field of functions.

If V is an affine variety, its coordinate ring is an integral domain and has thus a field of fractions which is denoted $k(V)$ and called the *field of the rational functions* on V or, shortly, the *function field* of V. Its elements are the restrictions to V of the rational functions over the affine space containing V. The domain of a rational function f is not V but the complement of the subvariety (a hypersurface) where the denominator of f vanishes.

Like for regular maps, one may define a *rational map* from a variety V to a variety V'. Like for the regular maps, the rational maps from V to V' may be identified to the field homomorphisms from $k(V')$ to $k(V)$.

Two affine varieties are *birationally equivalent* if there are two rational functions between them which are inverse one to the other in the regions where both are defined. Equivalently, they are birationally equivalent if their function fields are isomorphic.

An affine variety is a *rational variety* if it is birationally equivalent to an affine space. This means that the variety admits a rational parameterization. For example, the circle of equation $x^2 + y^2 - 1 = 0$ is a rational curve, as it has the parameterization

$$x = \frac{2t}{1+t^2}$$

$$y = \frac{1-t^2}{1+t^2},$$

which may also be viewed as a rational map from the line to the circle.

The problem of resolution of singularities is to know if every algebraic variety is birationally equivalent to a variety whose projective completion is nonsingular. It has been positively solved in characteristic 0 by Heisuke Hironaka in 1964 and is yet unsolved in finite characteristic.

Projective Variety

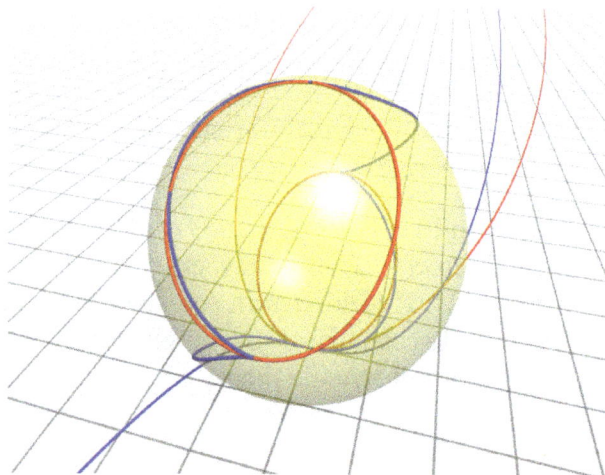

parabola ($y = x^2$, red) and cubic ($y = x^3$, blue) in projective space

Just as the formulas for the roots of 2nd, 3rd and 4th degree polynomials suggest extending real numbers to the more algebraically complete setting of the complex numbers, many properties of algebraic varieties suggest extending affine space to a more geometrically complete projective space. Whereas the complex numbers are obtained by adding the number i, a root of the polynomial x2 + 1, projective space is obtained by adding in appropriate points "at infinity", points where parallel lines may meet.

To see how this might come about, consider the variety $V(y - x^2)$. If we draw it, we get a parabola. As x goes to positive infinity, the slope of the line from the origin to the point (x, x^2) also goes to positive infinity. As x goes to negative infinity, the slope of the same line goes to negative infinity.

Compare this to the variety $V(y - x^3)$. This is a cubic curve. As x goes to positive infinity, the slope of the line from the origin to the point (x, x^3) goes to positive infinity just as before. But unlike before, as x goes to negative infinity, the slope of the same line goes to positive infinity as well; the exact opposite of the parabola. So the behavior "at infinity" of $V(y - x^3)$ is different from the behavior "at infinity" of $V(y - x^2)$.

The consideration of the *projective completion* of the two curves, which is their prolongation "at infinity" in the projective plane, allows to quantify this difference: the point at infinity of the parabola is a regular point, whose tangent is the line at infinity, while the point at infinity of the cubic curve is a cusp. Also, both curves are rational, as they are parameterized by x, and Riemann-Roch theorem implies that the cubic curve must have a singularity, which must be at infinity, as all its points in the affine space are regular.

Thus many of the properties of algebraic varieties, including birational equivalence and all the topological properties, depend on the behavior "at infinity" and so it is natural to study the varieties in projective space. Furthermore, the introduction of projective techniques made many theorems in algebraic geometry simpler and sharper: For example, Bézout's theorem on the number of intersection points between two varieties can be stated in its sharpest form only in projective space. For these reasons, projective space plays a fundamental role in algebraic geometry.

Nowadays, the *projective space* P^n of dimension n is usually defined as the set of the lines passing through a point, considered as the origin, in the affine space of dimension $n+1$, or equivalently to the set of the vector lines in a vector space of dimension $n+1$. When a coordinate system has been chosen in the space of dimension $n+1$, all the points of a line have the same set of coordinates, up to the multiplication by an element of k. This defines the homogeneous coordinates of a point of P^n as a sequence of $n+1$ elements of the base field k, defined up to the multiplication by a nonzero element of k (the same for the whole sequence).

Given a polynomial in $n+1$ variables, it vanishes at all the point of a line passing through the origin if and only if it is homogeneous. In this case, one says that the polynomial *vanishes* at the corresponding point of P^n. This allows to define a *projective algebraic set* in P^n as the set $V(f_1, ..., f_k)$ where a finite set of homogeneous polynomials $\{f_1, ..., f_k\}$ vanishes. Like for affine algebraic sets, there is a bijection between the projective algebraic sets and the reduced homogeneous ideals which define them. The *projective varieties* are the projective algebraic sets whose defining ideal is prime. In other words, a projective variety is a projective algebraic set, whose homogeneous coordinate ring is an integral domain, the *projective coordinates ring* being defined as the quotient of the graded ring or the polynomials in $n+1$ variables by the homogeneous (reduced) ideal defining the variety. Every projective algebraic set may be uniquely decomposed into a finite union of projective varieties.

The only regular functions which may be defined properly on a projective variety are the constant functions. Thus this notion is not used in projective situations. On the other hand, the *field of the rational functions* or *function field* is a useful notion, which, similarly as in the affine case, is defined as the set of the quotients of two homogeneous elements of the same degree in the homogeneous coordinate ring.

Real Algebraic Geometry

The real algebraic geometry is the study of the real points of the algebraic geometry.

The fact that the field of the reals number is an ordered field should not be ignored in such a study. For example, the curve of equation $x^2 + y^2 - a = 0$ is a circle if $a > 0$, , but does not have any real point if $a < 0$.. It follows that real algebraic geometry is not only the study of the real algebraic varieties, but has been generalized to the study of the *semi-algebraic sets*, which are the solutions of systems of polynomial equations and polynomial inequalities. For example, a branch of the hyperbola of equation $xy - 1 = 0$ is not an algebraic variety, but is a semi-algebraic set defined by $xy - 1 = 0$ and $x > 0$ or by $xy - 1 = 0$ and $x + y > 0$.

One of the challenging problems of real algebraic geometry is the unsolved Hilbert's sixteenth problem: Decide which respective positions are possible for the ovals of a nonsingular plane curve of degree 8.

Computational Algebraic Geometry

One may date the origin of computational algebraic geometry to meeting EUROSAM'79 (International Symposium on Symbolic and Algebraic Manipulation) held at Marseille, France in June 1979. At this meeting,

- Dennis S. Arnon showed that George E. Collins's Cylindrical algebraic decomposition (CAD) allows the computation of the topology of semi-algebraic sets,

- Bruno Buchberger presented the Gröbner bases and his algorithm to compute them,

- Daniel Lazard presented a new algorithm for solving systems of homogeneous polynomial equations with a computational complexity which is essentially polynomial in the expected number of solutions and thus simply exponential in the number of the unknowns. This algorithm is strongly related with Macaulay's multivariate resultant.

Since then, most results in this area are related to one or several of these items either by using or improving one of these algorithms, or by finding algorithms whose complexity is simply exponential in the number of the variables.

Gröbner Basis

A Gröbner basis is a system of generators of a polynomial ideal whose computation allows the deduction of many properties of the affine algebraic variety defined by the ideal.

Given an ideal I defining an algebraic set V:

- V is empty (over an algebraically closed extension of the basis field), if and only if the Gröbner basis for any monomial ordering is reduced to $\{1\}$.

- By means of the Hilbert series one may compute the dimension and the degree of V from any Gröbner basis of I for a monomial ordering refining the total degree.

- If the dimension of V is 0, one may compute the points (finite in number) of V from any Gröbner basis of I.

- A Gröbner basis computation allows to remove from V all irreducible components which are contained in a given hyper surface.

- A Gröbner basis computation allows to compute the Zariski closure of the image of V by the projection on the k first coordinates, and the subset of the image where the projection is not proper.

- More generally Gröbner basis computations allows to compute the Zariski closure of the image and the critical points of a rational function of V into another affine variety.

Gröbner basis computations do not allow to compute directly the primary decomposition of I nor the prime ideals defining the irreducible components of V, but most algorithms for this involve Gröbner basis computation. The algorithms which are not based on Gröbner bases use regular chains but may need Gröbner bases in some exceptional situations.

Gröbner base are deemed to be difficult to compute. In fact they may contain, in the worst case, polynomials whose degree is doubly exponential in the number of variables and a number of polynomials which is also doubly exponential. However, this is only a worst case complexity, and the complexity bound of Lazard's algorithm of 1979 may frequently apply. Faugère's F4 and F5 algorithms realize this complexity, as F5 algorithm may be viewed as an improvement of Lazard's 1979 algorithm. It follows that the best implementations allow to compute almost routinely with algebraic sets of degree more than 100. This means that, presently, the difficulty of computing a Gröbner basis is strongly related to the intrinsic difficulty of the problem.

Cylindrical Algebraic Decomposition (CAD)

CAD is an algorithm which had been introduced in 1973 by G. Collins to implement with an acceptable complexity the Tarski–Seidenberg theorem on quantifier elimination over the real numbers.

This theorem concerns the formulas of the first-order logic whose atomic formulas are polynomial equalities or inequalities between polynomials with real coefficients. These formulas are thus the formulas which may be constructed from the atomic formulas by the logical operators *and* ($\wedge$), *or* ($\vee$), *not* ($\neg$), *for all* ($\forall$) and *exists* ($\exists$). Tarski's theorem asserts that, from such a formula, one may compute an equivalent formula without quantifier ($\forall$, $\exists$).

The complexity of CAD is doubly exponential in the number of variables. This means that CAD allow, in theory, to solve every problem of real algebraic geometry which may be expressed by such a formula, that is almost every problem concerning explicitly given varieties and semi-algebraic sets.

While Gröbner basis computation has doubly exponential complexity only in rare cases, CAD has almost always this high complexity. This implies that, unless if most polynomials appearing in the input are linear, it may not solve problems with more than four variables.

Since 1973, most of the research on this subject is devoted either to improve CAD or to find alternate algorithms in special cases of general interest.

As an example of the state of art, there are efficient algorithms to find at least a point in every connected component of a semi-algebraic set, and thus to test if a semi-algebraic set is empty. On the other hand, CAD is yet, in practice, the best algorithm to count the number of connected components.

Asymptotic Complexity vs. Practical Efficiency

The basic general algorithms of computational geometry have a double exponential worst case complexity. More precisely, if d is the maximal degree of the input polynomials and n the number of variables, their complexity is at most $d^{2^{cn}}$ for some constant c, and, for some inputs, the complexity is at least $d^{2^{c'n}}$ for another constant c'.

During the last 20 years of 20th century, various algorithms have been introduced to solve specific subproblems with a better complexity. Most of these algorithms have a complexity $d^{O(n^2)}$.

Among these algorithms which solve a sub problem of the problems solved by Gröbner bases, one may cite *testing if an affine variety is empty* and *solving nonhomogeneous polynomial systems which have a finite number of solutions*. Such algorithms are rarely implemented because, on most entries Faugère's F4 and F5 algorithms have a better practical efficiency and probably a similar or better complexity (*probably* because the evaluation of the complexity of Gröbner basis algorithms on a particular class of entries is a difficult task which has been done only in a few special cases).

The main algorithms of real algebraic geometry which solve a problem solved by CAD are related to the topology of semi-algebraic sets. One may cite *counting the number of connected components, testing if two points are in the same components* or *computing a Whitney stratification of a real algebraic set*. They have a complexity of $d^{O(n^2)}$, but the constant involved by O notation is so high that using them to solve any nontrivial problem effectively solved by CAD, is impossible even if one could use all the existing computing power in the world. Therefore, these algorithms have never been implemented and this is an active research area to search for algorithms with have together a good asymptotic complexity and a good practical efficiency.

Abstract Modern Viewpoint

The modern approaches to algebraic geometry redefine and effectively extend the range of basic objects in various levels of generality to schemes, formal schemes, ind-schemes, algebraic spaces, algebraic stacks and so on. The need for this arises already from the useful ideas within theory of varieties, e.g. the formal functions of Zariski can be accommodated by introducing nilpotent elements in structure rings; considering spaces of loops and arcs, constructing quotients by group actions and developing formal grounds for natural intersection theory and deformation theory lead to some of the further extensions.

Most remarkably, in late 1950s, algebraic varieties were subsumed into Alexander Grothendieck's concept of a scheme. Their local objects are affine schemes or prime spectra which are locally ringed spaces which form a category which is antiequivalent to the category of commutative unital rings, extending the duality between the category of affine algebraic varieties over a field k, and the category of finitely generated reduced k-algebras. The gluing is along Zariski topology; one can glue within the category of locally ringed spaces, but also, using the Yoneda embedding, within the more abstract category of presheaves of sets over the category of affine schemes. The Zariski topology in the set theoretic sense is then replaced by a Grothendieck topology. Grothendieck introduced Grothendieck topologies having in mind more exotic but geometrically finer and more sensitive examples than the crude Zariski topology, namely the étale topology, and the two flat Grothendieck topologies: fppf and fpqc; nowadays some other examples became prominent including Nisnevich topology. Sheaves can be furthermore generalized to stacks in the sense of Grothendieck, usually with some additional representability conditions leading to Artin stacks and, even finer, Deligne-Mumford stacks, both often called algebraic stacks.

Sometimes other algebraic sites replace the category of affine schemes. For example, Nikolai Durov has introduced commutative algebraic monads as a generalization of local objects in a generalized algebraic geometry. Versions of a tropical geometry, of an absolute geometry over a field of one element and an algebraic analogue of Arakelov's geometry were realized in this setup.

Another formal generalization is possible to Universal algebraic geometry in which every variety of algebras has its own algebraic geometry. The term *variety of algebras* should not be confused with *algebraic variety*.

The language of schemes, stacks and generalizations has proved to be a valuable way of dealing with geometric concepts and became cornerstones of modern algebraic geometry.

Algebraic stacks can be further generalized and for many practical questions like deformation theory and intersection theory, this is often the most natural approach. One can extend the Grothendieck site of affine schemes to a higher categorical site of derived affine schemes, by replacing the commutative rings with an infinity category of differential graded commutative algebras, or of simplicial commutative rings or a similar category with an appropriate variant of a Grothendieck topology. One can also replace presheaves of sets by presheaves of simplicial sets (or of infinity groupoids). Then, in presence of an appropriate homotopic machinery one can develop a notion of derived stack as such a presheaf on the infinity category of derived affine schemes, which is satisfying certain infinite categorical version of a sheaf axiom (and to be algebraic, inductively a sequence of representability conditions). Quillen model categories, Segal categories and quasicategories are some of the most often used tools to formalize this yielding the *derived algebraic geometry*, introduced by the school of Carlos Simpson, including Andre Hirschowitz, Bertrand Toën, Gabrielle Vezzosi, Michel Vaquié and others; and developed further by Jacob Lurie, Bertrand Toën, and Gabrielle Vezzosi. Another (noncommutative) version of derived algebraic geometry, using A-infinity categories has been developed from early 1990s by Maxim Kontsevich and followers.

History

Prehistory: Before The 16th Century

Some of the roots of algebraic geometry date back to the work of the Hellenistic Greeks from the 5th century BC. The Delian problem, for instance, was to construct a length x so that the cube of side x contained the same volume as the rectangular box a^2b for given sides a and b. Menaechmus (circa 350 BC) considered the problem geometrically by intersecting the pair of plane conics $ay = x^2$ and $xy = ab$. The later work, in the 3rd century BC, of Archimedes and Apollonius studied more systematically problems on conic sections, and also involved the use of coordinates. The Arab mathematicians were able to solve by purely algebraic means certain cubic equations, and then to interpret the results geometrically. This was done, for instance, by Ibn al-Haytham in the 10th century AD. Subsequently, Persian mathematician Omar Khayyám (born 1048 A.D.) discovered the general method of solving cubic equations by intersecting a parabola with a circle. Each of these early developments in algebraic geometry dealt with questions of finding and describing the intersections of algebraic curves.

Renaissance

Such techniques of applying geometrical constructions to algebraic problems were also adopted by a number of Renaissance mathematicians such as Gerolamo Cardano and Niccolò Fontana "Tartaglia" on their studies of the cubic equation. The geometrical approach to construction problems, rather than the algebraic one, was favored by most 16th and 17th century mathematicians, notably Blaise Pascal who argued against the use of algebraic and analytical methods in geometry. The French mathematicians Franciscus Vieta and later René Descartes and Pierre de Fermat revolutionized the conventional way of thinking about construction problems through the introduction of coordinate geometry. They were interested primarily in the properties of *algebraic curves*, such as those defined by Diophantine equations (in the case of Fermat), and the algebraic reformulation of the classical Greek works on conics and cubics (in the case of Descartes).

During the same period, Blaise Pascal and Gérard Desargues approached geometry from a different perspective, developing the synthetic notions of projective geometry. Pascal and Desargues also studied curves, but from the purely geometrical point of view: the analog of the Greek *ruler and compass construction*. Ultimately, the analytic geometry of Descartes and Fermat won out, for it supplied the 18th century mathematicians with concrete quantitative tools needed to study physical problems using the new calculus of Newton and Leibniz. However, by the end of the 18th century, most of the algebraic character of coordinate geometry was subsumed by the *calculus of infinitesimals* of Lagrange and Euler.

19th and Early 20th Century

It took the simultaneous 19th century developments of non-Euclidean geometry and Abelian integrals in order to bring the old algebraic ideas back into the geometrical fold. The first of these new developments was seized up by Edmond Laguerre and Arthur Cayley, who attempted to ascertain the generalized metric properties of projective space. Cayley introduced the idea of *homogeneous polynomial forms*, and more specifically quadratic forms, on projective space. Subsequently, Felix Klein studied projective geometry (along with other types of geometry) from the viewpoint that the geometry on a space is encoded in a certain class of transformations on the space. By the end of the 19th century, projective geometers were studying more general kinds of transformations on figures in projective space. Rather than the projective linear transformations which were normally regarded as giving the fundamental Kleinian geometry on projective space, they concerned themselves also with the higher degree birational transformations. This weaker notion of congruence would later lead members of the 20th century Italian school of algebraic geometry to classify algebraic surfaces up to birational isomorphism.

The second early 19th century development, that of Abelian integrals, would lead Bernhard Riemann to the development of Riemann surfaces.

In the same period began the algebraization of the algebraic geometry through commutative algebra. The prominent results in this direction are Hilbert's basis theorem and Hilbert's Nullstellensatz, which are the basis of the connexion between algebraic geometry and commutative algebra, and Macaulay's multivariate resultant, which is the basis of elimination theory. Probably because of the size of the computation which is implied by multivariate resultants, elimination theory was forgotten during the middle of the 20th century until it was renewed by singularity theory and computational algebraic geometry.

20th Century

B. L. van der Waerden, Oscar Zariski and André Weil developed a foundation for algebraic geometry based on contemporary commutative algebra, including valuation theory and the theory of ideals. One of the goals was to give a rigorous framework for proving the results of Italian school of algebraic geometry. In particular, this school used systematically the notion of generic point without any precise definition, which was first given by these authors during the 1930s.

In the 1950s and 1960s Jean-Pierre Serre and Alexander Grothendieck recast the foundations making use of sheaf theory. Later, from about 1960, and largely led by Grothendieck, the idea of schemes was worked out, in conjunction with a very refined apparatus of homological techniques. After a decade of rapid development the field stabilized in the 1970s, and new applications were made, both to number theory and to more classical geometric questions on algebraic varieties, singularities and moduli.

An important class of varieties, not easily understood directly from their defining equations, are the abelian varieties, which are the projective varieties whose points form an abelian group. The prototypical examples are the elliptic curves, which have a rich theory. They were instrumental in the proof of Fermat's last theorem and are also used in elliptic curve cryptography.

In parallel with the abstract trend of the algebraic geometry, which is concerned with general statements about varieties, methods for effective computation with concretely-given varieties have also been developed, which lead to the new area of computational algebraic geometry. One of the founding methods of this area is the theory of Gröbner bases, introduced by Bruno Buchberger in 1965. Another founding method, more specially devoted to real algebraic geometry, is the cylindrical algebraic decomposition, introduced by George E. Collins in 1973.

Analytic Geometry

An analytic variety is defined locally as the set of common solutions of several equations involving analytic functions. It is analogous to the included concept of real or complex algebraic variety. Any complex manifold is an analytic variety. Since analytic varieties may have singular points, not all analytic varieties are manifolds.

Modern analytic geometry is essentially equivalent to real and complex algebraic geometry, as has been shown by Jean-Pierre Serre in his paper *GAGA*, the name of which is French for *Algebraic geometry and analytic geometry*. Nevertheless, the two fields remain distinct, as the methods of proof are quite different and algebraic geometry includes also geometry in finite characteristic.

Applications

Algebraic geometry now finds applications in statistics,control theory,robotics,error-correcting codes,phylogenetics and geometric modelling. There are also connections to string theory,game theory,graph matchings,solitons and integer programming.

References

- Martin J. Turner,Jonathan M. Blackledge,Patrick R. Andrews (1998). Fractal geometry in digital imaging. Academic Press. p. 1. ISBN 0-12-703970-8

- Otto Neugebauer (1975). A history of ancient mathematical astronomy. 1. Springer-Verlag. pp. 744–. ISBN 978-3-540-06995-9.

- Grattan-Guinness, Ivor (1997). The Rainbow of Mathematics: A History of the Mathematical Sciences. W.W. Norton. ISBN 0-393-32030-8.

- Ossendrijver, Mathieu (29 Jan 2016). "Ancient Babylonian astronomers calculated Jupiter's position from the area under a time-velocity graph". Science. 351 (6272): 482–484. doi:10.1126/science.aad8085. Retrieved 29 January 2016.

Processes of Mathematics

Mathematical proof is patterned according to the stages that a mathematical equation or postulate follows while being examined. There are many components to mathematical proof such as calculation, mathematical proposition and measurement. The aspects elucidated in this chapter are of vital importance, and provide a better understanding of mathematics.

Mathematical Proof

In mathematics, a proof is a deductive argument for a mathematical statement. In the argument, other previously established statements, such as theorems, can be used. In principle, a proof can be traced back to self-evident or assumed statements, known as axioms, along with accepted rules of inference. Axioms may be treated as conditions that must be met before the statement applies. Proofs are examples of deductive reasoning and are distinguished from inductive or empirical arguments; a proof must demonstrate that a statement is always true (occasionally by listing *all* possible cases and showing that it holds in each), rather than enumerate many confirmatory cases. An unproved proposition that is believed to be true is known as a conjecture.

P. Oxy. 29, one of the oldest surviving fragments of Euclid's *Elements*, a textbook used for millennia to teach proof-writing techniques. The diagram accompanies Book II, Proposition 5.

Proofs employ logic but usually include some amount of natural language which usually admits some ambiguity. In fact, the vast majority of proofs in written mathematics can be considered as applications of rigorous informal logic. Purely formal proofs, written in symbolic language instead of natural language, are considered in proof theory. The distinction between formal and informal proofs has led to much examination of current and historical mathematical practice, quasi-empiricism in mathematics, and so-called folk mathematics (in both senses of that term). The philosophy of mathematics is concerned with the role of language and logic in proofs, and mathematics as a language.

History and Etymology

The word "proof" comes from the Latin *probare* meaning "to test". Related modern words are the English "probe", "probation", and "probability", the Spanish *probar* (to smell or taste, or (lesser use) touch or test), Italian *provare* (to try), and the German *probieren* (to try). The early use of "probity" was in the presentation of legal evidence. A person of authority, such as a nobleman, was said to have probity, whereby the evidence was by his relative authority, which outweighed empirical testimony.

Plausibility arguments using heuristic devices such as pictures and analogies preceded strict mathematical proof. It is likely that the idea of demonstrating a conclusion first arose in connection with geometry, which originally meant the same as "land measurement". The development of mathematical proof is primarily the product of ancient Greek mathematics, and one of the greatest achievements thereof. Thales (624–546 BCE) proved some theorems in geometry. Eudoxus (408–355 BCE) and Theaetetus (417–369 BCE) formulated theorems but did not prove them. Aristotle (384–322 BCE) said definitions should describe the concept being defined in terms of other concepts already known. Mathematical proofs were revolutionized by Euclid (300 BCE), who introduced the axiomatic method still in use today, starting with undefined terms and axioms (propositions regarding the undefined terms assumed to be self-evidently true from the Greek "axios" meaning "something worthy"), and used these to prove theorems using deductive logic. His book, the *Elements*, was read by anyone who was considered educated in the West until the middle of the 20th century. In addition to theorems of geometry, such as the Pythagorean theorem, the *Elements* also covers number theory, including a proof that the square root of two is irrational and that there are infinitely many prime numbers.

Further advances took place in medieval Islamic mathematics. While earlier Greek proofs were largely geometric demonstrations, the development of arithmetic and algebra by Islamic mathematicians allowed more general proofs that no longer depended on geometry. In the 10th century CE, the Iraqi mathematician Al-Hashimi provided general proofs for numbers (rather than geometric demonstrations) as he considered multiplication, division, etc. for "lines." He used this method to provide a proof of the existence of irrational numbers. An inductive proof for arithmetic sequences was introduced in the *Al-Fakhri* (1000) by Al-Karaji, who used it to prove the binomial theorem and properties of Pascal's triangle. Alhazen also developed the method of proof by contradiction, as the first attempt at proving the Euclideanparallel postulate.

Modern proof theory treats proofs as inductively defined data structures. There is no longer an assumption that axioms are "true" in any sense; this allows for parallel mathematical theories built on alternate sets of axioms.

Nature and Purpose

As practiced, a proof is expressed in natural language and is a rigorous argument intended to convince the audience of the truth of a statement. The standard of rigor is not absolute and has varied throughout history. A proof can be presented differently depending on the intended audience. In order to gain acceptance, a proof has to meet communal statements of rigor; an argument considered vague or incomplete may be rejected.

The concept of a proof is formalized in the field of mathematical logic. A formal proof is written in a formal language instead of a natural language. A formal proof is defined as sequence of formulas in a formal language, in which each formula is a logical consequence of preceding formulas. Having a definition of formal proof makes the concept of proof amenable to study. Indeed, the field of proof theory studies formal proofs and their properties, for example, the property that a statement has a formal proof. An application of proof theory is to show that certain undecidable statements are not provable.

The definition of a formal proof is intended to capture the concept of proofs as written in the practice of mathematics. The soundness of this definition amounts to the belief that a published proof can, in principle, be converted into a formal proof. However, outside the field of automated proof assistants, this is rarely done in practice. A classic question in philosophy asks whether mathematical proofs are analytic or synthetic. Kant, who introduced the analytic-synthetic distinction, believed mathematical proofs are synthetic.

Proofs may be viewed as aesthetic objects, admired for their mathematical beauty. The mathematician Paul Erdős was known for describing proofs he found particularly elegant as coming from "The Book", a hypothetical tome containing the most beautiful method(s) of proving each theorem. The book *Proofs from THE BOOK*, published in 2003, is devoted to presenting 32 proofs its editors find particularly pleasing.

Methods

Direct Proof

In direct proof, the conclusion is established by logically combining the axioms, definitions, and earlier theorems. For example, direct proof can be used to establish that the sum of two evenintegers is always even:

Consider two even integers x and y. Since they are even, they can be written as $x = 2a$ and $y = 2b$, respectively, for integers a and b. Then the sum $x + y = 2a + 2b = 2(a+b)$. Therefore $x+y$ has 2 as a factor and, by definition, is even. Hence the sum of any two even integers is even.

This proof uses the definition of even integers, the integer properties of closure under addition and multiplication, and distributivity.

Proof By Mathematical Induction

Despite its name, mathematical induction is a method of deduction, not a form of inductive reasoning. In proof by mathematical induction, a single "base case" is proved, and an "induction rule" is proved that establishes that any arbitrary case implies the next case. Since in principle the induction rule can be applied repeatedly starting from the proved base case, we see that all (usually infinitely many) cases are provable. This avoids having to prove each case individually. A variant of mathematical induction is proof by infinite descent, which can be used, for example, to prove the irrationality of the square root of two.

A common application of proof by mathematical induction is to prove that a property known to hold for one number holds for all natural numbers: Let N = {1,2,3,4,...} be the set of natural numbers, and $P(n)$ be a mathematical statement involving the natural number n belonging to N such that

- (i)$P(1)$ is true, i.e., $P(n)$ is true for $n = 1$.

- (ii)$P(n+1)$ is true whenever $P(n)$ is true, i.e., $P(n)$ is true implies that $P(n+1)$ is true.

- Then $P(n)$ is true for all natural numbers n.

For example, we can prove by induction that all positive integers of the form $2n - 1$ are odd. Let $P(n)$ represent "$2n - 1$ is odd":

(i) For $n = 1$, $2n - 1 = 2(1) - 1 = 1$, and 1 is odd, since it leaves a remainder of 1 when divided by 2. Thus $P(1)$ is true.

(ii) For any n, if $2n - 1$ is odd ($P(n)$), then $(2n - 1) + 2$ must also be odd, because adding 2 to an odd number results in an odd number. But $(2n - 1) + 2 = 2n + 1 = 2(n+1) - 1$, so $2(n+1) - 1$ is odd ($P(n+1)$). So $P(n)$ implies $P(n+1)$.

Thus$2n - 1$ is odd, for all positive integers n.

The shorter phrase "proof by induction" is often used instead of "proof by mathematical induction".

Proof by Contraposition

Proof by contrapositioninfers the conclusion "if p then q" from the premise "if *not q* then *not p*". The statement "if *not q* then *not p*" is called the contrapositive of the statement "if p then q". For example, contraposition can be used to establish that, given an integer x, if x^2 is even, then x is even:

Suppose x is not even. Then x is odd. The product of two odd numbers is odd, hence $x^2 = x \cdot x$ is odd. Thus x^2 is not even. Thus, if x^2 *is* even, the supposition must be false, so x has to be even.

Proof by Contradiction

In proof by contradiction (also known as *reductio ad absurdum*, Latin for "by reduction to the absurd"), it is shown that if some statement were true, a logical contradiction occurs, hence the statement must be false. A famous example of proof by contradiction shows that $\sqrt{2}$ is an irrational number:

Suppose that $\sqrt{2}$ were a rational number, so by definition $\sqrt{2} = \dfrac{a}{b}$ where a and b are non-zero integers with no common factor. (If there is a common factor, divide both numerator and denominator by that factor to remove it, and repeat until no common factor remains. By the method of infinite descent, this process must terminate.) Thus, $b\sqrt{2} = a$. Squaring both sides yields $2b^2 = a^2$. Since 2 divides the left hand side, 2 must also divide the right hand side (otherwise an even number would equal an odd number). So a^2 is even, which implies that a must also be even. So we can write $a = 2c$, where c is also an integer. Substitution into the original equation yields $2b^2 = (2c)^2 = 4c^2$. Dividing both sides by 2 yields $b^2 = 2c^2$. But then, by the same argument as before, 2 divides b^2, so b must be even. However, if a and b are both even, they have a common factor, namely 2. This contradicts our initial supposition, so we are forced to conclude that $\sqrt{2}$ is an irrational number.

Proof by Construction

Proof by construction, or proof by example, is the construction of a concrete example with a property to show that something having that property exists. Joseph Liouville, for instance, proved the existence of transcendental numbers by constructing an explicit example. It can also be used to construct a counterexample to disprove a proposition that all elements have a certain property.

Proof by Exhaustion

In proof by exhaustion, the conclusion is established by dividing it into a finite number of cases and proving each one separately. The number of cases sometimes can become very large. For example, the first proof of the four color theorem was a proof by exhaustion with 1,936 cases. This proof was controversial because the majority of the cases were checked by a computer program, not by hand.

Probabilistic Proof

A probabilistic proof is one in which an example is shown to exist, with certainty, by using methods of probability theory. Probabilistic proof, like proof by construction, is one of many ways to show existence theorems.

This is not to be confused with an argument that a theorem is 'probably' true, a 'plausibility argument'. The work on the Collatz conjecture shows how far plausibility is from genuine proof.

Combinatorial Proof

A combinatorial proof establishes the equivalence of different expressions by showing that they count the same object in different ways. Often a bijection between two sets is used to show that the expressions for their two sizes are equal. Alternatively, a double counting argument provides two different expressions for the size of a single set, again showing that the two expressions are equal.

Nonconstructive Proof

A nonconstructive proof establishes that a mathematical object with a certain property exists without explaining how such an object can be found. Often, this takes the form of a proof by contradiction in which the nonexistence of the object is proved to be impossible. In contrast, a constructive proof establishes that a particular object exists by providing a method of finding it. A famous example of a nonconstructive proof shows that there exist two irrational numbers a and b such that a^b is a rational number:

Either $\sqrt{2}^{\sqrt{2}}$ is a rational number and we are done (take $a = b = \sqrt{2}$), or $\sqrt{2}^{\sqrt{2}}$ is irrational so we can write $a = \sqrt{2}^{\sqrt{2}}$ and $b = \sqrt{2}$. This then gives $\left(\sqrt{2}^{\sqrt{2}}\right)^{\sqrt{2}} = \sqrt{2}^2 = 2$ $\left(\sqrt{2}^{\sqrt{2}}\right)^{\sqrt{2}} = \sqrt{2}^2 = 2$, which is thus a rational of the form a^b.

Statistical Proofs in Pure Mathematics

The expression "statistical proof" may be used technically or colloquially in areas of pure mathematics, such as involving cryptography, chaotic series, and probabilistic or analytic number theory. It is less commonly used to refer to a mathematical proof in the branch of mathematics known as mathematical statistics.

Computer-assisted Proofs

Until the twentieth century it was assumed that any proof could, in principle, be checked by a competent mathematician to confirm its validity. However, computers are now used both to prove theorems and to carry out calculations that are too long for any human or team of humans to check; the first proof of the four color theorem is an example of a computer-assisted proof. Some mathematicians are concerned that the possibility of an error in a computer program or a run-time error in its calculations calls the validity of such computer-assisted proofs into question. In practice, the chances of an error invalidating a computer-assisted proof can be reduced by incorporating redundancy and self-checks into calculations, and by developing multiple independent approaches and programs. Errors can never be completely ruled out in case of verification of a proof by humans either, especially if the proof contains natural language and requires deep mathematical insight.

Undecidable Statements

A statement that is neither provable nor disprovable from a set of axioms is called undecidable (from those axioms). One example is the parallel postulate, which is neither provable nor refutable from the remaining axioms of Euclidean geometry.

Mathematicians have shown there are many statements that are neither provable nor disprovable in Zermelo-Fraenkel set theory with the axiom of choice (ZFC), the standard system of set theory in mathematics (assuming that ZFC is consistent).

Gödel's (first) incompleteness theorem shows that many axiom systems of mathematical interest will have undecidable statements.

Heuristic Mathematics and Experimental Mathematics

While early mathematicians such as Eudoxus of Cnidus did not use proofs, from Euclid to the foundational mathematics developments of the late 19th and 20th centuries, proofs were an essential part of mathematics. With the increase in computing power in the 1960s, significant work began to be done investigating mathematical objects outside of the proof-theorem framework, in experimental mathematics. Early pioneers of these methods intended the work ultimately to be embedded in a classical proof-theorem framework, e.g. the early development of fractal geometry, which was ultimately so embedded.

Related Concepts

Visual Proof

Although not a formal proof, a visual demonstration of a mathematical theorem is sometimes called a "proof without words". The left-hand picture below is an example of a historic visual proof of the Pythagorean theorem in the case of the (3,4,5) triangle.

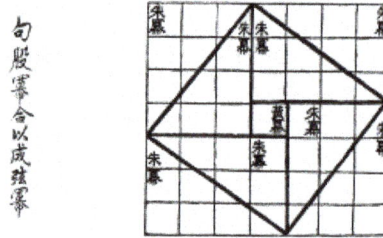

Visual proof for the (3, 4, 5) triangle as in the Chou Pei Suan Ching 500–200 BC.

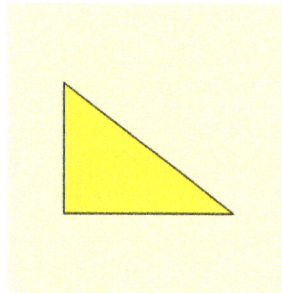

Animated visual proof for the Pythagorean theorem by rearrangement.

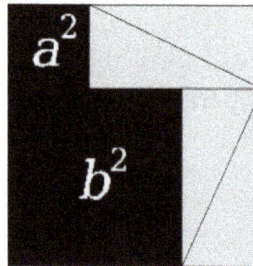

A second animated proof of the Pythagorean theorem.

Some illusory visual proofs, such as the missing square puzzle, can be constructed in a way which appear to prove a supposed mathematical fact but only do so under the presence of tiny errors (for example, supposedly straight lines which actually bend slightly) which are unnoticeable until the entire picture is closely examined, with lengths and angles precisely measured or calculated.

Elementary Proof

An elementary proof is a proof which only uses basic techniques. More specifically, the term is used in number theory to refer to proofs that make no use of complex analysis. For some time it was thought that certain theorems, like the prime number theorem, could only be proved using "higher" mathematics. However, over time, many of these results have been reproved using only elementary techniques.

Two-column Proof

A particular way of organising a proof using two parallel columns is often used in elementary geometry classes in the United States. The proof is written as a series of lines in two columns. In each line, the left-hand column contains a proposition, while the right-hand column contains a brief explanation of how the corresponding proposition in the left-hand column is either an axiom, a hypothesis, or can be logically derived from previous propositions. The left-hand column is typically headed "Statements" and the right-hand column is typically headed "Reasons".

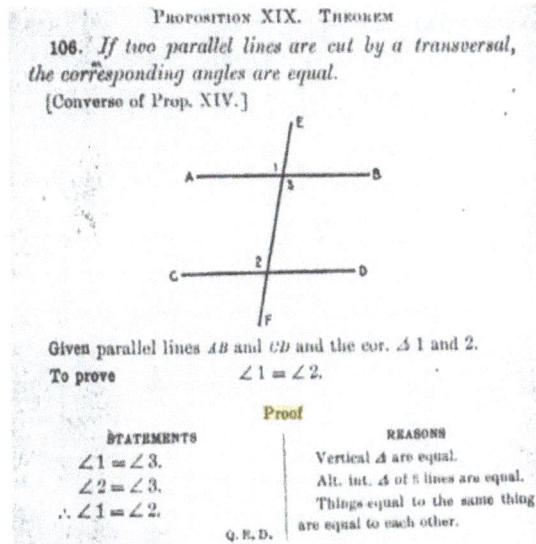

A two-column proof published in 1913

Colloquial Use of "Mathematical Proof"

The expression "mathematical proof" is used by lay people to refer to using mathematical methods or arguing with mathematical objects, such as numbers, to demonstrate something about everyday life, or when data used in an argument is numerical. It is sometimes also used to mean a "statistical proof" (below), especially when used to argue from data.

Statistical Proof Using Data

"Statistical proof" from data refers to the application of statistics, data analysis, or Bayesian analysis to infer propositions regarding the probability of data. While *using* mathematical proof to establish theorems in statistics, it is usually not a mathematical proof in that the *assumptions* from which probability statements are derived require empirical evidence from outside mathematics to verify. In physics, in addition to statistical methods, "statistical proof" can refer to the specialized *mathematical methods of physics* applied to analyze data in a particle physicsexperiment or observational study in cosmology. "Statistical proof" may also refer to raw data or a convincing diagram involving data, such as scatter plots, when the data or diagram is adequately convincing without further analysis.

Inductive Logic Proofs and Bayesian Analysis

Proofs using inductive logic, while considered mathematical in nature, seek to establish propositions with a degree of certainty, which acts in a similar manner to probability, and may be less than full certainty. Inductive logic should not be confused with mathematical induction.

Bayesian analysis uses Bayes' theorem to update a person's assessment of likelihoods of hypotheses when new evidence or information is acquired.

Proofs As Mental Objects

Psychologism views mathematical proofs as psychological or mental objects. Mathematician philosophers, such as Leibniz, Frege, and Carnap have variously criticized this view and attempted to develop a semantics for what they considered to be the language of thought, whereby standards of mathematical proof might be applied to empirical science.

Influence of Mathematical Proof Methods Outside Mathematics

Philosopher-mathematicians such as Spinoza have attempted to formulate philosophical arguments in an axiomatic manner, whereby mathematical proof standards could be applied to argumentation in general philosophy. Other mathematician-philosophers have tried to use standards of mathematical proof and reason, without empiricism, to arrive at statements outside of mathematics, but having the certainty of propositions deduced in a mathematical proof, such as Descartes' *cogito* argument.

Ending A Proof

Sometimes, the abbreviation *"Q.E.D."* is written to indicate the end of a proof. This abbreviation stands for *"Quod Erat Demonstrandum"*, which is Latin for *"that which was to be demonstrated"*. A more common alternative is to use a square or a rectangle, such as □ or ∎, known as a "tombstone" or "halmos" after its eponymPaul Halmos. Often, "which was to be shown" is verbally stated when writing "QED", "□", or "∎" during an oral presentation.

Calculation

A calculation is a deliberate process that transforms one or more inputs into one or more results, with variable change.

The term is used in a variety of senses, from the very definite arithmetical calculation of using an algorithm, to the vague heuristics of calculating a strategy in a competition, or calculating the chance of a successful relationship between two people.

For example, multiplying 7 by 6 is a simple algorithmic calculation. Estimating the fair price for financial instruments using the Black–Scholes model is a complex algorithmic calculation.

Statistical estimations of the likely election results from opinion polls also involve algorithmic calculations, but produces ranges of possibilities rather than exact answers.

To *calculate* means to ascertain by computing. The English word derives from the Latin*calculus*, which originally meant a small stone in the gall-bladder (from Latin *calx*). It also meant a pebble used for calculating, or a small stone used as a counter in an abacus. The abacus was an instrument used by Greeks and Romans for arithmetic calculations, preceding the slide-rule and the electronic calculator, and consisted of perforated pebbles sliding on an iron bars.

Comparison to Computation

Calculate comes from the Greek word *Κάχληκα* or gravel in English because Greeks used gravel for counting. Calculation is a prerequisite for computation. The difference in the meaning of *calculation* and *computation* appears to originate from the late medieval period.

Measurement

Measurement is the assignment of a number to a characteristic of an object or event, which can be compared with other objects or events. The scope and application of a measurement is dependent on the context and discipline. In the natural sciences and engineering, measurements do not apply to nominal properties of objects or events, which is consistent with the guidelines of the *International vocabulary of metrology* published by the International Bureau of Weights and Measures. However, in other fields such as statistics as well as the social and behavioral sciences, measurements can have multiple levels, which would include nominal, ordinal, interval, and ratio scales.

Measurement is a cornerstone of trade, science, technology, and quantitative research in many disciplines. Historically, many measurement systems existed for the varied fields of human existence to facilitate comparisons in these fields. Often these were achieved by local agreements between trading partners or collaborators. Since the 18th century, developments progressed towards unifying, widely accepted standards that resulted in the modern International System of Units (SI). This system reduces all physical measurements to a mathematical combination of seven base units. The science of measurement is pursued in the field of metrology.

A typical tape measure with both Metric and Imperial units and two US pennies for comparison

Methodology

The measurement of a property may be categorized by the following criteria: type, magnitude, unit, and uncertainty. They enable unambiguous comparisons between measurements.

The *type* or *level* of measurement is a taxonomy for the methodological character of a comparison.

For example, two states of a property may be compared by ratio, difference, or ordinal preference. The type is commonly not explicitly expressed, but implicit in the definition of a measurement procedure. The *magnitude* is the numerical value of the characterization, usually obtained with a suitably chosen measuring instrument. A *unit* assigns a mathematical weighting factor to the magnitude that is derived as a ratio to the property of an artifact used as standard or a natural physical quantity. An *uncertainty* represents the random and systemic errors of the measurement procedure; it indicates a confidence level in the measurement. Errors are evaluated by methodically repeating measurements and considering the accuracy and precision of the measuring instrument.

Standardization of Measurement Units

Measurements most commonly use the International System of Units (SI) as a comparison framework. The system defines seven fundamental units: kilogram, metre, candela, second, ampere, kelvin, and mole. Six of these units are defined without reference to a particular physical object which serves as a standard (artifact-free), with the exception of the kilogram which is still embodied in an artifact which rests at the BIPM outside Paris.

Artifact-free definitions fix measurements at an exact value related to a physical constant or other invariable phenomena in nature, in contrast to standard artifacts which are subject to deterioration or destruction. Instead, the measurement unit can only ever change through increased accuracy in determining the value of the constant it is tied to.

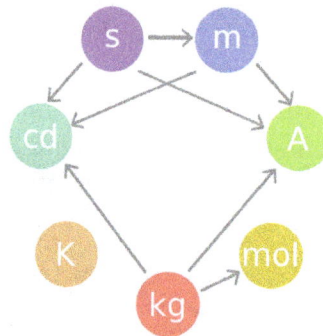

The seven base units in the SI system. Arrows point from units to those that depend on them.

The first proposal to tie an SI base unit to an experimental standard independent of fiat was by Charles Sanders Peirce (1839–1914), who proposed to define the metre in terms of the wavelength of a spectral line. This directly influenced the Michelson–Morley experiment; Michelson and Morley cite Peirce, and improve on his method.

Standards

With the exception of a few fundamental quantum constants, units of measurement are derived from historical agreements. Nothing inherent in nature dictates that an inch has to be a certain length, nor that a mile is a better measure of distance than a kilometre. Over the course of human history, however, first for convenience and then for necessity, standards of measurement evolved so that communities would have certain common benchmarks. Laws regulating measurement were originally developed to prevent fraud in commerce.

Units of measurement are generally defined on a scientific basis, overseen by governmental or independent agencies, and established in international treaties, pre-eminent of which is the General Conference on Weights and Measures (CGPM), established in 1875 by the Treaty of the metre and which oversees the International System of Units (SI) and which has custody of the International Prototype Kilogram. The metre, for example, was redefined in 1983 by the CGPM as the distance traveled by light in free space in 1⁄299,792,458 of a second while in 1960 the international yard was defined by the governments of the United States, United Kingdom, Australia and South Africa as being *exactly* 0.9144 metres.

In the United States, the National Institute of Standards and Technology (NIST), a division of the United States Department of Commerce, regulates commercial measurements. In the United Kingdom, the role is performed by the National Physical Laboratory (NPL), in Australia by the National Measurement Institute, in South Africa by the Council for Scientific and Industrial Research and in India the National Physical Laboratory of India.

Units and Systems

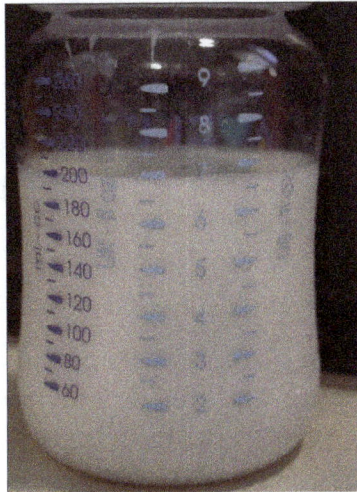

A baby bottle that measures in three measurement systems, Imperial (U.K.), U.S. customary, and metric.

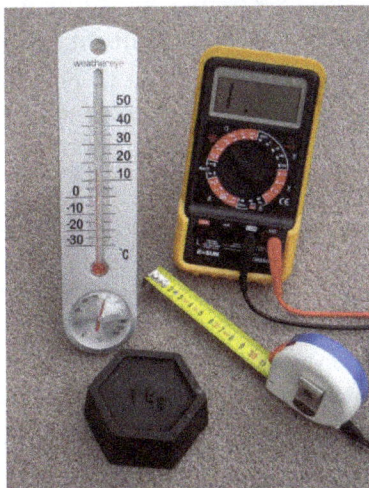

Four measuring devices having metric calibrations

Imperial and US Customary Systems

Before SI units were widely adopted around the world, the British systems of English units and later imperial units were used in Britain, the Commonwealth and the United States. The system came to be known as U.S. customary units in the United States and is still in use there and in a few Caribbean countries. These various systems of measurement have at times been called *foot-pound-second* systems after the Imperial units for length, weight and time even though the tons, hundredweights, gallons, and nautical miles, for example, are different for the U.S. units. Many Imperial units remain in use in Britain, which has officially switched to the SI system—with a few exceptions such as road signs, which are still in miles. Draught beer and cider must be sold by the imperial pint, and milk in returnable bottles can be sold by the imperial pint. Many people measure their height in feet and inches and their weight in stone and pounds, to give just a few examples. Imperial units are used in many other places, for example, in many Commonwealth countries that are considered metricated, land area is measured in acres and floor space in square feet, particularly for commercial transactions (rather than government statistics). Similarly, gasoline is sold by the gallon in many countries that are considered metricated.

Metric System

The metric system is a decimal systems of measurement based on its units for length, the metre and for mass, the kilogram. It exists in several variations, with different choices of base units, though these do not affect its day-to-day use. Since the 1960s, the International System of Units (SI) is the internationally recognised metric system. Metric units of mass, length, and electricity are widely used around the world for both everyday and scientific purposes.

The metric system features a single base unit for many physical quantities. Other quantities are derived from the standard SI units. Multiples and fractions of the units are expressed as Powers of 10 of each unit. Unit conversions are always simple because they are in the ratio of ten, one hundred, one thousand, etc., so that convenient magnitudes for measurements are achieved by simply moving the decimal place: 1.234 metres is 1234 millimetres or 0.001234 kilometres. The use of fractions, such as 2/5 of a metre, is not prohibited, but uncommon. All lengths and distances, for example, are measured in metres, or thousandths of a metre (millimetres), or thousands of metres (kilometres). There is no profusion of different units with different conversion factors as in the Imperial system which uses, for example, inches, feet, yards, fathoms, rods.

International System of Units

The International System of Units (abbreviated as SI from the French language name *Système International d'Unités*) is the modern revision of the metric system. It is the world's most widely used system of units, both in everyday commerce and in science. The SI was developed in 1960 from the metre-kilogram-second (MKS) system, rather than the centimetre-gram-second (CGS) system, which, in turn, had many variants. During its development the SI also introduced several newly named units that were previously not a part of the metric system. The original SI units for the seven basic physical quantities were:

Base quantity	Base unit	Symbol	Current SI constants	New SI constants (proposed)
time	second	s	hyperfine splitting in Cesium-133	same as current SI
length	metre	m	speed of light in vacuum, c	same as current SI
mass	kilogram	kg	mass of International Prototype Kilogram (IPK)	Planck's constant, h
electric current	Ampere	A	permeability of free space, permittivity of free space	charge of the electron, e
temperature	Kelvin	K	triple point of water, absolute zero	Boltzmann's constant, k
amount of substance	mole	mol	molar mass of Carbon-12	Avogadro constant N_A
luminous intensity	candela	cd	luminous efficacy of a 540 THz source	same as current SI

The mole was subsequently added to this list and the degree Kelvin renamed the kelvin.

There are two types of SI units, base units and derived units. Base units are the simple measurements for time, length, mass, temperature, amount of substance, electric current and light intensity. Derived units are constructed from the base units, for example, the Watt, i.e. the unit for power, is defined from the base units as $m^2{\cdot}kg{\cdot}s^{-3}$. Other physical properties may be measured in compound units, such as material density, measured in kg/m^3.

Converting Prefixes

The SI allows easy multiplication when switching among units having the same base but different prefixes. To convert from metres to centimetres it is only necessary to multiply the number of metres by 100, since there are 100 centimetres in a metre. Inversely, to switch from centimetres to metres one multiplies the number of centimetres by 0.01 or divide centimetres by 100.

Length

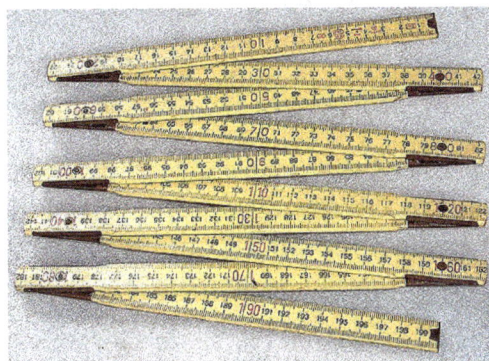

A 2-metre carpenter's ruler

A ruler or rule is a tool used in, for example, geometry, technical drawing, engineering, and carpentry, to measure lengths or distances or to draw straight lines. Strictly speaking, the *ruler* is the instrument used to rule straight lines and the calibrated instrument used for determining length is called a *measure*, however common usage calls both instruments *rulers* and the special name *straightedge* is used for an unmarked rule. The use of the word *measure*, in the sense of a measuring instrument, only survives in the phrase *tape measure*, an instrument that can be used to measure but cannot be used to draw straight lines. As can be seen in the photographs on this page, a two-metre carpenter's rule can be folded down to a length of only 20 centimetres, to easily fit in a pocket, and a five-metre-long tape measure easily retracts to fit within a small housing.

Some Special Names

Some non-systematic names are applied for some multiples of some units.

- 100 kilograms = 1 quintal; 1000 kilogram = 1 metric tonne;

- 10 years = 1 decade; 100 years = 1 century; 1000 years = 1 millennium

Building Trades

The Australian building trades adopted the metric system in 1966 and the units used for measurement of length are metres (m) and millimetres (mm). Centimetres (cm) are avoided as they cause confusion when reading plans. For example, the length two and a half metres is usually recorded as 2500 mm or 2.5 m; it would be considered non-standard to record this length as 250 cm.

Surveyor's Trade

American surveyors use a decimal-based system of measurement devised by Edmund Gunter in 1620. The base unit is Gunter's chain of 66 feet (20 m) which is subdivided into 4 rods, each of 16.5 ft or 100 links of 0.66 feet. A link is abbreviated "lk," and links "lks" in old deeds and Land Surveys done for the government.

Time

Time is an abstract measurement of elemental changes over a non spatial continuum. It is denoted by numbers and/or named periods such as hours, days, weeks, months and years. It is an apparently irreversible series of occurrences within this non spatial continuum. It is also used to denote an interval between two relative points on this continuum.

Mass

Mass refers to the intrinsic property of all material objects to resist changes in their momentum. *Weight*, on the other hand, refers to the downward force produced when a mass is in a gravitational field. In free fall, (no net gravitational forces) objects lack weight but retain their mass. The Imperial units of mass include the ounce, pound, and ton. The metric units gram and kilogram are units of mass.

One device for measuring weight or mass is called a weighing scale or, often, simply a *scale*. A spring scale measures force but not mass, a balance compares weight, both require a gravitational field to operate. Some of the most accurate instruments for measuring weight or mass are based on load cells with a digital read-out, but require a gravitational field to function and would not work in free fall.

Economics

The measures used in economics are physical measures, nominal price value measures and real price measures. These measures differ from one another by the variables they measure and by the variables excluded from measurements.

Difficulties

Since accurate measurement is essential in many fields, and since all measurements are necessarily approximations, a great deal of effort must be taken to make measurements as accurate as possible. For example, consider the problem of measuring the time it takes an object to fall a distance of one metre (about 39 in). Using physics, it can be shown that, in the gravitational field of the Earth, it should take any object about 0.45 second to fall one metre. However, the following are just some of the sources of error that arise:

- This computation used for the acceleration of gravity 9.8 metres per second squared (32 ft/s^2). But this measurement is not exact, but only precise to two significant digits.

- The Earth's gravitational field varies slightly depending on height above sea level and other factors.

- The computation of .45 seconds involved extracting a square root, a mathematical operation that required rounding off to some number of significant digits, in this case two significant digits.

Additionally, other sources of experimental error include:

- carelessness,

- determining of the exact time at which the object is released and the exact time it hits the ground,

- measurement of the height and the measurement of the time both involve some error,

- Air resistance.

Scientific experiments must be carried out with great care to eliminate as much error as possible, and to keep error estimates realistic.

Definitions and Theories

Classical Definition

In the classical definition, which is standard throughout the physical sciences, *measurement* is the determination or estimation of ratios of quantities. Quantity and measurement are mutually defined: quantitative attributes are those possible to measure, at least in principle. The classical concept of quantity can be traced back to John Wallis and Isaac Newton, and was foreshadowed in Euclid's Elements.

Representational Theory

In the representational theory, *measurement* is defined as "the correlation of numbers with entities that are not numbers". The most technically elaborate form of representational theory is also known as additive conjoint measurement. In this form of representational theory, numbers are assigned based on correspondences or similarities between the structure of number systems and the structure of qualitative systems. A property is quantitative if such structural similarities can be established. In weaker forms of representational theory, such as that implicit within the work of Stanley Smith Stevens, numbers need only be assigned according to a rule.

The concept of measurement is often misunderstood as merely the assignment of a value, but it is possible to assign a value in a way that is not a measurement in terms of the requirements of additive conjoint measurement. One may assign a value to a person's height, but unless it can be established that there is a correlation between measurements of height and empirical relations, it is not a measurement according to additive conjoint measurement theory. Likewise, computing and assigning arbitrary values, like the "book value" of an asset in accounting, is not a measurement because it does not satisfy the necessary criteria.

Information Theory

Information theory recognises that all data are inexact and statistical in nature. Thus the definition of measurement is: "A set of observations that reduce uncertainty where the result is expressed as a quantity." This definition is implied in what scientists actually do when they measure something and report both the mean and statistics of the measurements. In practical terms, one begins with an initial guess as to the value of a quantity, and then, using various methods and instruments, reduces the uncertainty in the value. Note that in this view, unlike the positivist representational theory, all measurements are uncertain, so instead of assigning one value, a range of values is assigned to a measurement. This also implies that there is not a clear or neat distinction between estimation and measurement.

Quantum Mechanics

In quantum mechanics, a measurement is an action that determines a particular property (position, momentum, energy, etc.) of a quantum system. Before a measurement is made, a quantum system is simultaneously described by all values in a spectrum, or range, of possible values, where the probability of measuring each value is determined by the wavefunction of the system. When a measurement is performed, the wavefunction of the quantum system "collapses" to a single, definite value. The unambiguous meaning of the measurement problem is an unresolved fundamental problem in quantum mechanics.

Shape

A shape is the form of an object or its external boundary, outline, or external surface, as opposed to other properties such as color, texture, or material composition.

Psychologists have theorized that humans mentally break down images into simple geometric shapes called geons. Examples of geons include cones and spheres.

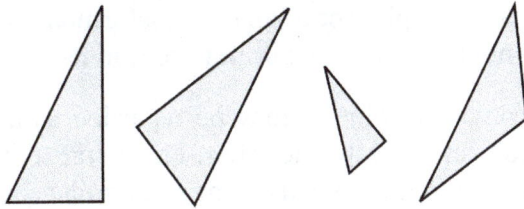

An example of the different definitions of **shape**. The two triangles on the left are congruent, while the third is similar to them. The last triangle is neither similar nor congruent to any of the others, but it is homeomorphic.

Classification of Simple Shapes

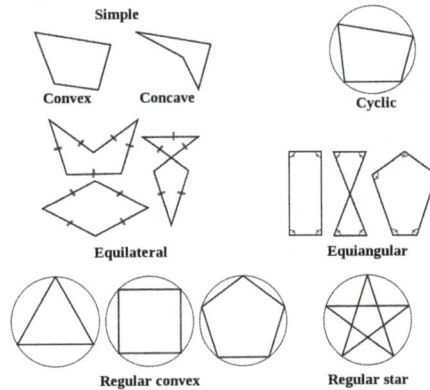

A variety of polygonal shapes.

Some simple shapes can be put into broad categories. For instance, polygons are classified according to their number of edges as triangles, quadrilaterals, pentagons, etc. Each of these is divided into smaller categories; triangles can be equilateral, isosceles, obtuse, acute, scalene, etc. while quadrilaterals can be rectangles, rhombi, trapezoids, squares, etc.

Other common shapes are points, lines, planes, and conic sections such as ellipses, circles, and parabolas.

Among the most common 3-dimensional shapes are polyhedra, which are shapes with flat faces; ellipsoids, which are egg-shaped or sphere-shaped objects; cylinders; and cones.

If an object falls into one of these categories exactly or even approximately, we can use it to describe the shape of the object. Thus, we say that the shape of a manhole cover is a disk, because it is approximately the same geometric object as an actual geometric disk.

Shape in Geometry

There are several ways to compare the shapes of two objects:

- Congruence: Two objects are congruent if one can be transformed into the other by a sequence of rotations, translations, and/or reflections.

- Similarity: Two objects are similar if one can be transformed into the other by a uniform scaling, together with a sequence of rotations, translations, and/or reflections.

- Isotopy: Two objects are isotopic if one can be transformed into the other by a sequence of deformations that do not tear the object or put holes in it.

Sometimes, two similar or congruent objects may be regarded as having a different shape if a reflection is required to transform one into the other. For instance, the letters "b" and "d" are a reflection of each other, and hence they are congruent and similar, but in some contexts they are not regarded as having the same shape. Sometimes, only the outline or external boundary of the object is considered to determine its shape. For instance, an hollow sphere may be considered to have the same shape as a solid sphere. Procrustes analysis is used in many sciences to determine whether or not two objects have the same shape, or to measure the difference between two shapes. In advanced mathematics, quasi-isometry can be used as a criterion to state that two shapes are approximately the same.

Simple shapes can often be classified into basic geometric objects such as a point, a line, a curve, a plane, a plane figure (e.g. square or circle), or a solid figure (e.g. cube or sphere). However, most shapes occurring in the physical world are complex. Some, such as plant structures and coastlines, may be so complicated as to defy traditional mathematical description – in which case they may be analyzed by differential geometry, or as fractals.

Equivalence of Shapes

In geometry, two subsets of a Euclidean space have the same shape if one can be transformed to the other by a combination of translations, rotations (together also called rigid transformations), and uniform scalings. In other words, the *shape* of a set of points is all the geometrical information that is invariant to translations, rotations, and size changes. Having the same shape is an equivalence relation, and accordingly a precise mathematical definition of the notion of shape can be given as being an equivalence class of subsets of a Euclidean space having the same shape.

Mathematician and statistician David George Kendall writes:

In this paper 'shape' is used in the vulgar sense, and means what one would normally expect it to mean. [...] We here define 'shape' informally as 'all the geometrical information that remains when location, scale and rotational effects are filtered out from an object.'

Shapes of physical objects are equal if the subsets of space these objects occupy satisfy the definition above. In particular, the shape does not depend on the size and placement in space of the object. For instance, a "d" and a "p" have the same shape, as they can be perfectly superimposed if the "d" is translated to the right by a given distance, rotated upside down and magnified by a given factor. However, a mirror image could be called a different shape. For instance, a "b" and a "p" have a different shape, at least when they are constrained to move within a two-dimensional space like the page on which they are written. Even though they have the same size, there's no way to perfectly superimpose them by translating and rotating them along the page. Similarly, within a three-dimensional space, a right hand and a left hand have a different shape, even if they are the mirror images of each other. Shapes may change if the object is scaled non-uniformly. For example, a sphere becomes an ellipsoid when scaled differently in the vertical and horizontal directions.

In other words, preserving axes of symmetry (if they exist) is important for preserving shapes. Also, shape is determined by only the outer boundary of an object.

Congruence and Similarity

Objects that can be transformed into each other by rigid transformations and mirroring (but not scaling) are congruent. An object is therefore congruent to its mirror image (even if it is not symmetric), but not to a scaled version. Two congruent objects always have either the same shape or mirror image shapes, and have the same size.

Objects that have the same shape or mirror image shapes are called geometrically similar, whether or not they have the same size. Thus, objects that can be transformed into each other by rigid transformations, mirroring, and uniform scaling are similar. Similarity is preserved when one of the objects is uniformly scaled, while congruence is not. Thus, congruent objects are always geometrically similar, but similar objects may not be congruent, as they may have different size.

Homeomorphism

A more flexible definition of shape takes into consideration the fact that realistic shapes are often deformable, e.g. a person in different postures, a tree bending in the wind or a hand with different finger positions.

One way of modeling non-rigid movements is by homeomorphisms. Roughly speaking, a homeomorphism is a continuous stretching and bending of an object into a new shape. Thus, a square and a circle are homeomorphic to each other, but a sphere and a donut are not. An often-repeated mathematical joke is that topologists can't tell their coffee cup from their donut, since a sufficiently pliable donut could be reshaped to the form of a coffee cup by creating a dimple and progressively enlarging it, while preserving the donut hole in a cup's handle.

Shape Analysis

The above-mentioned mathematical definitions of rigid and non-rigid shape have arisen in the field of statistical shape analysis. In particular Procrustes analysis, which is a technique used for comparing shapes of similar objects (e.g. bones of different animals), or measuring the deformation of a deformable object. Other methods are designed to work with non-rigid (bendable) objects, e.g. for posture independent shape retrieval.

Similarity Classes

All similar triangles have the same shape. These shapes can be classified using complex numbers in a method advanced by J.A. Lester and Rafael Artzy. For example, an equilateral triangle can be expressed by the complex numbers $0, 1, (1 + i\sqrt{3})/2$ representing its vertices. Lester and Artzy call the ratio

$$S(u, v, w) = \frac{u - w}{u - v}$$

the shape of triangle (u, v, w). Then the shape of the equilateral triangle is

$$(0-(1+\sqrt{3})/2)/(0-1) = (1+i\sqrt{3})/2 = \cos(60°) + i\sin(60°) = \exp(i\pi/3).$$

For any affine transformation of the complex plane, $z \mapsto az+b$, $a \neq 0$, a triangle is transformed but does not change its shape. Hence shape is an invariant of affine geometry. The shape $p = S(u,v,w)$ depends on the order of the arguments of function S, but permutations lead to related values. For instance,

$$1-p = 1-(u-w)/(u-v) = (w-v)/(u-v) = (v-w)/(v-u) = S(v,u,w)$$ Also

$$p^{-1} = S(u,w,v).$$

Combining these permutations gives $S(v,w,u) = (1-p)^{-1}$. Furthermore,

$$p(1-p)^{-1} = S(u,v,w)S(v,w,u) = (u-w)/(v-w) = S(w,v,u)$$. These relations are "conversion rules" for shape of a triangle.

The shape of a quadrilateral is associated with two complex numbers p,q. If the quadrilateral has vertices u,v,w,x, then $p = S(u,v,w)$ and $q = S(v,w,x)$. Artzy proves these propositions about quadrilateral shapes:

If $p = (1-q)^{-1}$, then the quadrilateral is a parallelogram.

If a parallelogram has $|\arg p| = |\arg q|$, then it is a rhombus.

When $p = 1+i$ and $q = (1+i)/2$, then the quadrilateral is square.

If $p = r(1-q^{-1})$ and sgn r = sgn(Im p), then the quadrilateral is a trapezoid.

A polygon $(z_1, z_2, ... z_n)$ has a shape defined by $n-2$ complex numbers $S(z_j, z_{j+1}, z_{j+2})$, $j = 1,...,n-2$. The polygon bounds a convex set when all these shape components have imaginary components of the same sign.

Mathematical Constant

A mathematical constant is a special number, usually a real number, that is "significantly interesting in some way". Constants arise in many areas of mathematics, with constants such as e and π occurring in such diverse contexts as geometry, number theory, and calculus.

What it means for a constant to arise "naturally", and what makes a constant "interesting", is ultimately a matter of taste, and some mathematical constants are notable more for historical reasons than for their intrinsic mathematical interest. The more popular constants have been studied throughout the ages and computed to many decimal places.

All mathematical constants are definable numbers and usually are also computable numbers (Chaitin's constant being a significant exception).

Common Mathematical Constants

These are constants which one is likely to encounter during pre-college education in many countries.

Archimedes' Constant Π

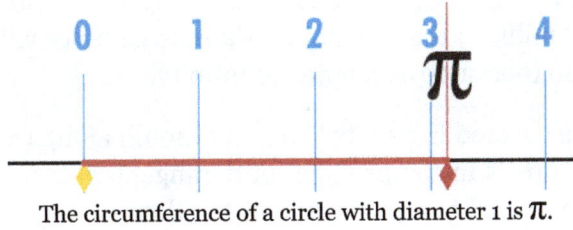

The circumference of a circle with diameter 1 is π.

The constant π (pi) has a natural definition in Euclidean geometry (the ratio between the circumference and diameter of a circle), but may be found in many places in mathematics: for example, the Gaussian integral in complex analysis, the roots of unity in number theory, and Cauchy distributions in probability. However, its universality is not limited to pure mathematics. Indeed, various formulae in physics, such as Heisenberg's uncertainty principle, and constants such as the cosmological constant include the constant π. The presence of π in physical principles, laws and formulae can have very simple explanations. For example, Coulomb's law, describing the inverse square proportionality of the magnitude of the electrostatic force between two electric charges and their distance, states that, in SI units,

$$F = \frac{1}{4\pi\varepsilon_0} \frac{|q_1 q_2|}{r^2}. \, [\,[\![$$

Besides ε_0 corresponding to the dielectric constant in vacuum, the $4\pi r^2$ factor in the above denominator expresses directly the surface of a sphere with radius r, having thus a very concrete meaning.

The numeric value of π is approximately 3.1415926535. Memorizing increasingly precise digits of π is a world record pursuit.

Euler's Number E

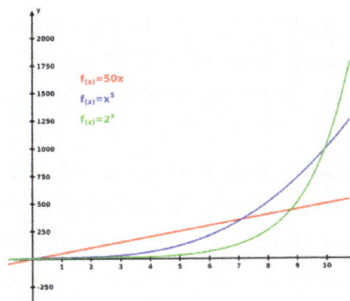

Exponential growth (green) describes many physical phenomena.

Euler's number e, also known as the exponential growth constant, appears in many areas of mathematics, and one possible definition of it is the value of the following expression:

$$e = \lim_{n \to \infty} \left(1 + \frac{1}{n}\right)^n$$

For example, the Swiss mathematician Jacob Bernoulli discovered that e arises in compound interest: An account that starts at \$1, and yields interest at annual rate R with continuous compounding, will accumulate to e^R dollars at the end of one year. The constant e also has applications to probability theory, where it arises in a way not obviously related to exponential growth. Suppose that a gambler plays a slot machine with a one in n probability of winning, and plays it n times. Then, for large n (such as a million) the probability that the gambler will win nothing at all is approximately $1/e$ and tends to this value as n tends to infinity.

Another application of e, discovered in part by Jacob Bernoulli along with French mathematician Pierre Raymond de Montmort, is in the problem of derangements, also known as the *hat check problem*. Here n guests are invited to a party, and at the door each guest checks his hat with the butler who then places them into labelled boxes. The butler does not know the name of the guests, and so must put them into boxes selected at random. The problem of de Montmort is: what is the probability that *none* of the hats gets put into the right box. The answer is

$$p_n = 1 - \frac{1}{1!} + \frac{1}{2!} - \frac{1}{3!} + \cdots + (-1)^n \frac{1}{n!}$$

and as n tends to infinity, p_n approaches $1/e$.

The numeric value of e is approximately 2.71828.

Pythagoras' Constant √2

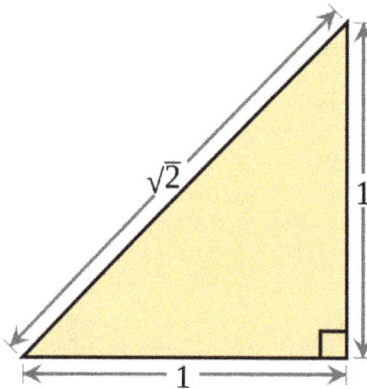

The square root of 2 is equal to the length of the hypotenuse of a right-angled triangle with legs of length 1.

The square root of 2, often known as root 2, radical 2, or Pythagoras's constant, and written as √2, is the positive algebraic number that, when multiplied by itself, gives the number 2. It is more precisely called the principal square root of 2, to distinguish it from the negative number with the same property.

Geometrically the square root of 2 is the length of a diagonal across a square with sides of one unit of length; this follows from the Pythagorean theorem. It was probably the first number known to be irrational. Its numerical value truncated to 65 decimal places is:

1.41421356237309504880168872420969807856967187537694807317667973799... (sequence A002193 in the OEIS).

The quick approximation 99/70 ($\approx$ 1.41429) for the square root of two is frequently used. Despite having a denominator of only 70, it differs from the correct value by less than 1/10,000 (approx. 7.2×10^{-5}).

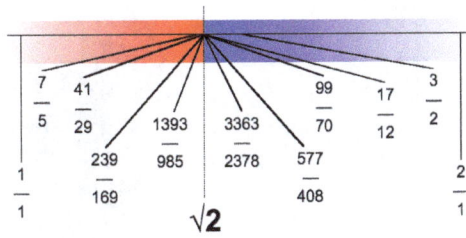

The square root of 2.

The Imaginary Unit *I*

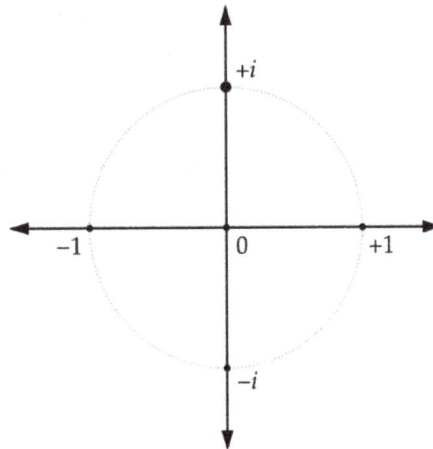

i in the complex or cartesian plane. Real numbers lie on the horizontal axis, and imaginary numbers lie on the vertical axis

The imaginary unit or unit imaginary number, denoted as *i*, is a mathematical concept which extends the real number system R to the complex number system C, which in turn provides at least one root for every polynomial $P(x)$. The imaginary unit's core property is that $i^2 = -1$. The term "imaginary" is used because there is no real number having a negative square.

There are in fact two complex square roots of −1, namely *i* and −*i*, just as there are two complex square roots of every other real number, except zero, which has one double square root.

In contexts where *i* is ambiguous or problematic, *j* or the Greek ι is sometimes used. In the disciplines of electrical engineering and control systems engineering, the imaginary unit is often denoted by *j* instead of *i*, because *i* is commonly used to denote electric current in these disciplines.

Constants in Advanced Mathematics

These are constants which are encountered frequently in higher mathematics.

The Feigenbaum Constants α and δ

Iterations of continuous maps serve as the simplest examples of models for dynamical systems. Named after mathematical physicist Mitchell Feigenbaum, the two Feigenbaum constants appear in such iterative processes: they are mathematical invariants of logistic maps with quadratic maximum points and their bifurcation diagrams.

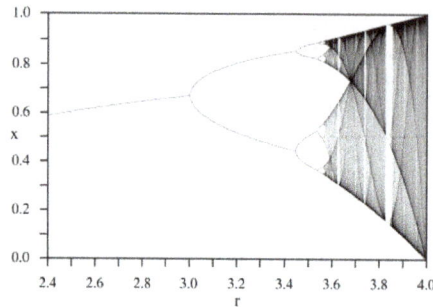

Bifurcation diagram of the logistic map.

The logistic map is a polynomial mapping, often cited as an archetypal example of how chaotic behaviour can arise from very simple non-linear dynamical equations. The map was popularized in a seminal 1976 paper by the Australian biologist Robert May, in part as a discrete-time demographic model analogous to the logistic equation first created by Pierre François Verhulst. The difference equation is intended to capture the two effects of reproduction and starvation.

The numeric value of α is approximately 2.5029. The numeric value of δ is approximately 4.6692.

Apéry's Constant $\zeta(3)$

$$\zeta(3) = 1 + \frac{1}{2^3} + \frac{1}{3^3} + \frac{1}{4^3} + \cdots$$

Despite being a special value of the Riemann zeta function, Apéry's constant arises naturally in a number of physical problems, including in the second- and third-order terms of the electron's gyromagnetic ratio, computed using quantum electrodynamics. The numeric value of $\zeta(3)$ is approximately 1.2020569.

The Golden Ratio Φ

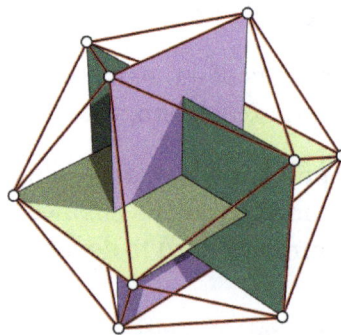

Golden rectangles in an icosahedron

$$F(n) = \frac{\varphi^n - (1-\varphi)^n}{\sqrt{5}}$$

An explicit formula for the nth Fibonacci number involving the golden ratio φ.

The number φ, also called the golden ratio, turns up frequently in geometry, particularly in figures with pentagonal symmetry. Indeed, the length of a regular pentagon's diagonal is φ times its side. The vertices of a regular icosahedron are those of three mutually orthogonalgolden rectangles.

Also, it appears in the Fibonacci sequence, related to growth by recursion. Kepler proved that it is the limit of the ratio of consecutive Fibonacci numbers. The golden ratio has the slowest convergence of any irrational number. It is, for that reason, one of the worst cases of Lagrange's approximation theorem and it is an extremal case of the Hurwitz inequality for Diophantine approximations. This may be why angles close to the golden ratio often show up in phyllotaxis (the growth of plants). It is approximately equal to 1.61803398874, or, more precisely

$$\frac{1+\sqrt{5}}{2}.$$

The Euler–Mascheroni Constant γ

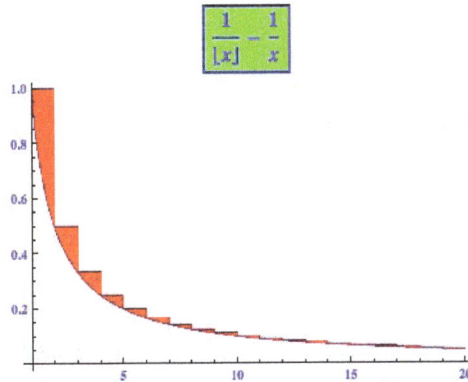

The area between the two curves (red) tends to a limit.

The Euler–Mascheroni constant is a recurring constant in number theory. The Belgian mathematician Charles Jean de la Vallée-Poussin proved in 1898 that when taking any positive integer n and dividing it by each positive integer m less than n, the average fraction by which the quotient n/m falls short of the next integer tends to γ as n tends to infinity. Surprisingly, this average doesn't tend to one half. The Euler–Mascheroni constant also appears in Merten's third theorem and has relations to the gamma function, the zeta function and many different integrals and series. The definition of the Euler–Mascheroni constant exhibits a close link between the discrete and the continuous.

The numeric value of γ is approximately 0.57721.

Conway's Constant λ

Conway's constant is the invariant growth rate of all derived strings similar to the look-and-say sequence (except for one trivial one).

It is given by the unique positive real root of a polynomial of degree 71 with integer coefficients.

The value of λ is approximately 1.30357.

Khinchin's Constant K

If a real number r is written as a simple continued fraction:

$$r = a_0 + \cfrac{1}{a_1 + \cfrac{1}{a_2 + \cfrac{1}{a_3 + \cdots}}},$$

where a_k are natural numbers for all k

then, as the Russian mathematician Aleksandr Khinchin proved in 1934, the limit as n tends to infinity of the geometric mean: $(a_1 a_2 ... a_n)^{1/n}$ exists and is a constant, Khinchin's constant, except for a set of measure 0.

The numeric value of K is approximately 2.6854520010.

The Glaisher–Kinkelin Constant A

The Glaisher–Kinkelin constant is defined as the limit:

$$A = \lim_{n \to \infty} \frac{\prod_{k=1}^{n} k^k}{n^{n^2/2 + n/2 + 1/12} e^{-n^2/4}}$$

It is an important constant which appears in many expressions for the derivative of the Riemann zeta function. It has a numerical value of approximately 1.2824271291.

Mathematical Curiosities and Unspecified Constants

Simple Representatives of Sets of Numbers

This Babylonian clay tablet gives an approximation of the square root of 2 in four sexagesimal figures: 1; 24, 51, 10, which is accurate to about six decimal figures.

$$c = \sum_{j=1}^{\infty} 10^{-j!} = 0.\overbrace{110001}^{3!\text{ digits}}\underbrace{000000000000000001}_{4!\text{ digits}}000\ldots$$

Liouville's constant is a simple example of a transcendental number.

Some constants, such as the square root of 2, Liouville's constant and Champernowne constant:

$$C_{10} = 0.12345678910111213141516\ldots$$

are not important mathematical invariants but retain interest being simple representatives of special sets of numbers, the irrational numbers, the transcendental numbers and the normal numbers (in base 10) respectively. The discovery of the irrational numbers is usually attributed to the PythagoreanHippasus of Metapontum who proved, most likely geometrically, the irrationality of the square root of 2. As for Liouville's constant, named after French mathematician Joseph Liouville, it was the first number to be proven transcendental.

Chaitin's Constant Ω

In the computer science subfield of algorithmic information theory, Chaitin's constant is the real number representing the probability that a randomly chosen Turing machine will halt, formed from a construction due to Argentine-American mathematician and computer scientistGregory Chaitin. Chaitin's constant, though not being computable, has been proven to be transcendental and normal. Chaitin's constant is not universal, depending heavily on the numerical encoding used for Turing machines; however, its interesting properties are independent of the encoding.

Unspecified Constants

When unspecified, constants indicate classes of similar objects, commonly functions, all equal up to a constant—technically speaking, this is may be viewed as 'similarity up to a constant'. Such constants appear frequently when dealing with integrals and differential equations. Though unspecified, they have a specific value, which often is not important.

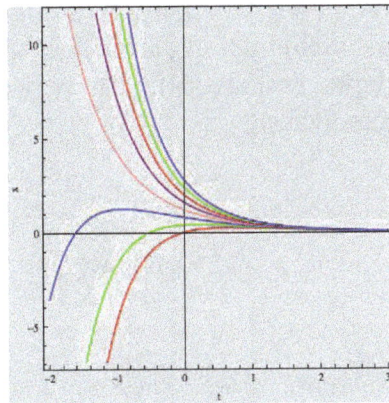

Solutions with different constants of integration of .

In Integrals

Indefinite integrals are called indefinite because their solutions are only unique up to a constant. For example, when working over the field of real numbers

$$\int \cos x \, dx = \sin x + C$$

where C, the constant of integration, is an arbitrary fixed real number. In other words, whatever the value of C, differentiating $\sin x + C$ with respect to x always yields $\cos x$.

In Differential Equations

In a similar fashion, constants appear in the solutions to differential equations where not enough initial values or boundary conditions are given. For example, the ordinary differential equation $y' = y(x)$ has solution Ce^x where C is an arbitrary constant.

When dealing with partial differential equations, the constants may be functions, constant with respect to some variables (but not necessarily all of them). For example, the PDE

$$\frac{\partial f(x, y)}{\partial x} = 0$$

has solutions $f(x,y) = C(y)$, where $C(y)$ is an arbitrary function in the variable y.

Notation

Representing Constants

It is common to express the numerical value of a constant by giving its decimal representation (or just the first few digits of it). For two reasons this representation may cause problems. First, even though rational numbers all have a finite or ever-repeating decimal expansion, irrational numbers don't have such an expression making them impossible to completely describe in this manner. Also, the decimal expansion of a number is not necessarily unique. For example, the two representations 0.999... and 1 are equivalent in the sense that they represent the same number.

Calculating digits of the decimal expansion of constants has been a common enterprise for many centuries. For example, German mathematician Ludolph van Ceulen of the 16th century spent a major part of his life calculating the first 35 digits of pi. Using computers and supercomputers, some of the mathematical constants, including π, e, and the square root of 2, have been computed to more than one hundred billion digits. Fast algorithms have been developed, some of which — as for Apéry's constant — are unexpectedly fast.

$$G = \left. \begin{array}{c} 3 \uparrow \underbrace{\ldots} \uparrow 3 \\ \vdots \\ 3 \uparrow\uparrow\uparrow\uparrow 3 \end{array} \right\} 64 \text{ layers}$$

Graham's number defined using Knuth's up-arrow notation.

Some constants differ so much from the usual kind that a new notation has been invented to represent them reasonably. Graham's number illustrates this as Knuth's up-arrow notation is used.

It may be of interest to represent them using continued fractions to perform various studies, including statistical analysis. Many mathematical constants have an analytic form, that is they can be constructed using well-known operations that lend themselves readily to calculation. Not all constants have known analytic forms, though; Grossman's constant and Foias' constant are examples.

Symbolizing and Naming of Constants

Symbolizing constants with letters is a frequent means of making the notation more concise. A standard convention, instigated by Leonhard Euler in the 18th century, is to use lower case letters from the beginning of the Latin alphabet $a, b, c, \ldots$ or the Greek alphabet $\alpha, \beta, \gamma, \ldots$ when dealing with constants in general.

Erdős–Borwein constant E_B
Embree–Trefethen constant β^*
Brun's constant for twin prime C_b
Champernowne constants C_b
cardinal numberaleph naught $\aleph_0$

Examples of different kinds of notation for constants.

However, for more important constants, the symbols may be more complex and have an extra letter, an asterisk, a number, a lemniscate or use different alphabets such as Hebrew, Cyrillic or Gothic.

$$\text{googol} = 10^{100}, \text{ googolplex} = 10^{\text{googol}} = 10^{10^{100}} a$$

Sometimes, the symbol representing a constant is a whole word. For example, American mathematician Edward Kasner's 9-year-old nephew coined the names googol and googolplex.

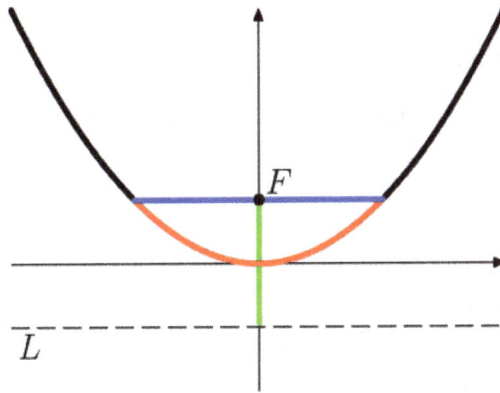

The universal parabolic constant is the ratio, for any parabola, of the arc length of the parabolic segment (red) formed by the latus rectum (blue) to the focal parameter (green).

The names are either related to the meaning of the constant (universal parabolic constant, twin prime constant, ...) or to a specific person (Sierpiński's constant, Josephson constant, ...).

Table of Selected Mathematical Constants

Abbreviations used:

> R – Rational number, I – Irrational number (may be algebraic or transcendental), A – Algebraic number (irrational), T – Transcendental number (irrational)

> Gen – General, NuT – Number theory, ChT – Chaos theory, Com – Combinatorics, Inf – Information theory, Ana – Mathematical analysis

Symbol	Value	Name	Field	N	First described	# of known digits
0	= 0	Zero	**Gen**	R	by c. 500 BC	N/A
1	= 1	One, Unity	**Gen**	R		N/A
i	= √−1	Imaginary unit, unit imaginary number	**Gen, Ana**	A	by c. 1500	N/A
π	≈ 3.14159 26535 89793 23846 26433 83279 50288	Pi, Archimedes' constant or Ludolph's number	**Gen, Ana**	T	by c. 2600 BC	12,100,000,000,000
e	≈ 2.71828 18284 59045 23536 02874 71352 66249	e, Napier's constant, or Euler's number	**Gen, Ana**	T	1618	100,000,000,000
√2	≈ 1.41421 35623 73095 04880 16887 24209 69807	Pythagoras' constant, square root of 2	**Gen**	A	by c. 800 BC	137,438,953,444
√3	≈ 1.73205 08075 68877 29352 74463 41505 87236	Theodorus' constant, square root of 3	**Gen**	A	by c. 800 BC	
√5	≈ 2.23606 79774 99789 69640 91736 68731 27623	square root of 5	**Gen**	A	by c. 800 BC	
	≈ 0.57721 56649 01532 86060 65120 90082 40243	Euler–Mascheroni constant	**Gen, NuT**		1735	14,922,244,771

Symbol	Value	Name	Field	N	First described	# of known digits
	≈ 1.61803 39887 49894 84820 45868 34365 63811	Golden ratio	**Gen**	*A*	by c. 200 BC	100,000,000,000
	≥ −1.1 • 10^{-12}	de Bruijn–Newman constant	**NuT**		1950 ?	none
M_1	≈ 0.26149 72128 47642 78375 54268 38608 69585	Meissel–Mertens constant	**NuT**		1866 1874	8,010
	≈ 0.28016 94990 23869 13303	Bern-stein's constant	**Ana**			
	≈ 0.30366 30028 98732 65859 74481 21901 55623	Gauss–Kuzmin–Wirsing constant	**Com**		1974	385
	≈ 0.35323 63718 54995 98454 35165 50432 68201	Hafner–Sarnak–McCurley constant	**NuT**		1993	
L	≈ 0.5	Landau's constant	**Ana**			1
Ω	≈ 0.56714 32904 09783 87299 99686 62210 35555	Omega constant	**Ana**	*T*		
,	≈ 0.62432 99885 43550 87099 29363 83100 83724	Golomb–Dickman constant	**Com, NuT**		1930 1964	

Symbol	Value	Name	Field	N	First described	# of known digits
	≈ 0.64341 05463	Cahen's constant		T	1891	4000
C_2	≈ 0.66016 18158 46869 57392 78121 10014 55577	Twin prime constant	**NuT**			5,020
	≈ 0.66274 34193 49181 58097 47420 97109 25290	Laplace limit				
*	≈ 0.70258	Embree–Trefethen constant	**NuT**			
K	≈ 0.76422 36535 89220 66299 06987 31250 09232	Lan-dau–Ra-manujan constant	**NuT**			30,010
	≈ 0.80939 40205	Alladi–Grinstead constant	**NuT**			
B_4	≈ 0.87058 83800	Brun's constant for prime quadru-plets	**NuT**			
K	≈ 0.91596 55941 77219 01505 46035 14932 38411	Catalan's constant	**Com**			15,510,000,000
B'_L	= 1	Legen-dre's constant	**NuT**	R		N/A
	≈ 1.09868 58055	Lengyel's constant	**Com**		1992	
K	≈ 1.13198 824	Viswa-nath's constant	**NuT**			8

Symbol	Value	Name	Field	N	First described	# of known digits
	≈ 1.20205 69031 59594 28539 97381 61511 44999	Apéry's constant		I	1979	15,510,000,000
	≈ 1.30357 72690 34296 39125 70991 12152 55189	Conway's constant	**NuT**	A		
	≈ 1.30637 78838 63080 69046 86144 92602 60571	Mills' constant	**NuT**		1947	6850
	≈ 1.32471 79572 44746 02596 09088 54478 09734	Plastic constant	**NuT**	A	1928	
	≈ 1.45136 92348 83381 05028 39684 85892 02744	Ramanu-jan–Soldner constant	**NuT**	I		75,500
	≈ 1.45607 49485 82689 67139 95953 51116 54356	Back-house's constant				
	≈ 1.46707 80794	Porter's constant	**NuT**		1975	
	≈ 1.53960 07178	Lieb's square ice constant	**Com**	A	1967	
E_B	≈ 1.60669 51524 15291 76378 33015 23190 92458	Erdős–Borwein constant	**NuT**	I		

Symbol	Value	Name	Field	N	First described	# of known digits
	≈ 1.70521 11401 05367 76428 85514 53434 50816	Niven's constant	**NuT**		1969	
B_2	≈ 1.90216 05823	Brun's constant for twin primes	**NuT**		1919	10
P_2	≈ 2.29558 71493 92638 07403 42980 49189 49039	Universal parabolic constant	**Gen**	T		
	≈ 2.50290 78750 95892 82228 39028 73218 21578	Feigen-baum constant	**ChT**			
K	≈ 2.58498 17595 79253 21706 58935 87383 17116	Sier-piński's constant				
	≈ 2.68545 20010 65306 44530 97148 35481 79569	Kh-inchin's constant	**NuT**		1934	7350
F	≈ 2.80777 02420 28519 36522 15011 86557 77293	Fransén–Robinson constant	**Ana**			
	≈ 3.27582 29187 21811 15978 76818 82453 84386	Lévy's constant	**NuT**			

Symbol	Value	Name	Field	N	First described	# of known digits
	≈ 3.35988 56662 43177 55317 20113 02918 92717	Recip-rocal Fibonacci constant		*I*		
	≈ 4.66920 16091 02990 67185 32038 20466 20161	Feigen-baum constant	**ChT**		1975	

Mathematical Object

A mathematical object is an abstract object arising in mathematics. The concept is studied in philosophy of mathematics.

In mathematical practice, an *object* is anything that has been (or could be) formally defined, and with which one may do deductive reasoning and mathematical proofs. Commonly encountered mathematical objects include numbers, permutations, partitions, matrices, sets, functions, and relations. Geometry as a branch of mathematics has such objects as hexagons, points, lines, triangles, circles, spheres, polyhedra, topological spaces and manifolds. Another branch—algebra—has groups, rings, fields, group-theoretic lattices, and order-theoretic lattices. Categories are simultaneously homes to mathematical objects and mathematical objects in their own right. In proof theory, proofs and theorems are also mathematical objects.

The ontological status of mathematical objects has been the subject of much investigation and debate by philosophers of mathematics.

Cantorian Framework

One view that emerged around the turn of the 20th century with the work of Cantor is that all mathematical objects can be defined as sets. The set {0,1} is a relatively clear-cut example. On the face of it the group Z_2 of integers mod 2 is also a set with two elements. However, it cannot simply be the set {0,1}, because this does not mention the additional structure imputed to Z_2 by the operations of addition and negation mod 2: how are we to tell which of 0 or 1 is the additive identity, for example? To organize this group as a set it can first be coded as the quadruple ({0,1},+,−,0), which in turn can be coded using one of several conventions as a set representing that quadruple, which in turn entails encoding the operations + and − and the constant 0 as sets.

Sets may include ordered denotation of the particular identities and operations that apply to them, indicating a group, abelian group, ring, field, or other mathematical object. These types of mathematical objects are commonly studied in abstract algebra.

Foundational Paradoxes

If, however, the goal of mathematical ontology is taken to be the internal consistency of mathematics, it is more important that mathematical objects be definable in some uniform way (for example, as sets) regardless of actual practice, in order to lay bare the essence of its paradoxes. This has been the viewpoint taken by foundations of mathematics, which has traditionally accorded the management of paradox higher priority than the faithful reflection of the details of mathematical practice as a justification for defining mathematical objects to be sets.

Much of the tension created by this foundational identification of mathematical objects with sets can be relieved without unduly compromising the goals of foundations by allowing two kinds of objects into the mathematical universe, sets and relations, without requiring that either be considered merely an instance of the other. These form the basis of model theory as the domain of discourse of predicate logic. From this viewpoint, mathematical objects are entities satisfying the axioms of a formal theory expressed in the language of predicate logic.

Category Theory

A variant of this approach replaces relations with operations, the basis of universal algebra. In this variant the axioms often take the form of equations, or implications between equations.

A more abstract variant is category theory, which abstracts sets as objects and the operations thereon as morphisms between those objects. At this level of abstraction mathematical objects reduce to mere vertices of a graph whose edges as the morphisms abstract the ways in which those objects can transform and whose structure is encoded in the composition law for morphisms. Categories may arise as the models of some axiomatic theory and the homomorphisms between them (in which case they are usually concrete, meaning equipped with a faithful forgetful functor to the category Set or more generally to a suitable topos), or they may be constructed from other more primitive categories, or they may be studied as abstract objects in their own right without regard for their provenance.

References

- Gossett, Eric. Discrete Mathematics with Proof. John Wiley and Sons, 2009. Definition 3.1 page 86. ISBN 0-470-45793-7

- Howard Eves, An Introduction to the History of Mathematics, Saunders, 1990, ISBN 0-03-029558-0 p. 141: "No work, except The Bible, has been more widely used...."

- Pedhazur, Elazar J.; Schmelkin, Liora Pedhazur (1991). Measurement, Design, and Analysis: An Integrated Approach (1st ed.). Hillsdale, NJ: Lawrence Erlbaum Associates. pp. 15–29. ISBN 0-8058-1063-3.

- Kirch, Wilhelm, ed. (2008). "Level of measurement". Encyclopedia of Public Health 2. Springer. p. 81. ISBN 0-321-02106-1.

- International Bureau of Weights and Measures (2006), The International System of Units (SI) (PDF) (8th ed.), p. 147, ISBN 92-822-2213-6

- Hubbard, John H.; West, Beverly H. (1995). Differential Equations: A Dynamical Systems Approach. Part II: Higher-Dimensional Systems. Texts in Applied Mathematics 18. Springer. p. 204. ISBN 978-0-387-94377-0.

- May, Robert (1976). Theoretical Ecology: Principles and Applications. Blackwell Scientific Publishers. ISBN 0-632-00768-0.

- Livio, Mario (2002). The Golden Ratio: The Story of Phi, The World's Most Astonishing Number. New York: Broadway Books. ISBN 0-7679-0815-5.

- Edwards, Henry; David Penney (1994). Calculus with analytic geometry (4e ed.). Prentice Hall. p. 269. ISBN 0-13-300575-5.

- Rudin, Walter (1976) [1953]. Principles of mathematical analysis (3e ed.). McGraw-Hill. p.61 theorem 3.26. ISBN 0-07-054235-X.

- Burgess, John, and Rosen, Gideon, 1997. A Subject with No Object: Strategies for Nominalistic Reconstrual of Mathematics. Oxford University Press. ISBN 0198236158

- Fibonacci Numbers and Nature - Part 2 : Why is the Golden section the "best" arrangement?, from Dr. Ron Knott's Fibonacci Numbers and the Golden Section, retrieved 2012-11-29.

5

Mathematical Theories and Models

Mathematics is applicable to all instances of life. Many a time, instances demand mathematical understanding and analysis. Theories listed in this chapter were developed with such analysis such as probability, graph theory, number theory, statistical models etc. The major categories of mathematics are dealt with great details in the chapter.

Probability Theory

Probability theory is the branch of mathematics concerned with probability, the analysis of random phenomena. The central objects of probability theory are random variables, stochastic processes, and events: mathematical abstractions of non-deterministic events or measured quantities that may either be single occurrences or evolve over time in an apparently random fashion.

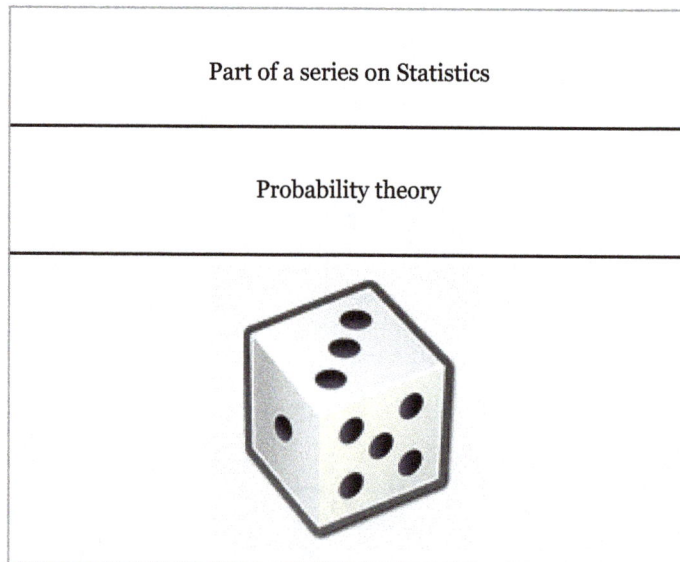

Part of a series on Statistics

Probability theory

It is not possible to predict precisely results of random events. However, if a sequence of individual events, such as coin flipping or the roll of dice, is influenced by other factors, such as friction, it will exhibit certain patterns, which can be studied and predicted. Two representative mathematical results describing such patterns are the law of large numbers and the central limit theorem.

As a mathematical foundation for statistics, probability theory is essential to many human activities that involve quantitative analysis of large sets of data. Methods of probability theory also apply to descriptions of complex systems given only partial knowledge of their state, as in statistical mechanics. A great discovery of twentieth century physics was the probabilistic nature of physical phenomena at atomic scales, described in quantum mechanics.

History

The mathematical theory of probability has its roots in attempts to analyze games of chance by Gerolamo Cardano in the sixteenth century, and by Pierre de Fermat and Blaise Pascal in the seventeenth century (for example the "problem of points"). Christiaan Huygens published a book on the subject in 1657 and in the 19th century a big work was done by Laplace in what can be considered today as the classic interpretation.

Initially, probability theory mainly considered discrete events, and its methods were mainly combinatorial. Eventually, analytical considerations compelled the incorporation of continuous variables into the theory.

This culminated in modern probability theory, on foundations laid by Andrey Nikolaevich Kolmogorov. Kolmogorov combined the notion of sample space, introduced by Richard von Mises, and measure theory and presented his axiom system for probability theory in 1933. Fairly quickly this became the mostly undisputed axiomatic basis for modern probability theory but alternatives exist, in particular the adoption of finite rather than countable additivity by Bruno de Finetti.

Treatment

Most introductions to probability theory treat discrete probability distributions and continuous probability distributions separately. The more mathematically advanced measure theory based treatment of probability covers both the discrete, the continuous, any mix of these two and more.

Motivation

Consider an experiment that can produce a number of outcomes. The set of all outcomes is called the *sample space* of the experiment. The *power set* of the sample space (or equivalently, the event space) is formed by considering all different collections of possible results. For example, rolling an honest die produces one of six possible results. One collection of possible results corresponds to getting an odd number. Thus, the subset {1,3,5} is an element of the power set of the sample space of die rolls. These collections are called *events*. In this case, {1,3,5} is the event that the die falls on some odd number. If the results that actually occur fall in a given event, that event is said to have occurred.

Probability is a way of assigning every "event" a value between zero and one, with the requirement that the event made up of all possible results (in our example, the event {1,2,3,4,5,6}) be assigned a value of one. To qualify as a probability distribution, the assignment of values must satisfy the requirement that if you look at a collection of mutually exclusive events (events that contain no common results, e.g., the events {1,6}, {3}, and {2,4} are all mutually exclusive), the probability that one of the events will occur is given by the sum of the probabilities of the individual events.

The probability that any one of the events {1,6}, {3}, or {2,4} will occur is 5/6. This is the same as saying that the probability of event {1,2,3,4,6} is 5/6. This event encompasses the possibility of any number except five being rolled. The mutually exclusive event {5} has a probability of 1/6, and the event {1,2,3,4,5,6} has a probability of 1, that is, absolute certainty.

Discrete Probability Distributions

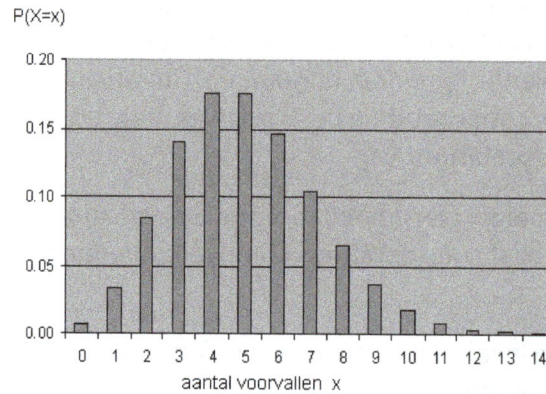

The Poisson distribution, a discrete probability distribution.

Discrete probability theory deals with events that occur in countable sample spaces.

Examples: Throwing dice, experiments with decks of cards, random walk, and tossing coins

Classical definition: Initially the probability of an event to occur was defined as number of cases favorable for the event, over the number of total outcomes possible in an equiprobable sample space.

For example, if the event is "occurrence of an even number when a die is rolled", the probability is given by $\frac{3}{6} = \frac{1}{2}$, since 3 faces out of the 6 have even numbers and each face has the same probability of appearing.

Modern definition: The modern definition starts with a finite or countable set called the sample space, which relates to the set of all *possible outcomes* in classical sense, denoted by $\Omega.$. It is then assumed that for each element $x \in \Omega,$, an intrinsic "probability" value $f(x)$ is attached, which satisfies the following properties:

$$f(x) \in [0,1] \text{ for all } x \in \Omega;$$

$$\sum_{x \in \Omega} f(x) = 1.$$

That is, the probability function $f(x)$ lies between zero and one for every value of x in the sample space Ω, and the sum of $f(x)$ over all values x in the sample space Ω is equal to 1. An event is defined as any subset E of the sample space Ω. The probability of the event E is defined as

$$P(E) = \sum_{x \in E} f(x).$$

So, the probability of the entire sample space is 1, and the probability of the null event is 0.

The function $f(x)$ mapping a point in the sample space to the "probability" value is called a probability mass function abbreviated as pmf. The modern definition does not try to answer how probability mass functions are obtained; instead it builds a theory that assumes their existence.

Continuous Probability Distributions

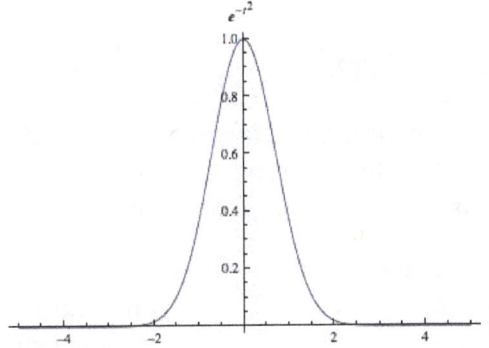

The normal distribution, a continuous probability distribution.

Continuous probability theory deals with events that occur in a continuous sample space.

Classical definition: The classical definition breaks down when confronted with the continuous case.

Modern definition: If the outcome space of a random variable X is the set of real numbers ($\mathbb{R}$) or a subset thereof, then a function called the cumulative distribution function (or cdf) F exists, defined by $F(x) = P(X \leq x)$.. That is, $F(x)$ returns the probability that X will be less than or equal to x.

The cdf necessarily satisfies the following properties.

1. F is a monotonically non-decreasing, right-continuous function;

2. $\lim_{x \to -\infty} F(x) = 0$;

3. $\lim_{x \to \infty} F(x) = 1$.

If F is absolutely continuous, i.e., its derivative exists and integrating the derivative gives us the cdf back again, then the random variable X is said to have a probability density function or pdf or

simply density $f(x) = \dfrac{dF(x)}{dx}$.

For a set $E \subseteq \mathbb{R}$, the probability of the random variable X being in E is

$$P(X \in E) = \int_{x \in E} dF(x).$$

In case the probability density function exists, this can be written as

$$P(X \in E) = \int_{x \in E} f(x)dx.$$

Whereas the *pdf* exists only for continuous random variables, the *cdf* exists for all random variables (including discrete random variables) that take values in $\mathbb{R}$.

These concepts can be generalized for multidimensional cases on $\mathbb{R}^n$ and other continuous sample spaces.

Measure-theoretic Probability Theory

The *raison d'être* of the measure-theoretic treatment of probability is that it unifies the discrete and the continuous cases, and makes the difference a question of which measure is used. Furthermore, it covers distributions that are neither discrete nor continuous nor mixtures of the two.

An example of such distributions could be a mix of discrete and continuous distributions—for example, a random variable that is 0 with probability 1/2, and takes a random value from a normal distribution with probability 1/2. It can still be studied to some extent by considering it to have a pdf of $(\delta[x] + \varphi(x))/2$, where $\delta[x]$ is the Dirac delta function.

Other distributions may not even be a mix, for example, the Cantor distribution has no positive probability for any single point, neither does it have a density. The modern approach to probability theory solves these problems using measure theory to define the probability space:

Given any set Ω, , (also called sample space) and a σ-algebra $\mathcal{F}$ on it, a measure P defined on $\mathcal{F}$ is called a probability measure if

If $P(\Omega) = 1$. is the Borel σ-algebra on the set of real numbers, then there is a unique probability measure on $\mathcal{F}$ for any cdf, and vice versa. The measure corresponding to a cdf is said to be induced by the cdf. This measure coincides with the pmf for discrete variables and pdf for continuous variables, making the measure-theoretic approach free of fallacies.

The *probability* of a set E in the σ-algebra $\mathcal{F}$ is defined as

$$P(E) = \int_{\omega \in E} \mu_F(d\omega)$$

where the integration is with respect to the measure μ_F induced by F.

Along with providing better understanding and unification of discrete and continuous probabilities, measure-theoretic treatment also allows us to work on probabilities outside $\mathbb{R}^n$, as in the theory of stochastic processes. For example, to study Brownian motion, probability is defined on a space of functions.

When it's convenient to work with a dominating measure, the Radon-Nikodym theorem is used to define a density as the Radon-Nikodym derivative of the probability distribution of interest with respect to this dominating measure. Discrete densities are usually defined as this derivative with respect to a counting measure over the set of all possible outcomes. Densities for absolutely continuous distributions are usually defined as this derivative with respect to the Lebesgue measure. If a theorem can be proved in this general setting, it holds for both discrete and continuous distributions as well as others; separate proofs are not required for discrete and continuous distributions.

Classical Probability Distributions

Certain random variables occur very often in probability theory because they well describe many natural or physical processes. Their distributions therefore have gained *special importance* in probability theory. Some fundamental *discrete distributions* are the discrete uniform, Bernoulli, binomial, negative binomial, Poisson and geometric distributions. Important *continuous distributions* include the continuous uniform, normal, exponential, gamma and beta distributions.

Convergence of Random Variables

In probability theory, there are several notions of convergence for random variables. They are listed below in the order of strength, i.e., any subsequent notion of convergence in the list implies convergence according to all of the preceding notions.

Weak convergence

A sequence of random variables $X_1, X_2, \ldots$, converges weakly to the random variable X if their respective cumulative *distribution functions* $F_1, F_2, \ldots$ converge to the cumulative distribution function F of X, wherever F is continuous. Weak convergence is also called convergence in distribution.

Most common shorthand notation: $X_n \xrightarrow{\mathcal{D}} X$

Convergence in probability

The sequence of random variables $X_1, X_2, \ldots$ is said to converge towards the random variable in probability if $\lim_{n \to \infty} P(|X_n - X| \geq \varepsilon) = 0$ for every $\varepsilon > 0$.

Most common shorthand notation: $X_n \xrightarrow{P} X$

Strong convergence

The sequence of random variables $X_1, X_2, \ldots$ is said to converge towards the random variable X strongly if $P(\lim_{n \to \infty} X_n = X) = 1$. Strong convergence is also known as almost sure convergence.

Most common shorthand notation: $X_n \xrightarrow{\text{a.s.}} X$

As the names indicate, weak convergence is weaker than strong convergence. In fact, strong convergence implies convergence in probability, and convergence in probability implies weak convergence. The reverse statements are not always true.

Law of Large Numbers

Common intuition suggests that if a fair coin is tossed many times, then *roughly* half of the time it will turn up *heads*, and the other half it will turn up *tails*. Furthermore, the more often the coin is tossed, the more likely it should be that the ratio of the number of *heads* to the number of *tails* will approach unity. Modern probability theory provides a formal version of this intuitive idea, known as the law of large numbers. This law is remarkable because it is not assumed in the foundations of probability theory, but instead emerges from these foundations as a theorem. Since it links theoretically derived probabilities to their actual frequency of occurrence in the real world, the law of large numbers is considered as a pillar in the history of statistical theory and has had widespread influence.

The law of large numbers (LLN) states that the sample average

$$\bar{X}_n = \frac{1}{n}\sum_{k=1}^{n} X_k$$

of a sequence of independent and identically distributed random variables X_k converges towards their common expectation μ, , provided that the expectation of $|X_k|$ is finite.

It is in the different forms of convergence of random variables that separates the *weak* and the *strong* law of large numbers

Weak law: $\bar{X}_n \xrightarrow{P} \mu$ for $n \to \infty$

Strong law: $\bar{X}_n \xrightarrow{a.s.} \mu$ for $n \to \infty$.

It follows from the LLN that if an event of probability p is observed repeatedly during independent experiments, the ratio of the observed frequency of that event to the total number of repetitions converges towards p.

For example, if $Y_1, Y_2,...$ are independent Bernoulli random variables taking values 1 with probability p and 0 with probability 1-p, then $E(Y_i) = p$ for all i, so that $\bar{Y}_n$ converges to p almost surely.

Central Limit Theorem

"The central limit theorem (CLT) is one of the great results of mathematics." (Chapter 18 in) It explains the ubiquitous occurrence of the normal distribution in nature.

The theorem states that the average of many independent and identically distributed random variables with finite variance tends towards a normal distribution *irrespective* of the distribution followed by the original random variables. Formally, let $X_1, X_2,...$ be independent random variables with mean μ and variance $\sigma^2 > 0$. Then the sequence of random variables

$$Z_n = \frac{\sum_{i=1}^{n}(X_i - \mu)}{\sigma\sqrt{n}}$$

converges in distribution to a standard normal random variable.

For some classes of random variables the classic central limit theorem works rather fast for example the distributions with finite first, second and third moment from the exponential family; on the other hand, for some random variables of the heavy tail and fat tail variety, it works very slowly or may not work at all: in such cases one may use the Generalized Central Limit Theorem (GCLT).

Graph Theory

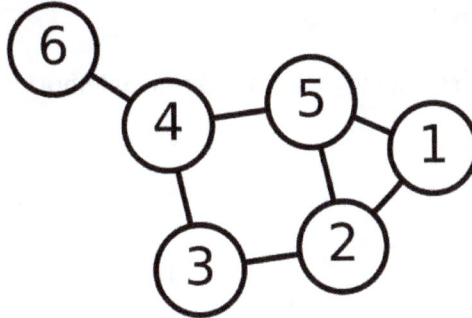

A drawing of a graph

In mathematics graph theory is the study of *graphs*, which are mathematical structures used to model pairwise relations between objects. A graph in this context is made up of *vertices, nodes,* or *points* which are connected by *edges, arcs,* or *lines*. A graph may be *undirected*, meaning that there is no distinction between the two vertices associated with each edge, or its edges may be *directed* from one vertex to another; Graph (discrete mathematics) for more detailed definitions and for other variations in the types of graph that are commonly considered. Graphs are one of the prime objects of study in discrete mathematics.

Refer to the glossary of graph theory for basic definitions in graph theory.

Definitions

Definitions in graph theory vary. The following are some of the more basic ways of defining graphs and related mathematical structures.

Graph

In the most common sense of the term, a graph is an ordered pair $G = (V, E)$ comprising a set V of *vertices* or *nodes* or *points* together with a set E of *edges* or *arcs* or *lines*, which are 2-element subsets of V (i.e. an edge is related with two vertices, and the relation is represented as an unordered pair of the vertices with respect to the particular edge). To avoid ambiguity, this type of graph may be described precisely as undirected and simple.

Other senses of *graph* stem from different conceptions of the edge set. In one more generalized notion, V is a set together with a relation of *incidence* that associates with each edge two vertices. In another generalized notion, E is a multiset of unordered pairs of (not necessarily distinct) vertices. Many authors call this type of object a multigraph or pseudograph.

All of these variants and others are described more fully below.

The vertices belonging to an edge are called the *ends* or *end vertices* of the edge. A vertex may exist in a graph and not belong to an edge.

V and E are usually taken to be finite, and many of the well-known results are not true (or are rather different) for infinite graphs because many of the arguments fail in the infinite case. The *order* of a graph is $|V|$, its number of vertices. The *size* of a graph is $|E|$, its number of edges. The *degree* or *valency* of a vertex is the number of edges that connect to it, where an edge that connects a vertex to itself (a loop) is counted twice.

For an edge $\{x, y\}$, graph theorists usually use the somewhat shorter notation xy.

Applications

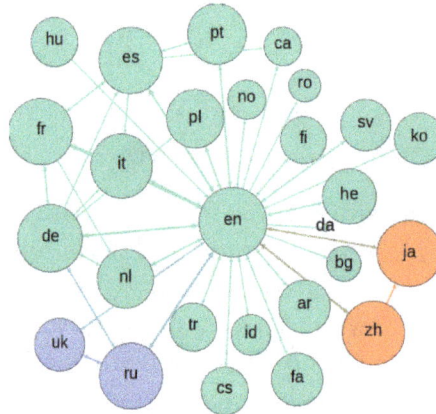

The network graph formed by Wikipedia editors (edges) contributing to different Wikipedia language versions (vertices) during one month in summer 2013

Graphs can be used to model many types of relations and processes in physical, biological, social and information systems. Many practical problems can be represented by graphs. Emphasizing their application to real-world systems, the term *network* is sometimes defined to mean a graph in which attributes (e.g. names) are associated with the nodes and/or edges.

In computer science, graphs are used to represent networks of communication, data organization, computational devices, the flow of computation, etc. For instance, the link structure of a website can be represented by a directed graph, in which the vertices represent web pages and directed edges represent links from one page to another. A similar approach can be taken to problems in social media, travel, biology, computer chip design, and many other fields. The development of algorithms to handle graphs is therefore of major interest in computer science. The transformation of graphs is often formalized and represented by graph rewrite systems. Complementary to graph transformation systems focusing on rule-based in-memory manipulation of graphs are graph databases geared towards transaction-safe, persistent storing and querying of graph-structured data.

Graph-theoretic methods, in various forms, have proven particularly useful in linguistics, since natural language often lends itself well to discrete structure. Traditionally, syntax and compositional semantics follow tree-based structures, whose expressive power lies in the principle of compositionality, modeled in a hierarchical graph. More contemporary approaches such as head-driven phrase structure grammar model the syntax of natural language using typed feature structures, which are directed acyclic graphs. Within lexical semantics, especially as applied to computers, modeling word meaning is easier when a given word is understood in terms of related words; semantic networks are therefore important in computational linguistics. Still other methods in phonology (e.g. optimality theory, which uses lattice graphs) and morphology (e.g. finite-state morphology, using finite-state transducers) are common in the analysis of language as a graph. Indeed, the usefulness of this area of mathematics to linguistics has borne organizations such as TextGraphs, as well as various 'Net' projects, such as WordNet, VerbNet, and others.

Graph theory is also used to study molecules in chemistry and physics. In condensed matter physics, the three-dimensional structure of complicated simulated atomic structures can be studied quantitatively by gathering statistics on graph-theoretic properties related to the topology of the atoms. In chemistry a graph makes a natural model for a molecule, where vertices represent atoms and edges bonds. This approach is especially used in computer processing of molecular structures, ranging from chemical editors to database searching. In statistical physics, graphs can represent local connections between interacting parts of a system, as well as the dynamics of a physical process on such systems. Similarly, in computational neuroscience graphs can be used to represent functional connections between brain areas that interact to give rise to various cognitive processes, where the vertices represent different areas of the brain and the edges represent the connections between those areas. Graphs are also used to represent the micro-scale channels of porous media, in which the vertices represent the pores and the edges represent the smaller channels connecting the pores.

Graph theory is also widely used in sociology as a way, for example, to measure actors' prestige or to explore rumor spreading, notably through the use of social network analysis software. Under the umbrella of social networks are many different types of graphs. Acquaintanceship and friendship graphs describe whether people know each other. Influence graphs model whether certain people can influence the behavior of others. Finally, collaboration graphs model whether two people work together in a particular way, such as acting in a movie together.

Likewise, graph theory is useful in biology and conservation efforts where a vertex can represent regions where certain species exist (or inhabit) and the edges represent migration paths, or movement between the regions. This information is important when looking at breeding patterns or tracking the spread of disease, parasites or how changes to the movement can affect other species.

In mathematics, graphs are useful in geometry and certain parts of topology such as knot theory. Algebraic graph theory has close links with group theory.

A graph structure can be extended by assigning a weight to each edge of the graph. Graphs with weights, or weighted graphs, are used to represent structures in which pairwise connections have some numerical values. For example, if a graph represents a road network, the weights could represent the length of each road.

History

The paper written by Leonhard Euler on the *Seven Bridges of Königsberg* and published in 1736 is regarded as the first paper in the history of graph theory. This paper, as well as the one written by Vandermonde on the *knight problem,* carried on with the *analysis situs* initiated by Leibniz. Euler's formula relating the number of edges, vertices, and faces of a convex polyhedron was studied and generalized by Cauchy and L'Huillier, and represents the beginning of the branch of mathematics known as topology.

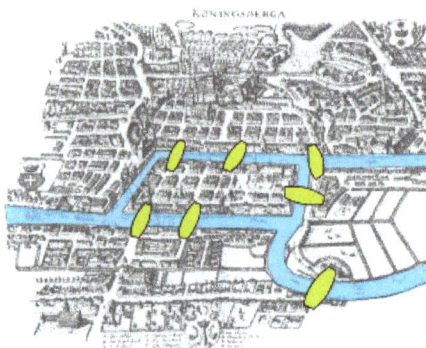

The Königsberg Bridge problem

More than one century after Euler's paper on the bridges of Königsberg and while Listing was introducing the concept of topology, Cayley was led by an interest in particular analytical forms arising from differential calculus to study a particular class of graphs, the *trees*. This study had many implications for theoretical chemistry. The techniques he used mainly concern the enumeration of graphs with particular properties. Enumerative graph theory then arose from the results of Cayley and the fundamental results published by Pólya between 1935 and 1937. These were generalized by De Bruijn in 1959. Cayley linked his results on trees with contemporary studies of chemical composition. The fusion of ideas from mathematics with those from chemistry began what has become part of the standard terminology of graph theory.

In particular, the term "graph" was introduced by Sylvester in a paper published in 1878 in *Nature*, where he draws an analogy between "quantic invariants" and "co-variants" of algebra and molecular diagrams:

> "[...] Every invariant and co-variant thus becomes expressible by a *graph* precisely identical with a Kekuléan diagram or chemicograph. [...] I give a rule for the geometrical multiplication of graphs, *i.e.* for constructing a *graph* to the product of in- or co-variants whose separate graphs are given. [...]" (italics as in the original).

The first textbook on graph theory was written by Dénes Kőnig, and published in 1936. Another book by Frank Harary, published in 1969, was "considered the world over to be the definitive textbook on the subject", and enabled mathematicians, chemists, electrical engineers and social scientists to talk to each other. Harary donated all of the royalties to fund the Pólya Prize.

One of the most famous and stimulating problems in graph theory is the four color problem: "Is it true that any map drawn in the plane may have its regions colored with four colors, in such a way that any two regions having a common border have different colors?" This problem was first posed by Francis Guthrie in 1852 and its first written record is in a letter of De Morgan addressed to Hamilton the same year. Many incorrect proofs have been proposed, including those by Cayley, Kempe, and others. The study and the generalization of this problem by Tait, Heawood, Ramsey and Hadwiger led to the study of the colorings of the graphs embedded on surfaces with arbitrary genus. Tait's reformulation generated a new class of problems, the *factorization problems*, particularly studied by Petersen and Kőnig. The works of Ramsey on colorations and more specially the results obtained by Turán in 1941 was at the origin of another branch of graph theory, *extremal graph theory*.

The four color problem remained unsolved for more than a century. In 1969 Heinrich Heesch published a method for solving the problem using computers. A computer-aided proof produced in 1976 by Kenneth Appel and Wolfgang Haken makes fundamental use of the notion of "discharging" developed by Heesch. The proof involved checking the properties of 1,936 configurations by computer, and was not fully accepted at the time due to its complexity. A simpler proof considering only 633 configurations was given twenty years later by Robertson, Seymour, Sanders and Thomas.

The autonomous development of topology from 1860 and 1930 fertilized graph theory back through the works of Jordan, Kuratowski and Whitney. Another important factor of common development of graph theory and topology came from the use of the techniques of modern algebra. The first example of such a use comes from the work of the physicist Gustav Kirchhoff, who published in 1845 his Kirchhoff's circuit laws for calculating the voltage and current in electric circuits.

The introduction of probabilistic methods in graph theory, especially in the study of Erdős and Rényi of the asymptotic probability of graph connectivity, gave rise to yet another branch, known as *random graph theory*, which has been a fruitful source of graph-theoretic results.

Graph Drawing

Graphs are represented visually by drawing a dot or circle for every vertex, and drawing an arc between two vertices if they are connected by an edge. If the graph is directed, the direction is indicated by drawing an arrow.

A graph drawing should not be confused with the graph itself (the abstract, non-visual structure) as there are several ways to structure the graph drawing. All that matters is which vertices are connected to which others by how many edges and not the exact layout. In practice it is often difficult to decide if two drawings represent the same graph. Depending on the problem domain some layouts may be better suited and easier to understand than others.

The pioneering work of W. T. Tutte was very influential in the subject of graph drawing. Among other achievements, he introduced the use of linear algebraic methods to obtain graph drawings.

Graph drawing also can be said to encompass problems that deal with the crossing number and its various generalizations. The crossing number of a graph is the minimum number of intersections between edges that a drawing of the graph in the plane must contain. For a planar graph, the crossing number is zero by definition.

Drawings on surfaces other than the plane are also studied.

Graph-theoretic Data Structures

There are different ways to store graphs in a computer system. The data structure used depends on both the graph structure and the algorithm used for manipulating the graph. Theoretically one can distinguish between list and matrix structures but in concrete applications the best structure is often a combination of both. List structures are often preferred for sparse graphs as they have smaller memory requirements. Matrix structures on the other hand provide faster access for some applications but can consume huge amounts of memory.

List structures include the incidence list, an array of pairs of vertices, and the adjacency list, which separately lists the neighbors of each vertex: Much like the incidence list, each vertex has a list of which vertices it is adjacent to.

Matrix structures include the incidence matrix, a matrix of 0's and 1's whose rows represent vertices and whose columns represent edges, and the adjacency matrix, in which both the rows and columns are indexed by vertices. In both cases a 1 indicates two adjacent objects and a 0 indicates two non-adjacent objects. The Laplacian matrix is a modified form of the adjacency matrix that incorporates information about the degrees of the vertices, and is useful in some calculations such as Kirchhoff's theorem on the number of spanning trees of a graph. The distance matrix, like the adjacency matrix, has both its rows and columns indexed by vertices, but rather than containing a 0 or a 1 in each cell it contains the length of a shortest path between two vertices.

Problems in Graph Theory

Enumeration

There is a large literature on graphical enumeration: the problem of counting graphs meeting specified conditions. Some of this work is found in Harary and Palmer (1973).

Subgraphs, Induced Subgraphs, and Minors

A common problem, called the subgraph isomorphism problem, is finding a fixed graph as a subgraph in a given graph. One reason to be interested in such a question is that many graph properties are *hereditary* for subgraphs, which means that a graph has the property if and only if all subgraphs have it too. Unfortunately, finding maximal subgraphs of a certain kind is often an NP-complete problem. For example:

- Finding the largest complete subgraph is called the clique problem (NP-complete).

- A similar problem is finding induced subgraphs in a given graph. Again, some important graph properties are hereditary with respect to induced subgraphs, which means that a graph has a property if and only if all induced subgraphs also have it. Finding maximal induced subgraphs of a certain kind is also often NP-complete. For example:

- Finding the largest edgeless induced subgraph or independent set is called the independent set problem (NP-complete).

- Still another such problem, the minor containment problem, is to find a fixed graph as a minor of a given graph. A minor or subcontraction of a graph is any graph obtained by taking a subgraph and contracting some (or no) edges. Many graph properties are hereditary for minors, which means that a graph has a property if and only if all minors have it too. For example, Wagner's Theorem states:

- A graph is planar if it contains as a minor neither the complete bipartite graph $K_{3,3}$ nor the complete graph K_5.

- A similar problem, the subdivision containment problem, is to find a fixed graph as a subdivision of a given graph. A subdivision or homeomorphism of a graph is any graph obtained by subdividing some (or no) edges. Subdivision containment is related to graph properties such as planarity. For example, Kuratowski's Theorem states:

- A graph is planar if it contains as a subdivision neither the complete bipartite graph $K_{3,3}$ nor the complete graph K_5.

- Another problem in subdivision containment is Kelmans-Seymour conjecture:

- Every 5-vertex-connected graph that is not planar contains a subdivision of the 5-vertex complete graph K_5.

Another class of problems has to do with the extent to which various species and generalizations of graphs are determined by their *point-deleted subgraphs*. For example:

- The reconstruction conjecture

Graph Coloring

Many problems have to do with various ways of coloring graphs, for example:

- Four-color theorem

- Strong perfect graph theorem

- Erdős–Faber–Lovász conjecture (unsolved)

- Total coloring conjecture, also called Behzad's conjecture (unsolved)

- List coloring conjecture (unsolved)

- Hadwiger conjecture (graph theory) (unsolved)

Subsumption and Unification

Constraint modeling theories concern families of directed graphs related by a partial order. In these applications, graphs are ordered by specificity, meaning that more constrained graphs—which are more specific and thus contain a greater amount of information—are subsumed by those that are more general. Operations between graphs include evaluating the direction of a subsumption relationship between two graphs, if any, and computing graph unification. The unification of two argument graphs is defined as the most general graph (or the computation thereof) that is consistent with (i.e. contains all of the information in) the inputs, if such a graph exists; efficient unification algorithms are known.

For constraint frameworks which are strictly compositional, graph unification is the sufficient satisfiability and combination function. Well-known applications include automatic theorem proving and modeling the elaboration of linguistic structure.

Route Problems

- Hamiltonian path problem

- Minimum spanning tree

- Route inspection problem (also called the "Chinese postman problem")
- Seven bridges of Königsberg
- Shortest path problem
- Steiner tree
- Three-cottage problem
- Traveling salesman problem (NP-hard)

Network Flow

There are numerous problems arising especially from applications that have to do with various notions of flows in networks, for example:

- Max flow min cut theorem

Visibility Problems

- Museum guard problem

Covering Problems

Covering problems in graphs are specific instances of subgraph-finding problems, and they tend to be closely related to the clique problem or the independent set problem.

- Set cover problem
- Vertex cover problem

Decomposition Problems

Decomposition, defined as partitioning the edge set of a graph (with as many vertices as necessary accompanying the edges of each part of the partition), has a wide variety of question. Often, it is required to decompose a graph into subgraphs isomorphic to a fixed graph; for instance, decomposing a complete graph into Hamiltonian cycles. Other problems specify a family of graphs into which a given graph should be decomposed, for instance, a family of cycles, or decomposing a complete graph K_n into $n - 1$ specified trees having, respectively, 1, 2, 3, ..., $n - 1$ edges.

Some specific decomposition problems that have been studied include:

- Arboricity, a decomposition into as few forests as possible
- Cycle double cover, a decomposition into a collection of cycles covering each edge exactly twice
- Edge coloring, a decomposition into as few matchings as possible
- Graph factorization, a decomposition of a regular graph into regular subgraphs of given degrees

Graph Classes

Many problems involve characterizing the members of various classes of graphs. Some examples of such questions are below:

- Enumerating the members of a class

- Characterizing a class in terms of forbidden substructures

- Ascertaining relationships among classes (e.g. does one property of graphs imply another)

- Finding efficient algorithms to decide membership in a class

- Finding representations for members of a class

Order Theory

Order theory is a branch of mathematics which investigates the intuitive notion of order using binary relations. It provides a formal framework for describing statements such as "this is less than that" or "this precedes that". This article introduces the field and provides basic definitions. A list of order-theoretic terms can be found in the order theory glossary.

Background and Motivation

Orders are everywhere in mathematics and related fields like computer science. The first order often discussed in primary school is the standard order on the natural numbers e.g. "2 is less than 3", "10 is greater than 5", or "Does Tom have fewer cookies than Sally?". This intuitive concept can be extended to orders on other sets of numbers, such as the integers and the reals. The idea of being greater than or less than another number is one of the basic intuitions of number systems (compare with numeral systems) in general (although one usually is also interested in the actual difference of two numbers, which is not given by the order). Other familiar examples of orderings are the alphabetical order of words in a dictionary and the genealogical property of lineal descent within a group of people.

The notion of order is very general, extending beyond contexts that have an immediate, intuitive feel of sequence or relative quantity. In other contexts orders may capture notions of containment or specialization. Abstractly, this type of order amounts to the subset relation, e.g., "Pediatricians are physicians," and "Circles are merely special-case ellipses."

Some orders, like "less-than" on the natural numbers and alphabetical order on words, have a special property: each element can be *compared* to any other element, i.e. it is smaller (earlier) than, larger (later) than, or identical to. However, many other orders do not. Consider for example the subset order on a collection of sets: though the set of birds and the set of dogs are both subsets of the set of animals, neither the birds nor the dogs constitutes a subset of the other. Those orders like the "subset-of" relation for which there exist *incomparable* elements are called *partial orders*; orders for which every pair of elements is comparable are *total orders*.

Order theory captures the intuition of orders that arises from such examples in a general setting.

This is achieved by specifying properties that a relation ≤ must have to be a mathematical order. This more abstract approach makes much sense, because one can derive numerous theorems in the general setting, without focusing on the details of any particular order. These insights can then be readily transferred to many less abstract applications.

Driven by the wide practical usage of orders, numerous special kinds of ordered sets have been defined, some of which have grown into mathematical fields of their own. In addition, order theory does not restrict itself to the various classes of ordering relations, but also considers appropriate functions between them. A simple example of an order theoretic property for functions comes from analysis where monotone functions are frequently found.

Basic Definitions

This section introduces ordered sets by building upon the concepts of set theory, arithmetic, and binary relations.

Partially Ordered Sets

Orders are special binary relations. Suppose that P is a set and that ≤ is a relation on P. Then ≤ is a partial order if it is reflexive, antisymmetric, and transitive, i.e., for all a, b and c in P, we have that:

$a ≤ a$ (reflexivity)

if $a ≤ b$ and $b ≤ a$ then $a = b$ (antisymmetry)

if $a ≤ b$ and $b ≤ c$ then $a ≤ c$ (transitivity).

A set with a partial order on it is called a partially ordered set, poset, or just an ordered set if the intended meaning is clear. By checking these properties, one immediately sees that the well-known orders on natural numbers, integers, rational numbers and reals are all orders in the above sense. However, they have the additional property of being total, i.e., for all a and b in P, we have that:

$a ≤ b$ or $b ≤ a$ (totality).

These orders can also be called linear orders or chains. While many classical orders are linear, the subset order on sets provides an example where this is not the case. Another example is given by the divisibility (or "*is-a-factor-of*") relation "|". For two natural numbers n and m, we write $n|m$ if n divides m without remainder. One easily sees that this yields a partial order. The identity relation = on any set is also a partial order in which every two distinct elements are incomparable. It is also the only relation that is both a partial order and an equivalence relation. Many advanced properties of posets are interesting mainly for non-linear orders.

Visualizing a Poset

Hasse diagrams can visually represent the elements and relations of a partial ordering. These are graph drawings where the vertices are the elements of the poset and the ordering relation is indicated by both the edges and the relative positioning of the vertices. Orders are drawn bottom-up: if an element x is smaller than (precedes) y then there exists a path from x to y that is directed upwards. It is often necessary for the edges connecting elements to cross each other, but elements must never be located within an edge. An instructive exercise is to draw the Hasse diagram for the set of natural numbers that are smaller than or equal to 13, ordered by | (the *divides* relation).

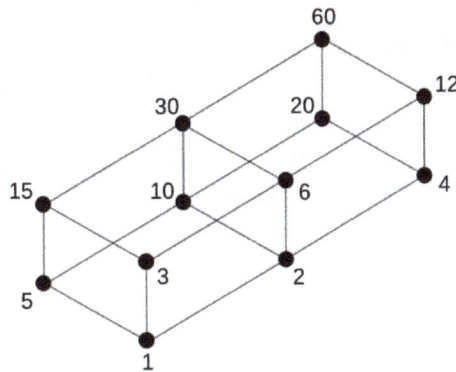

Hasse diagram of the set of all divisors of 60, partially ordered by divisibility

Even some infinite sets can be diagrammed by superimposing an ellipsis (...) on a finite sub-order. This works well for the natural numbers, but it fails for the reals, where there is no immediate successor above 0; however, quite often one can obtain an intuition related to diagrams of a similar kind.

Special Elements Within an Order

In a partially ordered set there may be some elements that play a special role. The most basic example is given by the least element of a poset. For example, 1 is the least element of the positive integers and the empty set is the least set under the subset order. Formally, an element m is a least element if:

$m \leq a$, for all elements a of the order.

The notation 0 is frequently found for the least element, even when no numbers are concerned. However, in orders on sets of numbers, this notation might be inappropriate or ambiguous, since the number 0 is not always least. An example is given by the above divisibility order |, where 1 is the least element since it divides all other numbers. In contrast, 0 is the number that is divided by all other numbers. Hence it is the greatest element of the order. Other frequent terms for the least and greatest elements is bottom and top or zero and unit.

Least and greatest elements may fail to exist, as the example of the real numbers shows. But if they exist, they are always unique. In contrast, consider the divisibility relation | on the set {2,3,4,5,6}. Although this set has neither top nor bottom, the elements 2, 3, and 5 have no elements below them, while 4, 5 and 6 have none above. Such elements are called minimal and maximal, respectively. Formally, an element m is minimal if:

$a \leq m$ implies $a = m$, for all elements a of the order.

Exchanging $\leq$ with $\geq$ yields the definition of maximality. As the example shows, there can be many maximal elements and some elements may be both maximal and minimal (e.g. 5 above). However, if there is a least element, then it is the only minimal element of the order. Again, in infinite posets maximal elements do not always exist - the set of all *finite* subsets of a given infinite set, ordered by subset inclusion, provides one of many counterexamples. An important tool to ensure the existence of maximal elements under certain conditions is Zorn's Lemma.

Subsets of partially ordered sets inherit the order. We already applied this by considering the subset {2,3,4,5,6} of the natural numbers with the induced divisibility ordering. Now there are also elements of a poset that are special with respect to some subset of the order. This leads to the definition of upper bounds. Given a subset S of some poset P, an upper bound of S is an element b of P that is above all elements of S. Formally, this means that

$$s \le b, \text{ for all } s \text{ in } S.$$

Lower bounds again are defined by inverting the order. For example, -5 is a lower bound of the natural numbers as a subset of the integers. Given a set of sets, an upper bound for these sets under the subset ordering is given by their union. In fact, this upper bound is quite special: it is the smallest set that contains all of the sets. Hence, we have found the least upper bound of a set of sets. This concept is also called supremum or join, and for a set S one writes $\sup(S)$ or $\bigvee$ for its least upper bound. Conversely, the greatest lower bound is known as infimum or meet and denoted $\inf(S)$ or $\bigwedge S$. These concepts play an important role in many applications of order theory. For two elements x and y, one also writes $x \vee y$ and $x \wedge y$ for $\sup(\{x,y\})$ and $\inf(\{x,y\})$, respectively. For example, 1 is the infimum of the positive integers as a subset of integers.

For another example, consider again the relation | on natural numbers. The least upper bound of two numbers is the smallest number that is divided by both of them, i.e. the least common multiple of the numbers. Greatest lower bounds in turn are given by the greatest common divisor.

Duality

In the previous definitions, we often noted that a concept can be defined by just inverting the ordering in a former definition. This is the case for "least" and "greatest", for "minimal" and "maximal", for "upper bound" and "lower bound", and so on. This is a general situation in order theory: A given order can be inverted by just exchanging its direction, pictorially flipping the Hasse diagram top-down. This yields the so-called dual, inverse, or opposite order.

Every order theoretic definition has its dual: it is the notion one obtains by applying the definition to the inverse order. Since all concepts are symmetric, this operation preserves the theorems of partial orders. For a given mathematical result, one can just invert the order and replace all definitions by their duals and one obtains another valid theorem. This is important and useful, since one obtains two theorems for the price of one. Some more details and examples can be found in the article on duality in order theory.

Constructing New Orders

There are many ways to construct orders out of given orders. The dual order is one example. Another important construction is the cartesian product of two partially ordered sets, taken together with the product order on pairs of elements. The ordering is defined by $(a, x) \le (b, y)$ if (and only if) $a \le b$ and $x \le y$. (Notice carefully that there are three distinct meanings for the relation symbol $\le$ in this definition.) The disjoint union of two posets is another typical example of order construction, where the order is just the (disjoint) union of the original orders.

Every partial order ≤ gives rise to a so-called strict order <, by defining $a < b$ if $a \leq b$ and not $b \leq a$. This transformation can be inverted by setting $a \leq b$ if $a < b$ or $a = b$. The two concepts are equivalent although in some circumstances one can be more convenient to work with than the other.

Functions Between Orders

It is reasonable to consider functions between partially ordered sets having certain additional properties that are related to the ordering relations of the two sets. The most fundamental condition that occurs in this context is monotonicity. A function f from a poset P to a poset Q is monotone, or order-preserving, if $a \leq b$ in P implies $f(a) \leq f(b)$ in Q (Noting that, strictly, the two relations here are different since they apply to different sets.). The converse of this implication leads to functions that are order-reflecting, i.e. functions f as above for which $f(a) \leq f(b)$ implies $a \leq b$. On the other hand, a function may also be order-reversing or antitone, if $a \leq b$ implies $f(b) \leq f(a)$.

An order-embedding is a function f between orders that is both order-preserving and order-reflecting. Examples for these definitions are found easily. For instance, the function that maps a natural number to its successor is clearly monotone with respect to the natural order. Any function from a discrete order, i.e. from a set ordered by the identity order "=", is also monotone. Mapping each natural number to the corresponding real number gives an example for an order embedding. The set complement on a powerset is an example of an antitone function.

An important question is when two orders are "essentially equal", i.e. when they are the same up to renaming of elements. Order isomorphisms are functions that define such a renaming. An order-isomorphism is a monotone bijective function that has a monotone inverse. This is equivalent to being a surjective order-embedding. Hence, the image $f(P)$ of an order-embedding is always isomorphic to P, which justifies the term "embedding".

A more elaborate type of functions is given by so-called Galois connections. Monotone Galois connections can be viewed as a generalization of order-isomorphisms, since they constitute of a pair of two functions in converse directions, which are "not quite" inverse to each other, but that still have close relationships.

Another special type of self-maps on a poset are closure operators, which are not only monotonic, but also idempotent, i.e. $f(x) = f(f(x))$), and extensive (or *inflationary*), i.e. $x \leq f(x)$. These have many applications in all kinds of "closures" that appear in mathematics.

Besides being compatible with the mere order relations, functions between posets may also behave well with respect to special elements and constructions. For example, when talking about posets with least element, it may seem reasonable to consider only monotonic functions that preserve this element, i.e. which map least elements to least elements. If binary infima ∧ exist, then a reasonable property might be to require that $f(x \wedge y) = f(x) \wedge f(y)$, for all x and y. All of these properties, and indeed many more, may be compiled under the label of limit-preserving functions.

Finally, one can invert the view, switching from *functions of orders* to *orders of functions*. Indeed, the functions between two posets P and Q can be ordered via the pointwise order. For two functions f and g, we have $f \leq g$ if $f(x) \leq g(x)$ for all elements x of P. This occurs for example in domain theory, where function spaces play an important role.

Special types of Orders

Many of the structures that are studied in order theory employ order relations with further properties. In fact, even some relations that are not partial orders are of special interest. Mainly the concept of a preorder has to be mentioned. A preorder is a relation that is reflexive and transitive, but not necessarily antisymmetric. Each preorder induces an equivalence relation between elements, where a is equivalent to b, if $a \leq b$ and $b \leq a$. Preorders can be turned into orders by identifying all elements that are equivalent with respect to this relation.

Several types of orders can be defined from numerical data on the items of the order: a total order results from attaching distinct real numbers to each item and using the numerical comparisons to order the items; instead, if distinct items are allowed to have equal numerical scores, one obtains a strict weak ordering. Requiring two scores to be separated by a fixed threshold before they may be compared leads to the concept of a semiorder, while allowing the threshold to vary on a per-item basis produces an interval order.

An additional simple but useful property leads to so-called well-orders, for which all non-empty subsets have a minimal element. Generalizing well-orders from linear to partial orders, a set is well partially ordered if all its non-empty subsets have a finite number of minimal elements.

Many other types of orders arise when the existence of infima and suprema of certain sets is guaranteed. Focusing on this aspect, usually referred to as completeness of orders, one obtains:

- Bounded posets, i.e. posets with a least and greatest element (which are just the supremum and infimum of the empty subset),

- Lattices, in which every non-empty finite set has a supremum and infimum,

- Complete lattices, where every set has a supremum and infimum, and

- Directed complete partial orders (dcpos), that guarantee the existence of suprema of all directed subsets and that are studied in domain theory.

- Partial orders with complements, or *poc sets*, are posets S having a unique bottom element $o \in S$, along with an order-reversing involution, such that $a \leq a^* \Rightarrow a = 0..$

However, one can go even further: if all finite non-empty infima exist, then $\wedge$ can be viewed as a total binary operation in the sense of universal algebra. Hence, in a lattice, two operations $\wedge$ and $\vee$ are available, and one can define new properties by giving identities, such as

$$x \wedge (y \vee z) = (x \wedge y) \vee (x \wedge z), \text{ for all } x, y, \text{ and } z.$$

This condition is called distributivity and gives rise to distributive lattices. There are some other important distributivity laws which are discussed in the article on distributivity in order theory. Some additional order structures that are often specified via algebraic operations and defining identities are

- Heyting algebras and

- Boolean algebras,

which both introduce a new operation ~ called negation. Both structures play a role in mathematical logic and especially Boolean algebras have major applications in computer science. Finally, various structures in mathematics combine orders with even more algebraic operations, as in the case of quantales, that allow for the definition of an addition operation.

Many other important properties of posets exist. For example, a poset is locally finite if every closed interval $[a, b]$ in it is finite. Locally finite posets give rise to incidence algebras which in turn can be used to define the Euler characteristic of finite bounded posets.

Subsets of Ordered Sets

In an ordered set, one can define many types of special subsets based on the given order. A simple example are upper sets; i.e. sets that contain all elements that are above them in the order. Formally, the upper closure of a set S in a poset P is given by the set $\{x$ in $P \mid$ there is some y in S with $y \leq x\}$. A set that is equal to its upper closure is called an upper set. Lower sets are defined dually.

More complicated lower subsets are ideals, which have the additional property that each two of their elements have an upper bound within the ideal. Their duals are given by filters. A related concept is that of a directed subset, which like an ideal contains upper bounds of finite subsets, but does not have to be a lower set. Furthermore, it is often generalized to preordered sets.

A subset which is - as a sub-poset - linearly ordered, is called a chain. The opposite notion, the antichain, is a subset that contains no two comparable elements; i.e. that is a discrete order.

Related Mathematical Areas

Although most mathematical areas *use* orders in one or the other way, there are also a few theories that have relationships which go far beyond mere application. Together with their major points of contact with order theory, some of these are to be presented below.

Universal Algebra

As already mentioned, the methods and formalisms of universal algebra are an important tool for many order theoretic considerations. Beside formalizing orders in terms of algebraic structures that satisfy certain identities, one can also establish other connections to algebra. An example is given by the correspondence between Boolean algebras and Boolean rings. Other issues are concerned with the existence of free constructions, such as *free lattices* based on a given set of generators. Furthermore, closure operators are important in the study of universal algebra.

Topology

In topology orders play a very prominent role. In fact, the set of open sets provides a classical example of a complete lattice, more precisely a complete Heyting algebra (or "frame" or "locale"). Filters and nets are notions closely related to order theory and the closure operator of sets can be used to define topology. Beyond these relations, topology can be looked at solely in terms of the open set lattices, which leads to the study of pointless topology. Furthermore, a natural preorder of elements of the underlying set of a topology is given by the so-called specialization order, that is actually a partial order if the topology is T_o.

Conversely, in order theory, one often makes use of topological results. There are various ways to define subsets of an order which can be considered as open sets of a topology. Especially, it is interesting to consider topologies on a poset $(X, \leq)$ that in turn induce $\leq$ as their specialization order. The *finest* such topology is the Alexandrov topology, given by taking all upper sets as opens. Conversely, the *coarsest* topology that induces the specialization order is the upper topology, having the complements of principal ideals (i.e. sets of the form $\{y$ in $X \mid y \leq x\}$ for some x) as a subbase. Additionally, a topology with specialization order $\leq$ may be order consistent, meaning that their open sets are "inaccessible by directed suprema" (with respect to $\leq$). The finest order consistent topology is the Scott topology, which is coarser than the Alexandrov topology. A third important topology in this spirit is the Lawson topology. There are close connections between these topologies and the concepts of order theory. For example, a function preserves directed suprema iff it is continuous with respect to the Scott topology (for this reason this order theoretic property is also called Scott-continuity).

Category Theory

The visualization of orders with Hasse diagrams has a straightforward generalization: instead of displaying lesser elements *below* greater ones, the direction of the order can also be depicted by giving directions to the edges of a graph. In this way, each order is seen to be equivalent to a directed acyclic graph, where the nodes are the elements of the poset and there is a directed path from a to b if and only if $a \leq b$. Dropping the requirement of being acyclic, one can also obtain all preorders.

When equipped with all transitive edges, these graphs in turn are just special categories, where elements are objects and each set of morphisms between two elements is at most singleton. Functions between orders become functors between categories. Interestingly, many ideas of order theory are just concepts of category theory in small. For example, an infimum is just a categorical product. More generally, one can capture infima and suprema under the abstract notion of a categorical limit (or *colimit*, respectively). Another place where categorical ideas occur is the concept of a (monotone) Galois connection, which is just the same as a pair of adjoint functors.

But category theory also has its impact on order theory on a larger scale. Classes of posets with appropriate functions as discussed above form interesting categories. Often one can also state constructions of orders, like the product order, in terms of categories. Further insights result when categories of orders are found categorically equivalent to other categories, for example of topological spaces. This line of research leads to various *representation theorems*, often collected under the label of Stone duality.

History

As explained before, orders are ubiquitous in mathematics. However, earliest explicit mentionings of partial orders are probably to be found not before the 19th century. In this context the works of George Boole are of great importance. Moreover, works of Charles Sanders Peirce, Richard Dedekind, and Ernst Schröder also consider concepts of order theory. Certainly, there are others to be named in this context and surely there exists more detailed material on the history of order theory.

The term *poset* as an abbreviation for partially ordered set was coined by Garrett Birkhoff in the second edition of his influential book *Lattice Theory*.

Number Theory

A Lehmer sieve, which is a primitive digital computer once used for finding primes and solving simple Diophantine equations.

Number theory or, in older usage, arithmetic is a branch of pure mathematics devoted primarily to the study of the integers. It is sometimes called "The Queen of Mathematics" because of its foundational place in the discipline. Number theorists study prime numbers as well as the properties of objects made out of integers (e.g., rational numbers) or defined as generalizations of the integers (e.g., algebraic integers).

Integers can be considered either in themselves or as solutions to equations (Diophantine geometry). Questions in number theory are often best understood through the study of analytical objects (e.g., the Riemann zeta function) that encode properties of the integers, primes or other number-theoretic objects in some fashion (analytic number theory). One may also study real numbers in relation to rational numbers, e.g., as approximated by the latter (Diophantine approximation).

The older term for number theory is *arithmetic*. By the early twentieth century, it had been superseded by "number theory". (The word "arithmetic" is used by the general public to mean "elementary calculations"; it has also acquired other meanings in mathematical logic, as in *Peano arithmetic*, and computer science, as in *floating point arithmetic*.) The use of the term *arithmetic* for *number theory* regained some ground in the second half of the 20th century, arguably in part due to French influence. In particular, *arithmetical* is preferred as an adjective to *number-theoretic*.

History

Origins

Dawn of Arithmetic

The first historical find of an arithmetical nature is a fragment of a table: the broken clay tablet

Plimpton 322 (Larsa, Mesopotamia, ca. 1800 BCE) contains a list of "Pythagorean triples", i.e., integers (a, b, c) such that $a^2 + b^2 = c^2$. The triples are too many and too large to have been obtained by brute force. The heading over the first column reads: "The *takiltum* of the diagonal which has been subtracted such that the width..."

The Plimpton 322 tablet

The table's layout suggests that it was constructed by means of what amounts, in modern language, to the identity

$$\left(\frac{1}{2}\left(x - \frac{1}{x} \right) \right)^2 + 1 = \left(\frac{1}{2}\left(x + \frac{1}{x} \right) \right)^2,$$

which is implicit in routine Old Babylonian exercises. If some other method was used, the triples were first constructed and then reordered by c / a, presumably for actual use as a "table", i.e., with a view to applications.

It is not known what these applications may have been, or whether there could have been any; Babylonian astronomy, for example, truly flowered only later. It has been suggested instead that the table was a source of numerical examples for school problems.

While Babylonian number theory—or what survives of Babylonian mathematics that can be called thus—consists of this single, striking fragment, Babylonian algebra (in the secondary-school sense of "algebra") was exceptionally well developed. Late Neoplatonic sources state that Pythagoras learned mathematics from the Babylonians. Much earlier sources state that Thales and Pythagoras traveled and studied in Egypt.

Euclid IX 21—34 is very probably Pythagorean; it is very simple material ("odd times even is even", "if an odd number measures [= divides] an even number, then it also measures [= divides] half of it"), but it is all that is needed to prove that $\sqrt{2}$ is irrational. Pythagorean mystics gave great importance to the odd and the even. The discovery that $\sqrt{2}$ is irrational is credited to the early Pythagoreans (pre-Theodorus). By revealing (in modern terms) that numbers could be irrational, this discovery seems to have provoked the first foundational crisis in mathematical history; its proof or its divulgation are sometimes credited to Hippasus, who was expelled or split from the Pythagorean sect. This forced a distinction between *numbers* (integers and the rationals—the subjects of arithmetic), on the one hand, and *lengths* and *proportions* (which we would identify with real numbers, whether rational or not), on the other hand.

The Pythagorean tradition spoke also of so-called polygonal or figurate numbers. While square numbers, cubic numbers, etc., are seen now as more natural than triangular numbers, pentagonal numbers, etc., the study of the sums of triangular and pentagonal numbers would prove fruitful in the early modern period (17th to early 19th century).

We know of no clearly arithmetical material in ancient Egyptian or Vedic sources, though there is some algebra in both. The Chinese remainder theorem appears as an exercise in Sun Zi's *Suan Ching*, also known as *The Mathematical Classic of Sun Zi* (3rd, 4th or 5th century CE.) (There is one important step glossed over in Sun Zi's solution)

There is also some numerical mysticism in Chinese mathematics,but, unlike that of the Pythagoreans, it seems to have led nowhere. Like the Pythagoreans' perfect numbers, magic squares have passed from superstition into recreation.

Classical Greece and the Early Hellenistic Period

Aside from a few fragments, the mathematics of Classical Greece is known to us either through the reports of contemporary non-mathematicians or through mathematical works from the early Hellenistic period. In the case of number theory, this means, by and large, *Plato* and *Euclid*, respectively.

Plato had a keen interest in mathematics, and distinguished clearly between arithmetic and calculation. (By *arithmetic* he meant, in part, theorising on number, rather than what *arithmetic* or *number theory* have come to mean.) It is through one of Plato's dialogues—namely, *Theaetetus*—that we know that Theodorus had proven that $\sqrt{3}, \sqrt{5}, \ldots, \sqrt{17}$ are irrational. Theaetetus was, like Plato, a disciple of Theodorus's; he worked on distinguishing different kinds of incommensurables, and was thus arguably a pioneer in the study of number systems. (Book X of Euclid's Elements is described by Pappus as being largely based on Theaetetus's work.)

Euclid devoted part of his *Elements* to prime numbers and divisibility, topics that belong unambiguously to number theory and are basic to it (Books VII to IX of Euclid's Elements). In particular, he gave an algorithm for computing the greatest common divisor of two numbers (the Euclidean algorithm; *Elements*, Prop. VII.2) and the first known proof of the infinitude of primes (*Elements*, Prop. IX.20).

In 1773, Lessing published an epigram he had found in a manuscript during his work as a librarian; it claimed to be a letter sent by Archimedes to Eratosthenes. The epigram proposed what has become known as Archimedes' cattle problem; its solution (absent from the manuscript) requires solving an indeterminate quadratic equation (which reduces to what would later be misnamed Pell's equation). As far as we know, such equations were first successfully treated by the Indian school. It is not known whether Archimedes himself had a method of solution.

Diophantus

Very little is known about Diophantus of Alexandria; he probably lived in the third century CE, that is, about five hundred years after Euclid. Six out of the thirteen books of Diophantus's *Arithmetica* survive in the original Greek; four more books survive in an Arabic translation. The *Arithmetica* is a collection of worked-out problems where the task is invariably to find rational solutions to a system of polynomial equations, usually of the form $f(x,y) = z^2$ or $f(x,y,z) = w^2$. Thus, nowadays, we speak of *Diophantine equations* when we speak of polynomial equations to which rational or integer solutions must be found.

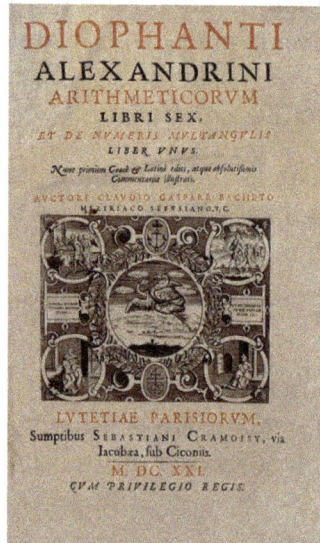

Title page of the 1621 edition of Diophantus' *Arithmetica*, translated into Latin by Claude Gaspard Bachet de Méziriac.

One may say that Diophantus was studying rational points — i.e., points whose coordinates are rational — on curves and algebraic varieties; however, unlike the Greeks of the Classical period, who did what we would now call basic algebra in geometrical terms, Diophantus did what we would now call basic algebraic geometry in purely algebraic terms. In modern language, what Diophantus did was to find rational parametrizations of varieties; that is, given an equation of the form (say) $f(x_1, x_2, x_3) = 0$, his aim was to find (in essence) three rational functions g_1, g_2, g_3 such that, for all values of r and s, setting $x_i = g_i(r,s)$ for $i = 1, 2, 3$ gives a solution to $f(x_1, x_2, x_3) = 0$.

Diophantus also studied the equations of some non-rational curves, for which no rational parametrisation is possible. He managed to find some rational points on these curves (elliptic curves, as it happens, in what seems to be their first known occurrence) by means of what amounts to a tangent construction: translated into coordinate geometry (which did not exist in Diophantus's time), his method would be visualised as drawing a tangent to a curve at a known rational point, and then finding the other point of intersection of the tangent with the curve; that other point is a new rational point. (Diophantus also resorted to what could be called a special case of a secant construction.)

While Diophantus was concerned largely with rational solutions, he assumed some results on integer numbers, in particular that every integer is the sum of four squares (though he never stated as much explicitly).

Āryabhaṭa, Brahmagupta, Bhāskara

While Greek astronomy probably influenced Indian learning, to the point of introducing trigonometry, it seems to be the case that Indian mathematics is otherwise an indigenous tradition; in particular, there is no evidence that Euclid's Elements reached India before the 18th century.

Āryabhata (476–550 CE) showed that pairs of simultaneous congruences $n \equiv a_1 \pmod{m}_1$, $n \equiv a_2 \pmod{m}_2$ could be solved by a method he called *kuṭṭaka*, or *pulveriser*; this is a procedure close to (a generalisation of) the Euclidean algorithm, which was probably discovered independently in India. Āryabhata seems to have had in mind applications to astronomical calculations.

Brahmagupta (628 CE) started the systematic study of indefinite quadratic equations—in particular, the misnamed Pell equation, in which Archimedes may have first been interested, and which did not start to be solved in the West until the time of Fermat and Euler. Later Sanskrit authors would follow, using Brahmagupta's technical terminology. A general procedure (the chakravala, or "cyclic method") for solving Pell's equation was finally found by Jayadeva (cited in the eleventh century; his work is otherwise lost); the earliest surviving exposition appears in Bhāskara II's Bīja-ganita (twelfth century).

Indian mathematics remained largely unknown in Europe until the late eighteenth century; Brahmagupta and Bhāskara's work was translated into English in 1817 by Henry Colebrooke.

Arithmetic in the Islamic Golden Age

Al-Haytham seen by the West: frontispice of *Selenographia*, showing Alhasen [*sic*] representing knowledge through reason, and Galileo representing knowledge through the senses.

In the early ninth century, the caliph Al-Ma'mun ordered translations of many Greek mathematical works and at least one Sanskrit work (the *Sindhind*, which may or may not be Brahmagupta's Brāhmasphuṭasiddhānta). Diophantus's main work, the *Arithmetica*, was translated into Arabic by Qusta ibn Luqa (820–912). Part of the treatise *al-Fakhri* (by al-Karajī, 953 – ca. 1029) builds on it to some extent. According to Rashed Roshdi, Al-Karajī's contemporary Ibn al-Haytham knew what would later be called Wilson's theorem.

Western Europe in the Middle Ages

Other than a treatise on squares in arithmetic progression by Fibonacci — who lived and studied in north Africa and Constantinople during his formative years, ca. 1175–1200 — no number theory to speak of was done in western Europe during the Middle Ages. Matters started to change in Europe in the late Renaissance, thanks to a renewed study of the works of Greek antiquity. A catalyst was the textual emendation and translation into Latin of Diophantus's *Arithmetica* (Bachet, 1621, following a first attempt by Xylander, 1575).

Early Modern Number Theory

Fermat

Pierre de Fermat

Pierre de Fermat (1601–1665) never published his writings; in particular, his work on number theory is contained almost entirely in letters to mathematicians and in private marginal notes. He wrote down nearly no proofs in number theory; he had no models in the area. He did make repeated use of mathematical induction, introducing the method of infinite descent.

One of Fermat's first interests was perfect numbers (which appear in Euclid, *Elements* IX) and amicable numbers; this led him to work on integer divisors, which were from the beginning among the subjects of the correspondence (1636 onwards) that put him in touch with the mathematical community of the day. He had already studied Bachet's edition of Diophantus carefully; by 1643, his interests had shifted largely to Diophantine problems and sums of squares (also treated by Diophantus).

Fermat's achievements in arithmetic include:

- Fermat's little theorem (1640), stating that, if a is not divisible by a prime p, then

- If a and b are coprime, then $a^2 + b^2$ is not divisible by any prime congruent to -1 modulo 4; and every prime congruent to 1 modulo 4 can be written in the form $a^2 + b^2$. These two statements also date from 1640; in 1659, Fermat stated to Huygens that he had proven the latter statement by the method of infinite descent. Fermat and Frenicle also did some work (some of it erroneous) on other quadratic forms.

- Fermat posed the problem of solving $x^2 - Ny^2 = 1$ as a challenge to English mathematicians (1657). The problem was solved in a few months by Wallis and Brouncker. Fermat considered their solution valid, but pointed out they had provided an algorithm without a proof (as had Jayadeva and Bhaskara, though Fermat would never know this.) He states that a proof can be found by descent.

- Fermat developed methods for (doing what in our terms amounts to) finding points on curves of genus 0 and 1. As in Diophantus, there are many special procedures and what amounts to a tangent construction, but no use of a secant construction.

- Fermat states and proves (by descent) in the appendix to *Observations on Diophantus* (Obs. XLV) that $x^4 + y^4 = z^4$ has no non-trivial solutions in the integers. Fermat also mentioned to his correspondents that $x^3 + y^3 = z^3$ has no non-trivial solutions, and that this could be proven by descent. The first known proof is due to Euler (1753; indeed by descent).

Fermat's claim ("Fermat's last theorem") to have shown there are no solutions to $x^n + y^n = z^n$ for all $n \geq 3$ (the only known proof of which is beyond his methods) appears only in his annotations on the margin of his copy of Diophantus; he never claimed this to others and thus would have had no need to retract it if he found any mistake in his supposed proof.

Euler

Leonhard Euler

The interest of Leonhard Euler (1707–1783) in number theory was first spurred in 1729, when a friend of his, the amateur Goldbach, pointed him towards some of Fermat's work on the subject. This has been called the "rebirth" of modern number theory, after Fermat's relative lack of success in getting his contemporaries' attention for the subject. Euler's work on number theory includes the following:

- *Proofs for Fermat's statements.* This includes Fermat's little theorem (generalised by Euler to non-prime moduli); the fact that $p = x^2 + y^2$ if and only if $p \equiv 1 \pmod 4$; initial work towards a proof that every integer is the sum of four squares (the first complete proof is by Joseph-Louis Lagrange (1770), soon improved by Euler himself); the lack of non-zero integer solutions to $x^4 + y^4 = z^2$ (implying the case *n=4* of Fermat's last theorem, the case *n=3* of which Euler also proved by a related method).

- *Pell's equation*, first misnamed by Euler. He wrote on the link between continued fractions and Pell's equation.

- *First steps towards analytic number theory.* In his work of sums of four squares, partitions, pentagonal numbers, and the distribution of prime numbers, Euler pioneered the use of what can be seen as analysis (in particular, infinite series) in number theory. Since he lived before the development of complex analysis, most of his work is restricted to the formal manipulation of power series. He did, however, do some very notable (though not fully rigorous) early work on what would later be called the Riemann zeta function.

- *Quadratic forms.* Following Fermat's lead, Euler did further research on the question of which primes can be expressed in the form $x^2 + Ny^2$, some of it prefiguring quadratic reciprocity.

- *Diophantine equations.* Euler worked on some Diophantine equations of genus 0 and 1. In particular, he studied Diophantus's work; he tried to systematise it, but the time was not yet ripe for such an endeavour – algebraic geometry was still in its infancy. He did notice there was a connection between Diophantine problems and elliptic integrals, whose study he had himself initiated.

Lagrange, Legendre and Gauss

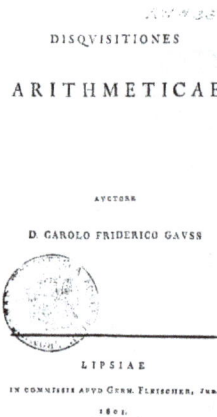

DISQVISITIONES

ARITHMETICAE

AVCTORE

D. CAROLO FRIDERICO GAVSS

LIPSIAE

IN COMMISSIS APVD GERM. FLEISCHER, Jun.

1801.

Carl Friedrich Gauss's Disquisitiones Arithmeticae, first edition

Joseph-Louis Lagrange (1736–1813) was the first to give full proofs of some of Fermat's and Euler's work and observations - for instance, the four-square theorem and the basic theory of the misnamed "Pell's equation" (for which an algorithmic solution was found by Fermat and his contemporaries, and also by Jayadeva and Bhaskara II before them.) He also studied quadratic forms in full generality (as opposed to $mX^2 + nY^2$) — defining their equivalence relation, showing how to put them in reduced form, etc.

Adrien-Marie Legendre (1752–1833) was the first to state the law of quadratic reciprocity. He also conjectured what amounts to the prime number theorem and Dirichlet's theorem on arithmetic progressions. He gave a full treatment of the equation $ax^2 + by^2 + cz^2 = 0$ and worked on quadratic forms along the lines later developed fully by Gauss. In his old age, he was the first to prove "Fermat's last theorem" for $n = 5$ (completing work by Peter Gustav Lejeune Dirichlet, and crediting both him and Sophie Germain).

In his *Disquisitiones Arithmeticae* (1798), Carl Friedrich Gauss (1777–1855) proved the law of quadratic reciprocity and developed the theory of quadratic forms (in particular, defining their composition). He also introduced some basic notation (congruences) and devoted a section to computational matters, including primality tests. The last section of the *Disquisitiones* established a link between roots of unity and number theory:

Carl Friedrich Gauss

The theory of the division of the circle...which is treated in sec. 7 does not belong by itself to arithmetic, but its principles can only be drawn from higher arithmetic.

In this way, Gauss arguably made a first foray towards both Évariste Galois's work and algebraic number theory.

Maturity and Division into Subfields

Ernst Kummer

Peter Gustav Lejeune Dirichlet

Starting early in the nineteenth century, the following developments gradually took place:

- The rise to self-consciousness of number theory (or *higher arithmetic*) as a field of study.

- The development of much of modern mathematics necessary for basic modern number theory: complex analysis, group theory, Galois theory—accompanied by greater rigor in analysis and abstraction in algebra.

- The rough subdivision of number theory into its modern subfields—in particular, analytic and algebraic number theory.

Algebraic number theory may be said to start with the study of reciprocity and cyclotomy, but truly came into its own with the development of abstract algebra and early ideal theory and valuation theory; A conventional starting point for analytic number theory is Dirichlet's theorem on arithmetic progressions (1837), whose proof introduced L-functions and involved some asymptotic analysis and a limiting process on a real variable. The first use of analytic ideas in number theory actually goes back to Euler (1730s), who used formal power series and non-rigorous (or implicit) limiting arguments. The use of *complex* analysis in number theory comes later: the work of Bernhard Riemann (1859) on the zeta function is the canonical starting point; Jacobi's four-square theorem (1839), which predates it, belongs to an initially different strand that has by now taken a leading role in analytic number theory (modular forms).

The history of each subfield is briefly addressed in its own section below; see the main article of each subfield for fuller treatments. Many of the most interesting questions in each area remain open and are being actively worked on.

Main Subdivisions

Elementary Tools

The term *elementary* generally denotes a method that does not use complex analysis. For example, the prime number theorem was first proven using complex analysis in 1896, but an elementary proof was found only in 1949 by Erdős and Selberg. The term is somewhat ambiguous: for example, proofs based on complex Tauberian theorems (e.g. Wiener–Ikehara) are often seen as quite enlightening but not elementary, in spite of using Fourier analysis, rather than complex analysis as such. Here as elsewhere, an *elementary* proof may be longer and more difficult for most readers than a non-elementary one.

Number theory has the reputation of being a field many of whose results can be stated to the layperson. At the same time, the proofs of these results are not particularly accessible, in part because the range of tools they use is, if anything, unusually broad within mathematics.

Analytic Number Theory

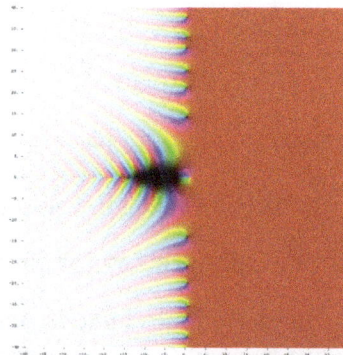

Riemann zeta function $\zeta(s)$ in the complex plane. The color of a point s gives the value of $\zeta(s)$: dark colors denote values close to zero and hue gives the value's argument.

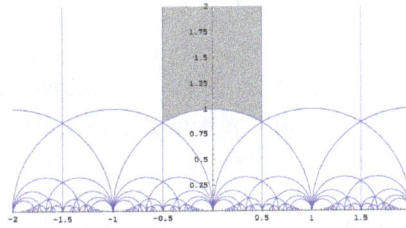

The action of the modular group on the upper half plane. The region in grey is the standard fundamental domain.

Analytic number theory may be defined

- in terms of its tools, as the study of the integers by means of tools from real and complex analysis; or

- in terms of its concerns, as the study within number theory of estimates on size and density, as opposed to identities.

Some subjects generally considered to be part of analytic number theory, e.g., sieve theory, are better covered by the second rather than the first definition: some of sieve theory, for instance, uses little analysis, yet it does belong to analytic number theory.

The following are examples of problems in analytic number theory: the prime number theorem, the Goldbach conjecture (or the twin prime conjecture, or the Hardy–Littlewood conjectures), the Waring problem and the Riemann Hypothesis. Some of the most important tools of analytic number theory are the circle method, sieve methods and L-functions (or, rather, the study of their properties). The theory of modular forms (and, more generally, automorphic forms) also occupies an increasingly central place in the toolbox of analytic number theory.

One may ask analytic questions about algebraic numbers, and use analytic means to answer such questions; it is thus that algebraic and analytic number theory intersect. For example, one may define prime ideals (generalizations of prime numbers in the field of algebraic numbers) and ask how many prime ideals there are up to a certain size. This question can be answered by means of an examination of Dedekind zeta functions, which are generalizations of the Riemann zeta function, a key analytic object at the roots of the subject. This is an example of a general procedure in analytic number theory: deriving information about the distribution of a sequence (here, prime ideals or prime numbers) from the analytic behavior of an appropriately constructed complex-valued function.

Algebraic Number Theory

An *algebraic number* is any complex number that is a solution to some polynomial equation $f(x) = 0$ with rational coefficients; for example, every solution x of $x^5 + (11/2)x^3 - 7x^2 + 9 = 0$ (say) is an algebraic number. Fields of algebraic numbers are also called *algebraic number fields*, or shortly *number fields*. Algebraic number theory studies algebraic number fields. Thus, analytic and algebraic number theory can and do overlap: the former is defined by its methods, the latter by its objects of study.

It could be argued that the simplest kind of number fields (viz., quadratic fields) were already studied by Gauss, as the discussion of quadratic forms in *Disquisitiones arithmeticae* can be restated in terms of ideals and norms in quadratic fields. (A *quadratic field* consists of all numbers of the form $a + b\sqrt{d}$, where a and b are rational numbers and d is a fixed rational number whose square root is not rational.) For that matter, the 11th-century chakravala method amounts—in modern terms—to an algorithm for finding the units of a real quadratic number field. However, neither Bhāskara nor Gauss knew of number fields as such.

The grounds of the subject as we know it were set in the late nineteenth century, when *ideal numbers*, the *theory of ideals* and *valuation theory* were developed; these are three complementary ways of dealing with the lack of unique factorisation in algebraic number fields. (For example, in the field generated by the rationals and $\sqrt{-5}$, the number **6** can be factorised both as $6 = 2 \cdot 3$ and $6 = (1+\sqrt{-5})(1-\sqrt{-5})$; all of 2, **3**, $1+\sqrt{-5}$, and $1-\sqrt{-5}$ are irreducible, and thus, in a naïve sense, analogous to primes among the integers.) The initial impetus for the development of ideal numbers (by Kummer) seems to have come from the study of higher reciprocity laws, i.e., generalisations of quadratic reciprocity.

Number fields are often studied as extensions of smaller number fields: a field L is said to be an *extension* of a field K if L contains K. (For example, the complex numbers C are an extension of the reals R, and the reals R are an extension of the rationals Q.) Classifying the possible extensions of a given number field is a difficult and partially open problem. Abelian extensions—that is, extensions L of K such that the Galois group $\mathrm{Gal}(L/K)$ of L over K is an abelian group—are relatively well understood. Their classification was the object of the programme of class field theory, which was initiated in the late 19th century (partly by Kronecker and Eisenstein) and carried out largely in 1900–1950.

An example of an active area of research in algebraic number theory is Iwasawa theory. The Langlands program, one of the main current large-scale research plans in mathematics, is sometimes described as an attempt to generalise class field theory to non-abelian extensions of number fields.

Diophantine Geometry

The central problem of *Diophantine geometry* is to determine when a Diophantine equation has solutions, and if it does, how many. The approach taken is to think of the solutions of an equation as a geometric object.

For example, an equation in two variables defines a curve in the plane. More generally, an equation, or system of equations, in two or more variables defines a curve, a surface or some other such object in n-dimensional space. In Diophantine geometry, one asks whether there are any *rational points* (points all of whose coordinates are rationals) or *integral points* (points all of whose coordinates are integers) on the curve or surface. If there are any such points, the next step is to ask how many there are and how they are distributed. A basic question in this direction is: are there finitely or infinitely many rational points on a given curve (or surface)? What about integer points?

An example here may be helpful. Consider the Pythagorean equation $x^2 + y^2 = 1$; we would like to study its rational solutions, i.e., its solutions (x, y) such that x and y are both rational. This is the same as asking for all integer solutions to $a^2 + b^2 = c^2$; any solution to the latter equation gives us a solution $x = a/c$, $y = b/c$ to the former. It is also the same as asking for all points with rational coordinates on the curve described by $x^2 + y^2 = 1$. (This curve happens to be a circle of radius 1 around the origin.)

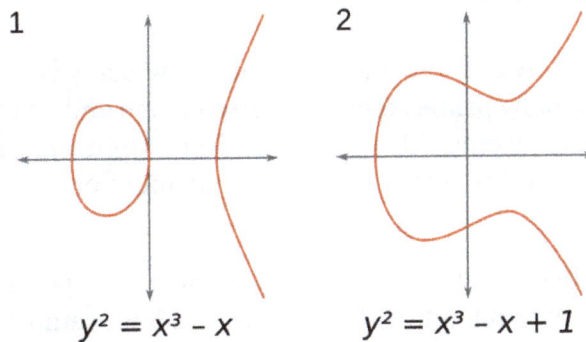

$$y^2 = x^3 - x \qquad\qquad y^2 = x^3 - x + 1$$

Two examples of an elliptic curve, i.e., a curve of genus 1 having at least one rational point. (Either graph can be seen as a slice of a torus in four-dimensional space.)

The rephrasing of questions on equations in terms of points on curves turns out to be felicitous. The finiteness or not of the number of rational or integer points on an algebraic curve—that is, rational or integer solutions to an equation $f(x,y)=0$, where f is a polynomial in two variables—turns out to depend crucially on the *genus* of the curve. The *genus* can be defined as follows: allow the variables in $f(x,y)=0$ to be complex numbers; then $f(x,y)=0$ defines a 2-dimensional surface in (projective) 4-dimensional space (since two complex variables can be decomposed into four real variables, i.e., four dimensions). Count the number of (doughnut) holes in the surface; call this number the *genus* of $f(x,y)=0$. Other geometrical notions turn out to be just as crucial.

There is also the closely linked area of Diophantine approximations: given a number x, how well can it be approximated by rationals? (We are looking for approximations that are good relative to the amount of space that it takes to write the rational: call a/q (with $\gcd(a,q)=1$) a good approximation to x if $|x-a/q|<\dfrac{1}{q^c}$, where c is large.) This question is of special interest if x is an algebraic number. If x cannot be well approximated, then some equations do not have integer or rational solutions. Moreover, several concepts (especially that of height) turn out to be crucial both in Diophantine geometry and in the study of Diophantine approximations. This question is also of special interest in transcendental number theory: if a number can be better approximated than any algebraic number, then it is a transcendental number. It is by this argument that π and e have been shown to be transcendental.

Diophantine geometry should not be confused with the geometry of numbers, which is a collection of graphical methods for answering certain questions in algebraic number theory. *Arithmetic geometry*, on the other hand, is a contemporary term for much the same domain as that covered by the term *Diophantine geometry*. The term *arithmetic geometry* is arguably used most often when one wishes to emphasise the connections to modern algebraic geometry (as in, for instance, Faltings' theorem) rather than to techniques in Diophantine approximations.

Recent Approaches and Subfields

The areas below date as such from no earlier than the mid-twentieth century, even if they are based on older material. For example, as is explained below, the matter of algorithms in number theory is very old, in some sense older than the concept of proof; at the same time, the modern study of computability dates only from the 1930s and 1940s, and computational complexity theory from the 1970s.

Probabilistic Number Theory

Take a number at random between one and a million. How likely is it to be prime? This is just another way of asking how many primes there are between one and a million. Further: how many prime divisors will it have, on average? How many divisors will it have altogether, and with what likelihood? What is the probability that it will have many more or many fewer divisors or prime divisors than the average?

Much of probabilistic number theory can be seen as an important special case of the study of variables that are almost, but not quite, mutually independent. For example, the event that a random integer between one and a million be divisible by two and the event that it be divisible by three are almost independent, but not quite.

It is sometimes said that probabilistic combinatorics uses the fact that whatever happens with probability greater than 0 must happen sometimes; one may say with equal justice that many applications of probabilistic number theory hinge on the fact that whatever is unusual must be rare. If certain algebraic objects (say, rational or integer solutions to certain equations) can be shown to be in the tail of certain sensibly defined distributions, it follows that there must be few of them; this is a very concrete non-probabilistic statement following from a probabilistic one.

At times, a non-rigorous, probabilistic approach leads to a number of heuristic algorithms and open problems, notably Cramér's conjecture.

Arithmetic Combinatorics

Let A be a set of N integers. Consider the set $A + A = \{ m + n \mid m, n \in A \}$ consisting of all sums of two elements of A. Is $A + A$ much larger than A? Barely larger? If $A + A$ is barely larger than A, must A have plenty of arithmetic structure, for example, does A resemble an arithmetic progression?

If we begin from a fairly "thick" infinite set A, does it contain many elements in arithmetic progression: $a, a+b, a+2b, a+3b, \ldots, a+10b$, say? Should it be possible to write large integers as sums of elements of A?

These questions are characteristic of *arithmetic combinatorics*. This is a presently coalescing field; it subsumes *additive number theory* (which concerns itself with certain very specific sets A of arithmetic significance, such as the primes or the squares) and, arguably, some of the *geometry of numbers*, together with some rapidly developing new material. Its focus on issues of growth and distribution accounts in part for its developing links with ergodic theory, finite group theory, model theory, and other fields. The term *additive combinatorics* is also used; however, the sets A being studied need not be sets of integers, but rather subsets of non-commutative groups, for which the multiplication symbol, not the addition symbol, is traditionally used; they can also be subsets of rings, in which case the growth of $A + A$ and $A \cdot A$ may be compared.

Computations in Number Theory

While the word *algorithm* goes back only to certain readers of al-Khwārizmī, careful descriptions of methods of solution are older than proofs: such methods (that is, algorithms) are as old as any recognisable mathematics—ancient Egyptian, Babylonian, Vedic, Chinese—whereas proofs appeared only with the Greeks of the classical period. An interesting early case is that of what we now call the Euclidean algorithm. In its basic form (namely, as an algorithm for computing the greatest common divisor) it appears as Proposition 2 of Book VII in *Elements*, together with a proof of correctness. However, in the form that is often used in number theory (namely, as an algorithm for finding integer solutions to an equation $ax + by = c,$, or, what is the same, for finding the quantities whose existence is assured by the Chinese remainder theorem) it first appears in the works of Āryabhaṭa (5th–6th century CE) as an algorithm called *kuṭṭaka* ("pulveriser"), without a proof of correctness.

There are two main questions: "can we compute this?" and "can we compute it rapidly?". Anybody can test whether a number is prime or, if it is not, split it into prime factors; doing so rapidly is another matter. We now know fast algorithms for testing primality, but, in spite of much work (both theoretical and practical), no truly fast algorithm for factoring.

The difficulty of a computation can be useful: modern protocols for encrypting messages (e.g., RSA) depend on functions that are known to all, but whose inverses (a) are known only to a chosen few, and (b) would take one too long a time to figure out on one's own. For example, these functions can be such that their inverses can be computed only if certain large integers are factorized. While many difficult computational problems outside number theory are known, most working encryption protocols nowadays are based on the difficulty of a few number-theoretical problems.

On a different note — some things may not be computable at all; in fact, this can be proven in some instances. For instance, in 1970, it was proven, as a solution to Hilbert's 10th problem, that there is no Turing machine which can solve all Diophantine equations. In particular, this means that, given a computably enumerable set of axioms, there are Diophantine equations for which there is no proof, starting from the axioms, of whether the set of equations has or does not have integer solutions. (We would necessarily be speaking of Diophantine equations for which there are no integer solutions, since, given a Diophantine equation with at least one solution, the solution itself provides a proof of the fact that a solution exists. We cannot prove, of course, that a particular Diophantine equation is of this kind, since this would imply that it has no solutions.)

Applications

The number-theorist Leonard Dickson (1874-1954) said "Thank God that number theory is unsullied by any application". Such a view is no longer applicable to number theory. In 1974, Donald Knuth said "...virtually every theorem in elementary number theory arises in a natural, motivated way in connection with the problem of making computers do high-speed numerical calculations". Elementary number theory is taught in discrete mathematics courses for computer scientists; and, on the other hand, number theory also has applications to the continuous in numerical analysis. As well as the well-known applications to cryptography, there are also applications to many other areas of mathematics.

Literature

Two of the most popular introductions to the subject are:

- *G. H. Hardy; E. M. Wright (2008) [1938]. An introduction to the theory of numbers (rev.*

by D. R. Heath-Brown and J. H. Silverman, 6th ed.). Oxford University Press. ISBN 978-0-19-921986-5. Retrieved 2016-03-02.

- *Vinogradov, I. M. (2003) [1954]. Elements of Number Theory (reprint of the 1954 ed.). Mineola, NY: Dover Publications.*

- Hardy and Wright's book is a comprehensive classic, though its clarity sometimes suffers due to the authors' insistence on elementary methods. Vinogradov's main attraction consists in its set of problems, which quickly lead to Vinogradov's own research interests; the text itself is very basic and close to minimal. Other popular first introductions are:

- *Ivan M. Niven; Herbert S. Zuckerman; Hugh L. Montgomery (2008) [1960]. An introduction to the theory of numbers (reprint of the 5th edition 1991 ed.). John Wiley & Sons. ISBN 978-8-12-651811-1. Retrieved 2016-02-28.*

- *Kenneth H. Rosen (2010). Elementary Number Theory (6th ed.). Pearson Education. ISBN 978-0-32-171775-7. Retrieved 2016-02-28.*

- Popular choices for a second textbook include:

- *Borevich, A. I.; Shafarevich, Igor R. (1966). Number theory. Pure and Applied Mathematics. **20**. Boston, MA: Academic Press. ISBN 978-0-12-117850-5. MR 0195803.*

- *Serre, Jean-Pierre (1996) [1973]. A course in arithmetic. Graduate texts in mathematics. 7. Springer. ISBN 978-0-387-90040-7.*

Prizes

The American Mathematical Society awards the *Cole Prize in Number Theory*. Moreover number theory is one of the three mathematical subdisciplines rewarded by the *Fermat Prize*.

Pythagorean Theorem

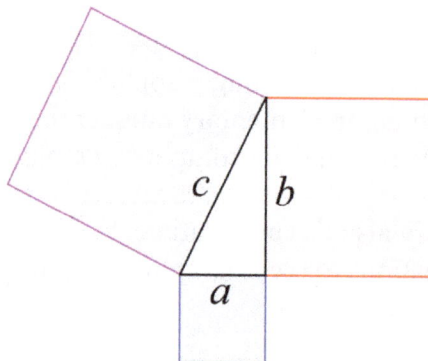

Pythagorean theorem
The sum of the areas of the two squares on the legs (a and b) equals the area of the square on the hypotenuse (c).

Geometry

Projecting a sphere to a plane.

In mathematics, the Pythagorean theorem, also known as Pythagoras' theorem, is a fundamental relation in Euclidean geometry among the three sides of a right triangle. It states that the square of the hypotenuse (the side opposite the right angle) is equal to the sum of the squares of the other two sides. The theorem can be written as an equation relating the lengths of the sides a, b and c, often called the "Pythagorean equation":

$$a^2 + b^2 = c^2,$$

where c represents the length of the hypotenuse and a and b the lengths of the triangle's other two sides.

Although it is often argued that knowledge of the theorem predates him, the theorem is named after the ancient Greek mathematician Pythagoras (c. 570 – c. 495 BC) as it is he who, by tradition, is credited with its first recorded proof. There is some evidence that Babylonian mathematicians understood the formula, although little of it indicates an application within a mathematical framework. Mesopotamian, Indian and Chinese mathematicians all discovered the theorem independently and, in some cases, provided proofs for special cases.

The theorem has been given numerous proofs – possibly the most for any mathematical theorem. They are very diverse, including both geometric proofs and algebraic proofs, with some dating back thousands of years. The theorem can be generalized in various ways, including higher-dimensional spaces, to spaces that are not Euclidean, to objects that are not right triangles, and indeed, to objects that are not triangles at all, but n-dimensional solids. The Pythagorean theorem has attracted interest outside mathematics as a symbol of mathematical abstruseness, mystique, or intellectual power; popular references in literature, plays, musicals, songs, stamps and cartoons abound.

Pythagorean Proof

$$c^2 = a^2 + b^2$$

The Pythagorean proof (click to view animation)

The Pythagorean Theorem was known long before Pythagoras, but he may well have been the first to prove it. In any event, the proof attributed to him is very simple, and is called a proof by rearrangement.

The two large squares shown in the figure each contain four identical triangles, and the only difference between the two large squares is that the triangles are arranged differently. Therefore, the white space within each of the two large squares must have equal area. Equating the area of the white space yields the Pythagorean Theorem, Q.E.D.

That Pythagoras originated this very simple proof is sometimes inferred from the writings of the later Greek philosopher and mathematician Proclus. Several other proofs of this theorem are described below, but this is known as the Pythagorean one.

Other forms of the Theorem

As pointed out in the introduction, if c denotes the length of the hypotenuse and a and b denote the lengths of the other two sides, the Pythagorean theorem can be expressed as the Pythagorean equation:

$$a^2 + b^2 = c^2.$$

If the length of both a and b are known, then c can be calculated as

$$c = \sqrt{a^2 + b^2}.$$

If the length of the hypotenuse c and of one side (a or b) are known, then the length of the other side can be calculated as

$$a = \sqrt{c^2 - b^2}$$

or

$$b = \sqrt{c^2 - a^2}.$$

The Pythagorean equation relates the sides of a right triangle in a simple way, so that if the lengths of any two sides are known the length of the third side can be found. Another corollary of the theorem is that in any right triangle, the hypotenuse is greater than any one of the other sides, but less than their sum.

A generalization of this theorem is the law of cosines, which allows the computation of the length of any side of any triangle, given the lengths of the other two sides and the angle between them. If the angle between the other sides is a right angle, the law of cosines reduces to the Pythagorean equation.

Other Proofs of the Theorem

This theorem may have more known proofs than any other (the law of quadratic reciprocity being another contender for that distinction); the book *The Pythagorean Proposition* contains 370 proofs.

Proof Using Similar Triangles

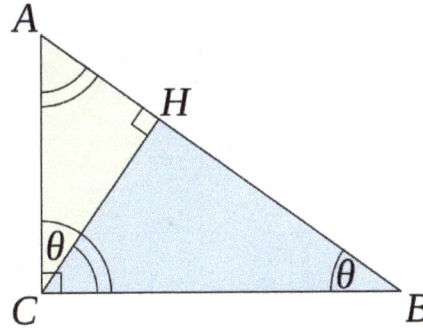

Proof using similar triangles

This proof is based on the proportionality of the sides of two similar triangles, that is, upon the fact that the ratio of any two corresponding sides of similar triangles is the same regardless of the size of the triangles.

Let ABC represent a right triangle, with the right angle located at C, as shown on the figure. Draw the altitude from point C, and call H its intersection with the side AB. Point H divides the length of the hypotenuse c into parts d and e. The new triangle ACH is similar to triangle ABC, because they both have a right angle (by definition of the altitude), and they share the angle at A, meaning that the third angle will be the same in both triangles as well, marked as θ in the figure. By a similar reasoning, the triangle CBH is also similar to ABC. The proof of similarity of the triangles requires the triangle postulate: the sum of the angles in a triangle is two right angles, and is equivalent to the parallel postulate. Similarity of the triangles leads to the equality of ratios of corresponding sides:

$$\frac{BC}{AB} = \frac{BH}{BC} \text{ and } \frac{AC}{AB} = \frac{AH}{AC}.$$

The first result equates the cosines of the angles θ, whereas the second result equates their sines.

These ratios can be written as

$$BC^2 = AB \times BH \text{ and } AC^2 = AB \times AH.$$

Summing these two equalities results in

$$BC^2 + AC^2 = AB \times BH + AB \times AH = AB \times (AH + BH) = AB^2,$$

which, after simplification, expresses the Pythagorean theorem:

$$BC^2 + AC^2 = AB^2.$$

The role of this proof in history is the subject of much speculation. The underlying question is why Euclid did not use this proof, but invented another. One conjecture is that the proof by similar triangles involved a theory of proportions, a topic not discussed until later in the *Elements*, and that the theory of proportions needed further development at that time.

Euclid's Proof

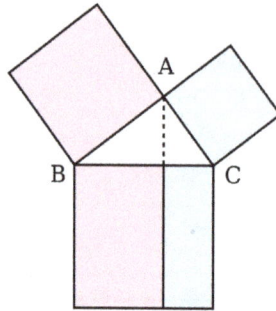

Proof in Euclid's *Elements*

In outline, here is how the proof in Euclid's *Elements* proceeds. The large square is divided into a left and right rectangle. A triangle is constructed that has half the area of the left rectangle. Then another triangle is constructed that has half the area of the square on the left-most side. These two triangles are shown to be congruent, proving this square has the same area as the left rectangle. This argument is followed by a similar version for the right rectangle and the remaining square. Putting the two rectangles together to reform the square on the hypotenuse, its area is the same as the sum of the area of the other two squares. The details follow.

Let *A*, *B*, *C* be the vertices of a right triangle, with a right angle at *A*. Drop a perpendicular from *A* to the side opposite the hypotenuse in the square on the hypotenuse. That line divides the square on the hypotenuse into two rectangles, each having the same area as one of the two squares on the legs.

For the formal proof, we require four elementary lemmata:

1. If two triangles have two sides of the one equal to two sides of the other, each to each, and the angles included by those sides equal, then the triangles are congruent (side-angle-side).

2. The area of a triangle is half the area of any parallelogram on the same base and having the same altitude.

3. The area of a rectangle is equal to the product of two adjacent sides.

4. The area of a square is equal to the product of two of its sides (follows from 3).

Next, each top square is related to a triangle congruent with another triangle related in turn to one of two rectangles making up the lower square.

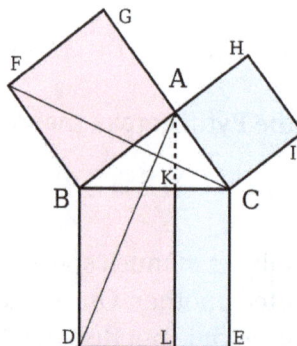

Illustration including the new lines

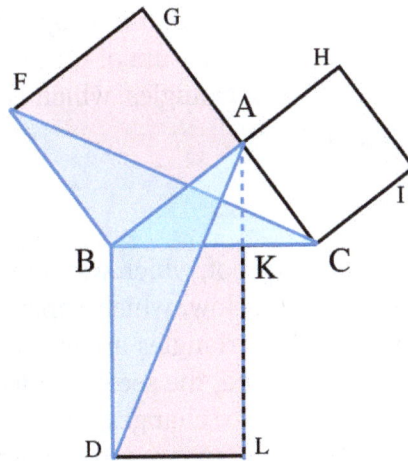

Showing the two congruent triangles of half the area of rectangle BDLK and square BAGF

The proof is as follows:

1. Let ACB be a right-angled triangle with right angle CAB.

2. On each of the sides BC, AB, and CA, squares are drawn, CBDE, BAGF, and ACIH, in that order. The construction of squares requires the immediately preceding theorems in Euclid, and depends upon the parallel postulate.

3. From A, draw a line parallel to BD and CE. It will perpendicularly intersect BC and DE at K and L, respectively.

4. Join CF and AD, to form the triangles BCF and BDA.

5. Angles CAB and BAG are both right angles; therefore C, A, and G are collinear. Similarly for B, A, and H.

6. Angles CBD and FBA are both right angles; therefore angle ABD equals angle FBC, since both are the sum of a right angle and angle ABC.

7. Since AB is equal to FB and BD is equal to BC, triangle ABD must be congruent to triangle FBC.

8. Since A-K-L is a straight line, parallel to BD, then rectangle BDLK has twice the area of triangle ABD because they share the base BD and have the same altitude BK, i.e., a line normal to their common base, connecting the parallel lines BD and AL. (lemma 2)

9. Since C is collinear with A and G, square BAGF must be twice in area to triangle FBC.

10. Therefore, rectangle BDLK must have the same area as square BAGF = AB^2.

11. Similarly, it can be shown that rectangle CKLE must have the same area as square ACIH = AC^2.

12. Adding these two results, $AB^2 + AC^2 = BD \times BK + KL \times KC$

13. Since BD = KL, $BD \times BK + KL \times KC = BD(BK + KC) = BD \times BC$

14. Therefore, $AB^2 + AC^2 = BC^2$, since CBDE is a square.

This proof, which appears in Euclid's *Elements* as that of Proposition 47 in Book 1, demonstrates that the area of the square on the hypotenuse is the sum of the areas of the other two squares. This is quite distinct from the proof by similarity of triangles, which is conjectured to be the proof that Pythagoras used.

Proofs by Dissection and Rearrangement

We have already discussed the Pythagorean proof, which was a proof by rearrangement. The same idea is conveyed by the leftmost animation below, which consists of a large square, side $a + b$, containing four identical right triangles. The triangles are shown in two arrangements, the first of which leaves two squares a^2 and b^2 uncovered, the second of which leaves square c^2 uncovered. The area encompassed by the outer square never changes, and the area of the four triangles is the same at the beginning and the end, so the black square areas must be equal, therefore $a^2 + b^2 = c^2$.

A second proof by rearrangement is given by the middle animation. A large square is formed with area c^2, from four identical right triangles with sides a, b and c, fitted around a small central square. Then two rectangles are formed with sides a and b by moving the triangles. Combining the smaller square with these rectangles produces two squares of areas a^2 and b^2, which must have the same area as the initial large square.

The third, rightmost image also gives a proof. The upper two squares are divided as shown by the blue and green shading, into pieces that when rearranged can be made to fit in the lower square on the hypotenuse – or conversely the large square can be divided as shown into pieces that fill the other two. This way of cutting one figure into pieces and rearranging them to get another figure is called dissection. This shows the area of the large square equals that of the two smaller ones.

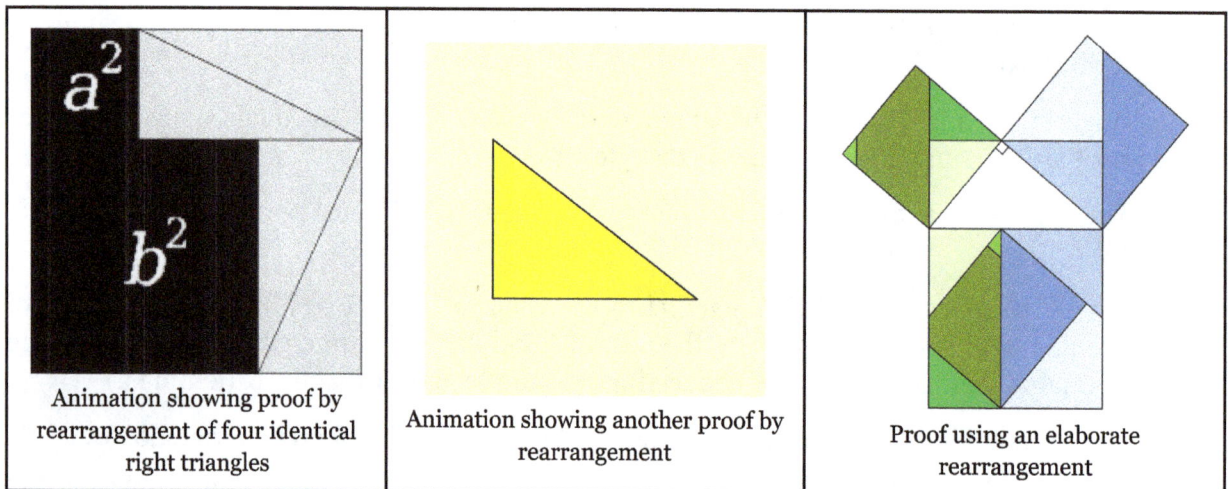

| Animation showing proof by rearrangement of four identical right triangles | Animation showing another proof by rearrangement | Proof using an elaborate rearrangement |

Einstein's Proof by Dissection without Rearrangement

Albert Einstein gave a proof by dissection in which the pieces need not get moved. Instead of using a square on the hypotenuse and two squares on the legs, one can use any other shape that includes the hypotenuse, and two similar shapes that each include one of two legs instead of the hypotenuse. In Einstein's proof, the shape that includes the hypotenuse is the right triangle itself. The dissection consists of dropping a perpendicular from the vertex of the right angle of the triangle to the hypotenuse, thus splitting the whole triangle into two parts. Those two parts have the same shape as the original right triangle, and have the legs of the original triangle as their hypotenuses, and the sum of their areas is that of the original triangle. Because the ratio of the area of a right triangle to the square of its hypotenuse is the same for similar triangles, the relationship between the areas of the three triangles holds for the squares of the sides of the large triangle as well.

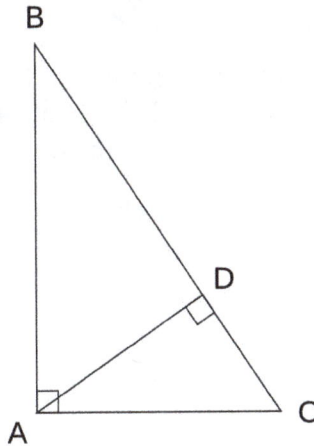

Right triangle on the hypotenuse dissected into two similar right triangles on the legs, according to Einstein's proof

Algebraic Proofs

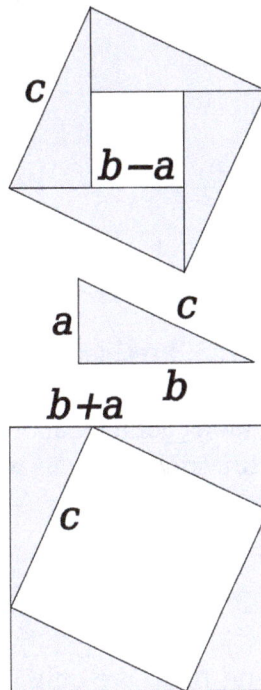

Diagram of the two algebraic proofs

The theorem can be proved algebraically using four copies of a right triangle with sides a, b and c, arranged inside a square with side c as in the top half of the diagram. The triangles are similar with area $\frac{1}{2}ab$, while the small square has side $b - a$ and area $(b - a)^2$. The area of the large square is therefore

$$(b-a)^2 + 4\frac{ab}{2} = (b-a)^2 + 2ab = a^2 + b^2.$$

But this is a square with side c and area c^2, so

$$c^2 = a^2 + b^2.$$

A similar proof uses four copies of the same triangle arranged symmetrically around a square with side c, as shown in the lower part of the diagram. This results in a larger square, with side $a + b$ and area $(a + b)^2$. The four triangles and the square side c must have the same area as the larger square,

$$(b+a)^2 = c^2 + 4\frac{ab}{2} = c^2 + 2ab,$$

giving

$$c^2 = (b+a)^2 - 2ab = a^2 + b^2.$$

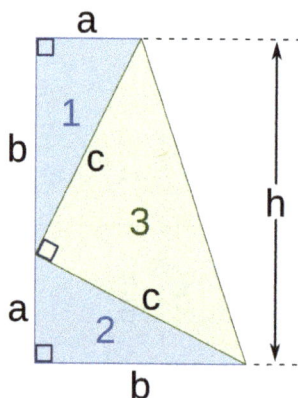

Diagram of Garfield's proof

A related proof was published by future U.S. President James A. Garfield (then a U.S. Representative). Instead of a square it uses a trapezoid, which can be constructed from the square in the second of the above proofs by bisecting along a diagonal of the inner square, to give the trapezoid as shown in the diagram. The area of the trapezoid can be calculated to be half the area of the square, that is

$$\frac{1}{2}(b+a)^2.$$

The inner square is similarly halved, and there are only two triangles so the proof proceeds as above except for a factor of $\frac{1}{2}$, which is removed by multiplying by two to give the result.

Proof using Differentials

One can arrive at the Pythagorean theorem by studying how changes in a side produce a change in the hypotenuse and employing calculus.

The triangle ABC is a right triangle, as shown in the upper part of the diagram, with BC the hypotenuse. At the same time the triangle lengths are measured as shown, with the hypotenuse of length y, the side AC of length x and the side AB of length a, as seen in the lower diagram part.

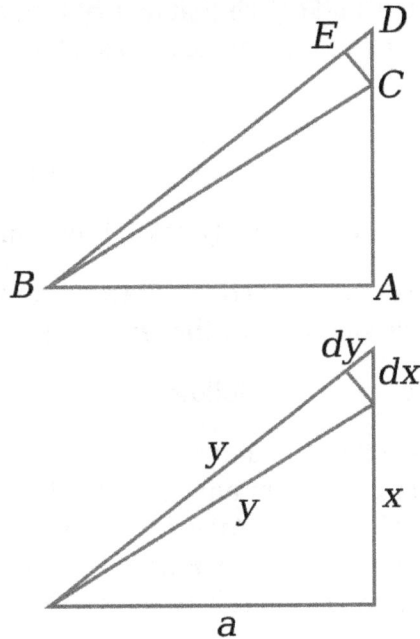

Diagram for differential proof

If x is increased by a small amount dx by extending the side AC slightly to D, then y also increases by dy. These form two sides of a triangle, CDE, which (with E chosen so CE is perpendicular to the hypotenuse) is a right triangle approximately similar to ABC. Therefore, the ratios of their sides must be the same, that is:

$$\frac{dy}{dx} = \frac{x}{y}.$$

This can be rewritten as $y\,dy = x\,dx$, which is a differential equation that can be solved by direct integration:

$$\int y\,dy = \int x\,dx,$$

giving

$$y^2 = x^2 + C.$$

The constant can be deduced from $x = 0$, $y = a$ to give the equation

$$y^2 = x^2 + a^2.$$

This is more of an intuitive proof than a formal one: it can be made more rigorous if proper limits are used in place of dx and dy.

Converse

The converse of the theorem is also true:

For any three positive numbers a, b, and c such that $a^2 + b^2 = c^2$, there exists a triangle with sides a, b and c, and every such triangle has a right angle between the sides of lengths a and b.

An alternative statement is:

For any triangle with sides a, b, c, if $a^2 + b^2 = c^2$, then the angle between a and b measures 90°.

This converse also appears in Euclid's *Elements* (Book I, Proposition 48):

"If in a triangle the square on one of the sides equals the sum of the squares on the remaining two sides of the triangle, then the angle contained by the remaining two sides of the triangle is right."

It can be proven using the law of cosines or as follows:

Let ABC be a triangle with side lengths a, b, and c, with $a^2 + b^2 = c^2$. Construct a second triangle with sides of length a and b containing a right angle. By the Pythagorean theorem, it follows that the hypotenuse of this triangle has length $c = \sqrt{a^2 + b^2}$, the same as the hypotenuse of the first triangle. Since both triangles' sides are the same lengths a, b and c, the triangles are congruent and must have the same angles. Therefore, the angle between the side of lengths a and b in the original triangle is a right angle.

The above proof of the converse makes use of the Pythagorean Theorem itself. The converse can also be proven without assuming the Pythagorean Theorem.

A corollary of the Pythagorean theorem's converse is a simple means of determining whether a triangle is right, obtuse, or acute, as follows. Let c be chosen to be the longest of the three sides and $a + b > c$ (otherwise there is no triangle according to the triangle inequality). The following statements apply:

- If $a^2 + b^2 = c^2$, then the triangle is right.

- If $a^2 + b^2 > c^2$, then the triangle is acute.

- If $a^2 + b^2 < c^2$, then the triangle is obtuse.

Edsger Dijkstra has stated this proposition about acute, right, and obtuse triangles in this language:

$$\mathrm{sgn}(\alpha + \beta - \gamma) = \mathrm{sgn}(a^2 + b^2 - c^2),$$

where α is the angle opposite to side a, β is the angle opposite to side b, γ is the angle opposite to side c, and sgn is the sign function.

Consequences and uses of the Theorem

Pythagorean Triples

A Pythagorean triple has three positive integers a, b, and c, such that $a^2 + b^2 = c^2$. In other words, a

Pythagorean triple represents the lengths of the sides of a right triangle where all three sides have integer lengths. Evidence from megalithic monuments in Northern Europe shows that such triples were known before the discovery of writing. Such a triple is commonly written (a, b, c). Some well-known examples are $(3, 4, 5)$ and $(5, 12, 13)$.

A primitive Pythagorean triple is one in which a, b and c are coprime (the greatest common divisor of a, b and c is 1).

The following is a list of primitive Pythagorean triples with values less than 100:

$(3, 4, 5)$, $(5, 12, 13)$, $(7, 24, 25)$, $(8, 15, 17)$, $(9, 40, 41)$, $(11, 60, 61)$, $(12, 35, 37)$, $(13, 84, 85)$, $(16, 63, 65)$, $(20, 21, 29)$, $(28, 45, 53)$, $(33, 56, 65)$, $(36, 77, 85)$, $(39, 80, 89)$, $(48, 55, 73)$, $(65, 72, 97)$

Incommensurable Lengths

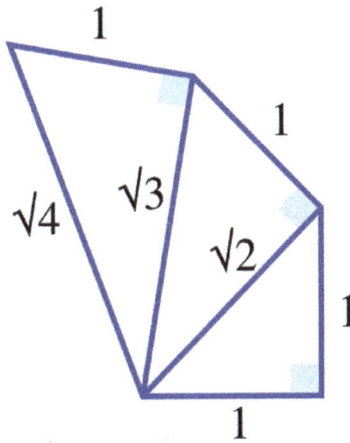

The spiral of Theodorus: A construction for line segments with lengths whose ratios are the square root of a positive integer

One of the consequences of the Pythagorean theorem is that line segments whose lengths are incommensurable (so the ratio of which is not a rational number) can be constructed using a straight-edge and compass. Pythagoras's theorem enables construction of incommensurable lengths because the hypotenuse of a triangle is related to the sides by the square root operation.

The figure on the right shows how to construct line segments whose lengths are in the ratio of the square root of any positive integer. Each triangle has a side (labeled "1") that is the chosen unit for measurement. In each right triangle, Pythagoras's theorem establishes the length of the hypotenuse in terms of this unit. If a hypotenuse is related to the unit by the square root of a positive integer that is not a perfect square, it is a realization of a length incommensurable with the unit, such as $\sqrt{2}$, $\sqrt{3}$, $\sqrt{5}$. For more detail.

Incommensurable lengths conflicted with the Pythagorean school's concept of numbers as only whole numbers. The Pythagorean school dealt with proportions by comparison of integer multiples of a common subunit. According to one legend, Hippasus of Metapontum (ca. 470 B.C.) was drowned at sea for making known the existence of the irrational or incommensurable.

Complex Numbers

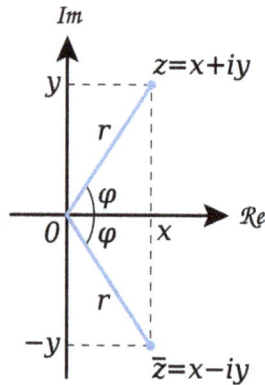

The absolute value of a complex number z is the distance r from z to the origin

For any complex number

$$z = x + iy,$$

the absolute value or modulus is given by

$$r = |z| = \sqrt{x^2 + y^2}.$$

So the three quantities, r, x and y are related by the Pythagorean equation,

$$r^2 = x^2 + y^2.$$

Note that r is defined to be a positive number or zero but x and y can be negative as well as positive. Geometrically r is the distance of the z from zero or the origin O in the complex plane.

This can be generalised to find the distance between two points, z_1 and z_2 say. The required distance is given by

$$|z_1 - z_2| = \sqrt{(x_1 - x_2)^2 + (y_1 - y_2)^2},$$

so again they are related by a version of the Pythagorean equation,

$$|z_1 - z_2|^2 = (x_1 - x_2)^2 + (y_1 - y_2)^2.$$

Euclidean Distance in Various Coordinate Systems

The distance formula in Cartesian coordinates is derived from the Pythagorean theorem. If (x_1, y_1) and (x_2, y_2) are points in the plane, then the distance between them, also called the Euclidean distance, is given by

$$\sqrt{(x_1 - x_2)^2 + (y_1 - y_2)^2}.$$

More generally, in Euclidean n-space, the Euclidean distance between two points, $A = (a_1, a_2, \ldots, a_n)$ and $B = (b_1, b_2, \ldots, b_n)$, , is defined, by generalization of the Pythagorean theorem, as:

$$\sqrt{(a_1 - b_1)^2 + (a_2 - b_2)^2 + \cdots + (a_n - b_n)^2} = \sqrt{\sum_{i=1}^{n} (a_i - b_i)^2}.$$

If Cartesian coordinates are not used, for example, if polar coordinates are used in two dimensions or, in more general terms, if curvilinear coordinates are used, the formulas expressing the Euclidean distance are more complicated than the Pythagorean theorem, but can be derived from it. A typical example where the straight-line distance between two points is converted to curvilinear coordinates can be found in the applications of Legendre polynomials in physics. The formulas can be discovered by using Pythagoras's theorem with the equations relating the curvilinear coordinates to Cartesian coordinates. For example, the polar coordinates (r, θ) can be introduced as:

$$x = r \cos \theta, \; y = r \sin \theta.$$

Then two points with locations (r_1, θ_1) and (r_2, θ_2) are separated by a distance s:

$$s^2 = (x_1 - x_2)^2 + (y_1 - y_2)^2 = (r_1 \cos \theta_1 - r_2 \cos \theta_2)^2 + (r_1 \sin \theta_1 - r_2 \sin \theta_2)^2.$$

Performing the squares and combining terms, the Pythagorean formula for distance in Cartesian coordinates produces the separation in polar coordinates as:

$$s^2 = r_1^2 + r_2^2 - 2r_1 r_2 \left(\cos \theta_1 \cos \theta_2 + \sin \theta_1 \sin \theta_2 \right)$$

$$= r_1^2 + r_2^2 - 2r_1 r_2 \cos \left(\theta_1 - \theta_2 \right)$$

$$= r_1^2 + r_2^2 - 2r_1 r_2 \cos \Delta \theta,$$

using the trigonometric product-to-sum formulas. This formula is the law of cosines, sometimes called the Generalized Pythagorean Theorem. From this result, for the case where the radii to the two locations are at right angles, the enclosed angle $\Delta \theta = \pi/2$, and the form corresponding to Pythagoras's theorem is regained: $s^2 = r_1^2 + r_2^2$. The Pythagorean theorem, valid for right triangles, therefore is a special case of the more general law of cosines, valid for arbitrary triangles.

Pythagorean Trigonometric Identity

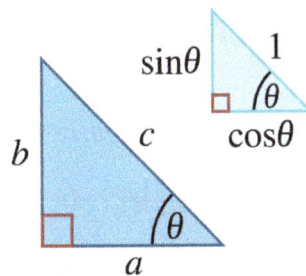

Similar right triangles showing sine and cosine of angle θ

In a right triangle with sides a, b and hypotenuse c, trigonometry determines the sine and cosine of the angle θ between side a and the hypotenuse as:

$$\sin\theta = \frac{b}{c}, \quad \cos\theta = \frac{a}{c}.$$

From that it follows:

$$\cos^2\theta + \sin^2\theta = \frac{a^2 + b^2}{c^2} = 1,$$

where the last step applies Pythagoras's theorem. This relation between sine and cosine is sometimes called the fundamental Pythagorean trigonometric identity. In similar triangles, the ratios of the sides are the same regardless of the size of the triangles, and depend upon the angles. Consequently, in the figure, the triangle with hypotenuse of unit size has opposite side of size $\sin\theta$ and adjacent side of size $\cos\theta$ in units of the hypotenuse.

Relation to the Cross Product

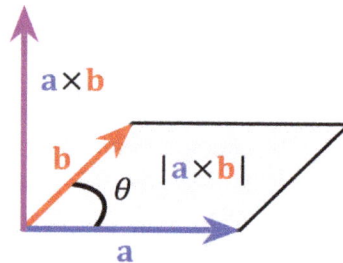

The area of a parallelogram as a cross product; vectors **a** and **b** identify a plane and **a** × **b** is normal to this plane.

The Pythagorean theorem relates the cross product and dot product in a similar way:

$$\| \mathbf{a} \times \mathbf{b} \|^2 + (\mathbf{a} \cdot \mathbf{b})^2 = \| \mathbf{a} \|^2 \| \mathbf{b} \|^2 .$$

This can be seen from the definitions of the cross product and dot product, as

$$\mathbf{a} \times \mathbf{b} = ab\mathbf{n}\sin\theta$$

$$\mathbf{a} \cdot \mathbf{b} = ab\cos\theta,$$

with n a unit vector normal to both a and b. The relationship follows from these definitions and the Pythagorean trigonometric identity.

This can also be used to define the cross product. By rearranging the following equation is obtained

$$\| \mathbf{a} \times \mathbf{b} \|^2 = \| \mathbf{a} \|^2 \| \mathbf{b} \|^2 - (\mathbf{a} \cdot \mathbf{b})^2 .$$

This can be considered as a condition on the cross product and so part of its definition, for example in seven dimensions.

Generalizations

Similar Figures on the Three Sides

A generalization of the Pythagorean theorem extending beyond the areas of squares on the three sides to similar figures was known by Hippocrates of Chios in the 5th century BC, and was included by Euclid in his *Elements*:

If one erects similar figures with corresponding sides on the sides of a right triangle, then the sum of the areas of the ones on the two smaller sides equals the area of the one on the larger side.

This extension assumes that the sides of the original triangle are the corresponding sides of the three congruent figures (so the common ratios of sides between the similar figures are $a{:}b{:}c$). While Euclid's proof only applied to convex polygons, the theorem also applies to concave polygons and even to similar figures that have curved boundaries (but still with part of a figure's boundary being the side of the original triangle).

The basic idea behind this generalization is that the area of a plane figure is proportional to the square of any linear dimension, and in particular is proportional to the square of the length of any side. Thus, if similar figures with areas A, B and C are erected on sides with corresponding lengths a, b and c then:

$$\frac{A}{a^2} = \frac{B}{b^2} = \frac{C}{c^2},$$

$$\Rightarrow A + B = \frac{a^2}{c^2}C + \frac{b^2}{c^2}C.$$

But, by the Pythagorean theorem, $a^2 + b^2 = c^2$, so $A + B = C$.

Conversely, if we can prove that $A + B = C$ for three similar figures without using the Pythagorean theorem, then we can work backwards to construct a proof of the theorem. For example, the starting center triangle can be replicated and used as a triangle C on its hypotenuse, and two similar right triangles (A and B) constructed on the other two sides, formed by dividing the central triangle by its altitude. The sum of the areas of the two smaller triangles therefore is that of the third, thus $A + B = C$ and reversing the above logic leads to the Pythagorean theorem $a^2 + b^2 = c^2$.

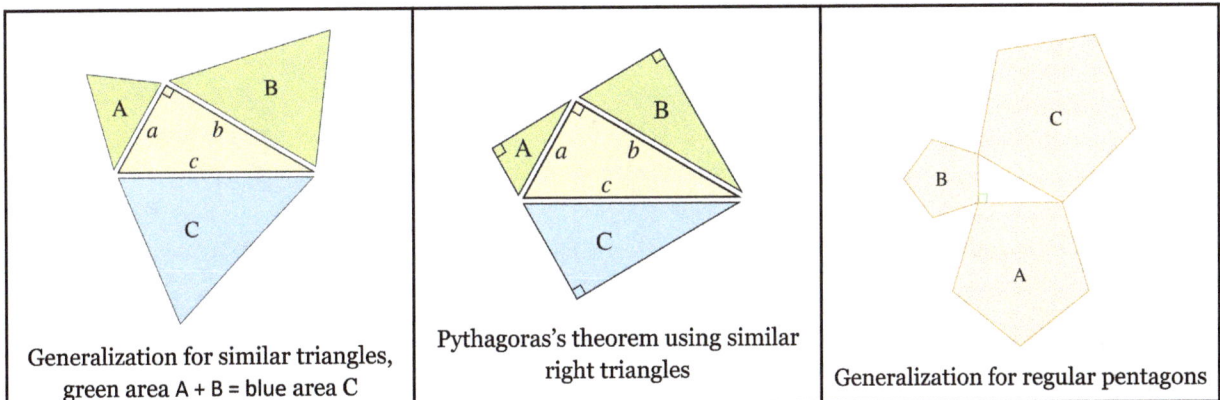

| Generalization for similar triangles, green area A + B = blue area C | Pythagoras's theorem using similar right triangles | Generalization for regular pentagons |

Law of Cosines

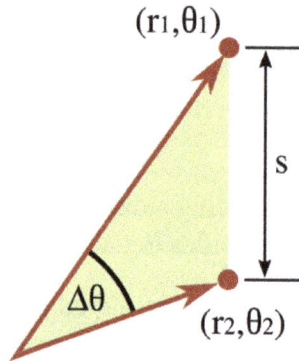

The separation s of two points (r_1, θ_1) and (r_2, θ_2) in polar coordinates is given by the law of cosines. Interior angle $\Delta\theta = \theta_1 - \theta_2$.

The Pythagorean theorem is a special case of the more general theorem relating the lengths of sides in any triangle, the law of cosines:

$$a^2 + b^2 - 2ab\cos\theta = c^2,$$

where θ is the angle between sides a and b.

When θ is 90 degrees ($\pi/2$ radians), then $\cos\theta = 0$, and the formula reduces to the usual Pythagorean theorem.

Arbitrary Triangle

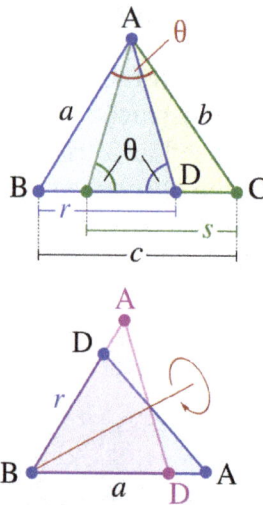

Generalization of Pythagoras's theorem by Tâbit ibn Qorra. Lower panel: reflection of triangle ABD (top) to form triangle DBA, similar to triangle ABC (top).

At any selected angle of a general triangle of sides a, b, c, inscribe an isosceles triangle such that the equal angles at its base θ are the same as the selected angle. Suppose the selected angle θ is opposite the side labeled c. Inscribing the isosceles triangle forms triangle ABD with angle θ opposite side a and with side r along c. A second triangle is formed with angle θ opposite side b and a side with length s along c, as shown in the figure. Thābit ibn Qurra stated that the sides of the three triangles were related as:

$$a^2 + b^2 = c(r + s).$$

As the angle θ approaches $\pi/2$, the base of the isosceles triangle narrows, and lengths r and s overlap less and less. When $\theta = \pi/2$, ADB becomes a right triangle, $r + s = c$, and the original Pythagorean theorem is regained.

One proof observes that triangle ABC has the same angles as triangle ABD, but in opposite order. (The two triangles share the angle at vertex B, both contain the angle θ, and so also have the same third angle by the triangle postulate.) Consequently, ABC is similar to the reflection of ABD, the triangle DBA in the lower panel. Taking the ratio of sides opposite and adjacent to θ,

$$\frac{c}{a} = \frac{a}{r}.$$

Likewise, for the reflection of the other triangle,

$$\frac{c}{b} = \frac{b}{s}.$$

Clearing fractions and adding these two relations:

$$cr + cs = a^2 + b^2 ,$$

the required result.

The theorem remains valid if the angle θ is obtuse so the lengths r and s are non-overlapping.

General Triangles Using Parallelograms

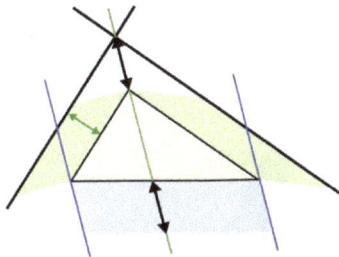

Generalization for arbitrary triangles,
green area = blue area

Pappus's area theorem is a further generalization, that applies to triangles that are not right triangles, using parallelograms on the three sides in place of squares (squares are a special case, of course). The upper figure shows that for a scalene triangle, the area of the parallelogram on the longest side is the sum of the areas of the parallelograms on the other two sides, provided the parallelogram on the long side is constructed as indicated (the dimensions labeled with arrows are the same, and determine the sides of the bottom parallelogram). This replacement of squares with parallelograms bears a clear resemblance to the original Pythagoras's theorem, and was considered a generalization by Pappus of Alexandria in 4 A.D.

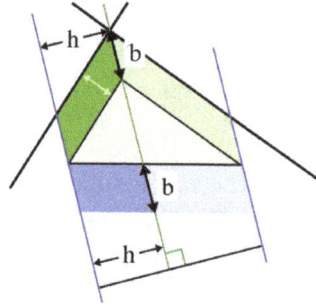

Construction for proof of parallelogram generalization

The lower figure shows the elements of the proof. Focus on the left side of the figure. The left green parallelogram has the same area as the left, blue portion of the bottom parallelogram because both have the same base b and height h. However, the left green parallelogram also has the same area as the left green parallelogram of the upper figure, because they have the same base (the upper left side of the triangle) and the same height normal to that side of the triangle. Repeating the argument for the right side of the figure, the bottom parallelogram has the same area as the sum of the two green parallelograms.

Solid Geometry

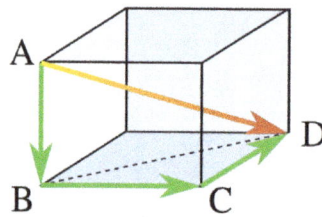

Pythagoras's theorem in three dimensions relates the diagonal AD to the three sides.

A tetrahedron with outward facing right-angle corner

In terms of solid geometry, Pythagoras's theorem can be applied to three dimensions as follows. Consider a rectangular solid as shown in the figure. The length of diagonal BD is found from Pythagoras's theorem as:

$$\overline{BD}^2 = \overline{BC}^2 + \overline{CD}^2,$$

where these three sides form a right triangle. Using horizontal diagonal BD and the vertical edge AB, the length of diagonal AD then is found by a second application of Pythagoras's theorem as:

$$\overline{AD}^2 = \overline{AB}^2 + \overline{BD}^2,$$

or, doing it all in one step:

$$\overline{AD}^2 = \overline{AB}^2 + \overline{BC}^2 + \overline{CD}^2.$$

This result is the three-dimensional expression for the magnitude of a vector v (the diagonal AD) in terms of its orthogonal components $\{v_k\}$ (the three mutually perpendicular sides):

$$\| \mathbf{v} \|^2 = \sum_{k=1}^{3} \| \mathbf{v}_k \|^2.$$

This one-step formulation may be viewed as a generalization of Pythagoras's theorem to higher dimensions. However, this result is really just the repeated application of the original Pythagoras's theorem to a succession of right triangles in a sequence of orthogonal planes.

A substantial generalization of the Pythagorean theorem to three dimensions is de Gua's theorem, named for Jean Paul de Gua de Malves: If a tetrahedron has a right angle corner (like a corner of a cube), then the square of the area of the face opposite the right angle corner is the sum of the squares of the areas of the other three faces. This result can be generalized as in the "n-dimensional Pythagorean theorem":

Let $x_1, x_2, \ldots, x_n$ be orthogonal vectors in $\mathbb{R}^n$. Consider the n-dimensional simplex S with vertices $0, x_1, \ldots, x_n$. (Think of the $(n-1)$-dimensional simplex with vertices $0, x_1, \ldots, x_n$. not including the origin as the "hypotenuse" of S and the remaining $(n-1)$-dimensional faces of S as its "legs".) Then the square of the volume of the hypotenuse of S is the sum of the squares of the volumes of the n legs.

This statement is illustrated in three dimensions by the tetrahedron in the figure. The "hypotenuse" is the base of the tetrahedron at the back of the figure, and the "legs" are the three sides emanating from the vertex in the foreground. As the depth of the base from the vertex increases, the area of the "legs" increases, while that of the base is fixed. The theorem suggests that when this depth is at the value creating a right vertex, the generalization of Pythagoras's theorem applies. In a different wording:

Given an n-rectangular n-dimensional simplex, the square of the $(n-1)$-content of the facet opposing the right vertex will equal the sum of the squares of the $(n-1)$-contents of the remaining facets.

Inner Product Spaces

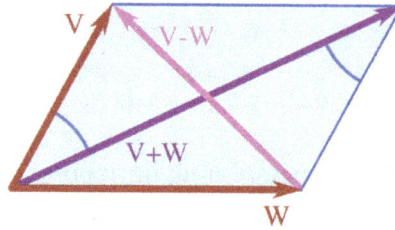

Vectors involved in the parallelogram law

The Pythagorean theorem can be generalized to inner product spaces, which are generalizations of the familiar 2-dimensional and 3-dimensional Euclidean spaces. For example, a function may be considered as a vector with infinitely many components in an inner product space, as in functional analysis.

In an inner product space, the concept of perpendicularity is replaced by the concept of orthogonality: two vectors v and w are orthogonal if their inner product $\langle \mathbf{v}, \mathbf{w} \rangle$ is zero. The inner product is a generalization of the dot product of vectors. The dot product is called the *standard* inner product or the *Euclidean* inner product. However, other inner products are possible.

The concept of length is replaced by the concept of the norm $||v||$ of a vector v, defined as:

$$\| \mathbf{v} \| \equiv \sqrt{\langle \mathbf{v}, \mathbf{v} \rangle}.$$

In an inner-product space, the Pythagorean theorem states that for any two orthogonal vectors v and w we have

$$\left\| \mathbf{v} + \mathbf{w} \right\|^2 = \left\| \mathbf{v} \right\|^2 + \left\| \mathbf{w} \right\|^2.$$

Here the vectors v and w are akin to the sides of a right triangle with hypotenuse given by the vector sum v + w. This form of the Pythagorean theorem is a consequence of the properties of the inner product:

$$\left\| \mathbf{v} + \mathbf{w} \right\|^2 = \langle \mathbf{v} + \mathbf{w}, \mathbf{v} + \mathbf{w} \rangle = \langle \mathbf{v}, \mathbf{v} \rangle + \langle \mathbf{w}, \mathbf{w} \rangle + \langle \mathbf{v}, \mathbf{w} \rangle + \langle \mathbf{w}, \mathbf{v} \rangle = \left\| \mathbf{v} \right\|^2 + \left\| \mathbf{w} \right\|^2,$$

where the inner products of the cross terms are zero, because of orthogonality.

A further generalization of the Pythagorean theorem in an inner product space to non-orthogonal vectors is the *parallelogram law* :

$$2 \| \mathbf{v} \|^2 + 2 \| \mathbf{w} \|^2 = \| \mathbf{v} + \mathbf{w} \|^2 + \| \mathbf{v} - \mathbf{w} \|^2,$$

which says that twice the sum of the squares of the lengths of the sides of a parallelogram is the sum of the squares of the lengths of the diagonals. Any norm that satisfies this equality is *ipso facto* a norm corresponding to an inner product.

The Pythagorean identity can be extended to sums of more than two orthogonal vectors. If $v_1, v_2, ...,$

v_n are pairwise-orthogonal vectors in an inner-product space, then application of the Pythagorean theorem to successive pairs of these vectors (as described for 3-dimensions in the section on solid geometry) results in the equation

$$\left\| \sum_{k=1}^{n} \mathbf{v}_k \right\|^2 = \sum_{k=1}^{n} \left\| \mathbf{v}_k \right\|^2.$$

Sets of m-dimensional Objects in n-dimensional Space

Another generalization of the Pythagorean theorem, introduced by Donald R. Conant and William A. Beyer, applies to a wide range of objects and sets of objects in any number of dimensions. Specifically, the square of the measure of an m-dimensional set of objects in one or more parallel m-dimensional flats in n-dimensional Euclidean space is equal to the sum of the squares of the measures of the orthogonal projections of the object(s) onto all m-dimensional coordinate subspaces.

In mathematical terms:

$$\mu_{ms}^2 = \sum_{i=1}^{x} \mu_{mp_i}^2$$

where:

- μ_m is a measure in m-dimensions (a length in one dimension, an area in two dimensions, a volume in three dimensions, etc.).

- s is a set of one or more non-overlapping m-dimensional objects in one or more parallel m-dimensional flats in n-dimensional Euclidean space.

- μ_{ms} is the total measure (sum) of the set of m-dimensional objects.

- p represents an m-dimensional projection of the original set onto an orthogonal coordinate subspace.

- μ_{mp_i} is the measure of the m-dimensional set projection onto m-dimensional coordinate subspace i. Because object projections can overlap on a coordinate subspace, the measure of each object projection in the set must be calculated individually, then measures of all projections added together to provide the total measure for the set of projections on the given coordinate subspace.

- x is the number of orthogonal, m-dimensional coordinate subspaces in n-dimensional space ($\mathbf{R}^n$) onto which the m-dimensional objects are projected ($m \leq n$):

$$x = \binom{n}{m} = \frac{n!}{m!(n-m)!}$$

For example, for a set of one or more two-dimensional parallel objects in three-dimensional space, $m = 2$ and $n = 3$. Therefore, the coordinate subspace calculation for this scenario is: $x = 3!/2!(3-2)! = 3*2*1/2*1*1 = 6/2 = 3$

Thus, three coordinate planes (*xy*-plane, *xz*-plane, and *yz*-plane) are required to capture the necessary projections for calculating the area of the set. If the set contained one-dimensional parallel line segments instead, three coordinate *axes* (*x*, *y*, and *z*), rather than planes, would be needed to capture the projections for calculating the length of the set.

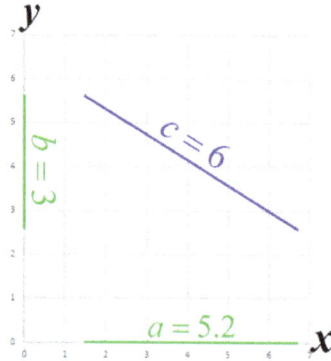

The Conant-Beyer generalization applied to a one-dimensional object in two dimensions of space:
$a^2 + b^2 = c^2$.

Applied to Sets Containing a Single Object

This generalized formula can be applied in the simplest case to a single one-dimensional object, a line segment, in two-dimensional space. The animation illustrates this case with a line segment shown in blue and its projections onto the *x*- and *y*- axes shown in green. The lengths of the projections squared and added together are equal to the length of the original line segment squared. This produces the familiar Pythagorean theorem formula:

$$a^2 + b^2 = c^2$$

where *c* is the length of the original line segment, *a* is the length of the segment projected onto the *x*-axis, and *b* is the length of the segment projected onto the *y*-axis. In the animation, $a^2 = 27$, $b^2 = 9$, and $c^2 = 36$. Bringing the line segment together with its coordinate projections forms the traditional right triangle.

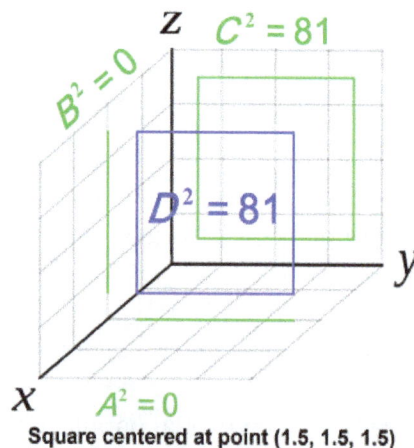

Square centered at point (1.5, 1.5, 1.5)

The Conant-Beyer generalization applied to a two-dimensional object in three dimensions of space:
$A^2 + B^2 + C^2 = D^2$.

Similarly, for any two-dimensional object in three-dimensional space, the formula can be stated as:

$$A^2 + B^2 + C^2 = D^2$$

where D is the area of a specified two-dimensional object, A is the area of the object's projection onto the xy-coordinate plane, B is the area of the object's projection onto the xz-coordinate plane, and C is the area of the object's projection onto the yz-coordinate plane.

The animation showing a blue three-by-three square object in three dimensions of space illustrates this application of the generalization to an object of more than one dimension. As the orientation of the object changes, the proportions of the green coordinate plane projections adjust accordingly, so the squares of the areas of the projections always add up to the same value: the square of the area of the original object. In this case, the sum of the squares of the projection areas always add up to 81.

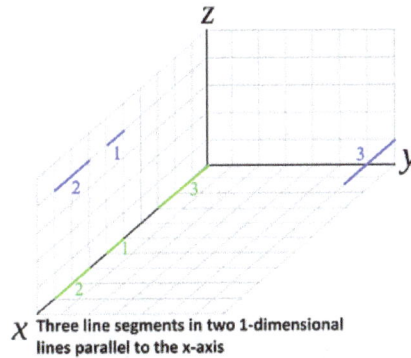

Three line segments in two 1-dimensional lines parallel to the x-axis

The Conant-Beyer generalization applied to a set of one-dimensional objects in three dimensions of space:
$$a^2 + b^2 + c^2 = d^2$$
where a, b, and c represent total lengths of the line segment sets on the x-, y-, and z- axes, and d the total length of the original line segment set.

Applied to Sets Containing Multiple Objects

The generalization applies equally to sets of multiple objects, as long as they are in the same plane or parallel planes. The measures of the objects in such a set can be added together and essentially treated as a single object. The multiple line-segment animation illustrates the generalization applied to a set of three one-dimensional objects in three dimensions of space. In this case, two sequential line segments exist in parallel to a third line segment. Because lines are one-dimensional, the coordinate subspaces onto which they are projected must also be one-dimensional. Thus, projections appear on the coordinate axes rather than on the coordinate planes. The lengths of the projected line segments on a given axis are summed, then squared, then added to the total lengths squared on the other axes. The result is the squared sum of the lengths of the original line segments. For the sake of simplicity, when projections are single points of zero length, they are not shown, since they do not affect the calculations.

The generalization applies to flat objects of any shape, regular or irregular. The multi-object animation illustrates the use of the generalization on a set of several different objects in different planes – in this case, a triangle and a circle on one plane, and a flat cat on a parallel plane (shown in blue). Projections of the set are shown in green on the coordinate plane subspaces. Objects shown initially upright in the yz-plane are subsequently tilted in parallel. Again, regardless of set orientation, the result remains the same. On each coordinate plane subspace, the areas of object projections are calculated individually (to avoid miscalculations due to projection overlap), then added together to produce the total projection area of the set on that plane. The projection set area is then squared for each coordinate plane. The sum of all projection set areas squared is always equal to the original set area squared.

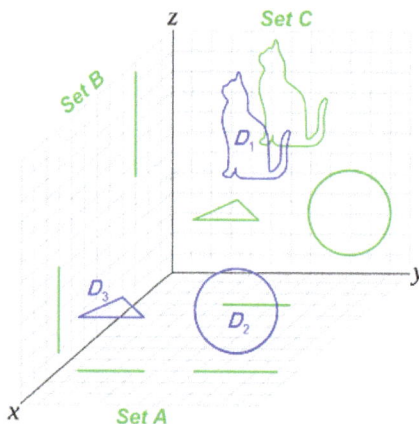

The Conant-Beyer generalization applied to a two-dimensional set of objects in three dimensions of space:
$$A^2 + B^2 + C^2 = D^2.$$

Applied in Any Number of Dimensions

This generalization holds regardless of the number of dimensions involved. The volume squared for a three-dimensional object or set can be calculated by summing the squares of the volumes of the associated three-dimensional projections onto three-dimensional subspaces. Any number of dimensions is valid for the set as long as one uses the same number of dimensions for the coordinate subspaces and projections.

It is the built-in symmetry of the Cartesian coordinate system where coordinates are orthogonal vectors of unit length in flat Euclidean space that allows this generalization to apply so broadly.

Non-Euclidean Geometry

The Pythagorean theorem is derived from the axioms of Euclidean geometry, and in fact, the Pythagorean theorem given above does not hold in a non-Euclidean geometry. (The Pythagorean theorem has been shown, in fact, to be equivalent to Euclid's Parallel (Fifth) Postulate.) In other words, in non-Euclidean geometry, the relation between the sides of a triangle must necessarily take a non-Pythagorean form. For example, in spherical geometry, all three sides of the right triangle (say a, b, and c) bounding an octant of the unit sphere have length equal to $\pi/2$, and all its angles are right angles, which violates the Pythagorean theorem because $a^2 + b^2 \neq c^2$.

Here two cases of non-Euclidean geometry are considered—spherical geometry and hyperbolic plane geometry; in each case, as in the Euclidean case for non-right triangles, the result replacing the Pythagorean theorem follows from the appropriate law of cosines.

However, the Pythagorean theorem remains true in hyperbolic geometry and elliptic geometry if the condition that the triangle be right is replaced with the condition that two of the angles sum to the third, say $A+B = C$. The sides are then related as follows: the sum of the areas of the circles with diameters a and b equals the area of the circle with diameter c.

Spherical Geometry

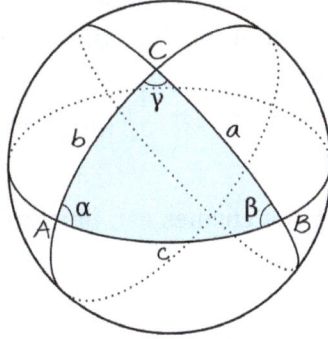

Spherical triangle

For any right triangle on a sphere of radius R (for example, if γ in the figure is a right angle), with sides a, b, c, the relation between the sides takes the form:

$$\cos\left(\frac{c}{R}\right) = \cos\left(\frac{a}{R}\right)\cos\left(\frac{b}{R}\right).$$

This equation can be derived as a special case of the spherical law of cosines that applies to all spherical triangles:

$$\cos\left(\frac{c}{R}\right) = \cos\left(\frac{a}{R}\right)\cos\left(\frac{b}{R}\right) + \sin\left(\frac{a}{R}\right)\sin\left(\frac{b}{R}\right)\cos\gamma.$$

By expressing the Maclaurin series for the cosine function as an asymptotic expansion with the remainder term in big O notation,

$$\cos x = 1 - \frac{x^2}{2} + O(x^4) \text{ as } x \to 0,$$

it can be shown that as the radius R approaches infinity and the arguments a/R, b/R, and c/R tend to zero, the spherical relation between the sides of a right triangle approaches the Euclidean form of the Pythagorean theorem. Substituting the asymptotic expansion for each of the cosines into the spherical relation for a right triangle yields

$$1 - \frac{1}{2}\left(\frac{c}{R}\right)^2 + O\left(\frac{1}{R^4}\right) = \left[1 - \frac{1}{2}\left(\frac{a}{R}\right)^2 + O\left(\frac{1}{R^4}\right)\right]\left[1 - \frac{1}{2}\left(\frac{b}{R}\right)^2 + O\left(\frac{1}{R^4}\right)\right] \text{ as } R \to \infty.$$

The constants a^4, b^4, and c^4 have been absorbed into the big O remainder terms since they are independent of the radius R. This asymptotic relationship can be further simplified by multiplying out the bracketed quantities, cancelling the ones, multiplying through by -2, and collecting all the error terms together:

$$\left(\frac{c}{R}\right)^2 = \left(\frac{a}{R}\right)^2 + \left(\frac{b}{R}\right)^2 + O\left(\frac{1}{R^4}\right) \text{ as } R \to \infty.$$

After multiplying through by R^2, the Euclidean Pythagorean relationship $c^2 = a^2 + b^2$ is recovered in the limit as the radius R approaches infinity (since the remainder term tends to zero):

$$c^2 = a^2 + b^2 + O\left(\frac{1}{R^2}\right) \text{ as } R \to \infty.$$

For small right triangles ($a, b << R$), the cosines can be eliminated to avoid loss of significance, giving

$$\sin^2 \frac{c}{2R} = \sin^2 \frac{a}{2R} + \sin^2 \frac{b}{2R} - 2\sin^2 \frac{a}{2R} \sin^2 \frac{b}{2R}.$$

Hyperbolic Geometry

In a hyperbolic space with uniform curvature $-1/R^2$, for a right triangle with legs a, b, and hypotenuse c, the relation between the sides takes the form:

$$\cosh \frac{c}{R} = \cosh \frac{a}{R} \cosh \frac{b}{R}$$

where cosh is the hyperbolic cosine. This formula is a special form of the hyperbolic law of cosines that applies to all hyperbolic triangles:

$$\cosh \frac{c}{R} = \cosh \frac{a}{R} \cosh \frac{b}{R} - \sinh \frac{a}{R} \sinh \frac{b}{R} \cos \gamma,$$

with γ the angle at the vertex opposite the side c.

By using the Maclaurin series for the hyperbolic cosine, $\cosh x \approx 1 + x^2/2$, it can be shown that as a hyperbolic triangle becomes very small (that is, as a, b, and c all approach zero), the hyperbolic relation for a right triangle approaches the form of Pythagoras's theorem.

For small right triangles ($a, b << R$), the hyperbolic cosines can be eliminated to avoid loss of significance, giving

$$\sinh^2 \frac{c}{2R} = \sinh^2 \frac{a}{2R} + \sinh^2 \frac{b}{2R} + 2\sinh^2 \frac{a}{2R} \sinh^2 \frac{b}{2R}.$$

Very Small Triangles

For any uniform curvature K (positive, zero, or negative), in very small right triangles ($|K|a^2$, $|K|b^2$ $<< 1$) with hypotenuse c, it can be shown that

$$c^2 = a^2 + b^2 - \frac{K}{3}a^2b^2 - \frac{K^2}{45}a^2b^2(a^2+b^2) - \frac{2K^3}{945}a^2b^2(a^2-b^2)^2 + O(K^4c^{10}).$$

Differential Geometry

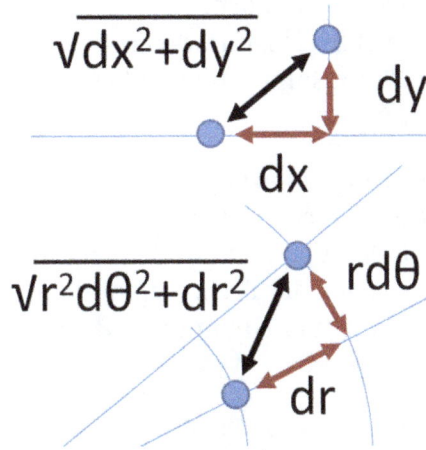

Distance between infinitesimally separated points in Cartesian coordinates (top) and polar coordinates (bottom), as given by Pythagoras's theorem

On an infinitesimal level, in three dimensional space, Pythagoras's theorem describes the distance between two infinitesimally separated points as:

$$ds^2 = dx^2 + dy^2 + dz^2,$$

with ds the element of distance and (dx, dy, dz) the components of the vector separating the two points. Such a space is called a Euclidean space. However, in Riemannian geometry, a generalization of this expression useful for general coordinates (not just Cartesian) and general spaces (not just Euclidean) takes the form:

$$ds^2 = \sum_{i,j}^{n} g_{ij}\, dx_i\, dx_j$$

which is called the metric tensor. (Sometimes, by abuse of language, the same term is applied to the set of coefficients g_{ij}.) It may be a function of position, and often describes curved space. A simple example is Euclidean (flat) space expressed in curvilinear coordinates. For example, in polar coordinates:

$$ds^2 = dr^2 + r^2 d\theta^2 .$$

History

There is debate whether the Pythagorean theorem was discovered once, or many times in many places, and the date of first discovery is uncertain, as is the date of the first proof. According to Joran Friberg, a historian of mathematics, evidence indicates that the Pythagorean Theorem was well-known to the mathematicians of the First Babylonian Dynasty (20th to 16th centuries BC), which would have been over a thousand years before Pythagoras was born, thus an example of Stigler's law of eponymy. (Yale's Institute for the Preservation of Cultural Heritage's 3-D scan of a cuneiform tablet depicting the proof is one of their mostly widely used images.) Other sources, such as a book by Leon Lederman and Dick Teresi, mention that Pythagoras discovered the theorem, although Teresi subsequently stated that the Babylonians developed the theorem "at least fifteen hundred years before Pythagoras was born." The history of the theorem can be divided into four parts: knowledge of Pythagorean triples, knowledge of the relationship among the sides of a right triangle, knowledge of the relationships among adjacent angles, and proofs of the theorem within some deductive system.

Bartel Leendert van der Waerden (1903–1996) conjectured that Pythagorean triples were discovered algebraically by the Babylonians. Written between 2000 and 1786 BC, the Middle Kingdom Egyptian *Berlin Papyrus 6619* includes a problem whose solution is the Pythagorean triple 6:8:10, but the problem does not mention a triangle. The Mesopotamian tablet *Plimpton 322*, written between 1790 and 1750 BC during the reign of Hammurabi the Great, contains many entries closely related to Pythagorean triples.

In India, the *Baudhayana Sulba Sutra*, the dates of which are given variously as between the 8th and 5th century BC, contains a list of Pythagorean triples discovered algebraically, a statement of the Pythagorean theorem, and a geometrical proof of the Pythagorean theorem for an isosceles right triangle. The *Apastamba Sulba Sutra* (c. 600 BC) contains a numerical proof of the general Pythagorean theorem, using an area computation. Van der Waerden believed that "it was certainly based on earlier traditions". Carl Boyer states that the Pythagorean theorem in *Śulba-sũtram* may have been influenced by ancient Mesopotamian math, but there is no conclusive evidence in favor or opposition of this possibility.

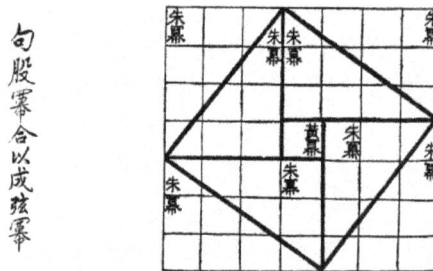

Geometric proof of the Pythagorean theorem from the *Zhou Bi Suan Jing*.

With contents known much earlier, but in surviving texts dating from roughly the 1st century BC, the Chinese text *Zhou Bi Suan Jing*, (*The Arithmetical Classic of the Gnomon and the Circular Paths of Heaven*) gives a reasoning for the Pythagorean theorem for the (3, 4, 5) triangle—in China it is called the "Gougu Theorem". During the Han Dynasty (202 BC to 220 AD), Pythagorean triples appear in *The Nine Chapters on the Mathematical Art*, together with a mention of right triangles. Some believe the theorem arose first in China, where it is alternatively known as the "Shang Gao Theorem", named after the Duke of Zhou's astronomer and mathematician, whose reasoning composed most of what was in the *Zhou Bi Suan Jing*.

Pythagoras, whose dates are commonly given as 569–475 BC, used algebraic methods to construct Pythagorean triples, according to Proclus's commentary on Euclid. Proclus, however, wrote between 410 and 485 AD. According to Thomas L. Heath (1861–1940), no specific attribution of the theorem to Pythagoras exists in the surviving Greek literature from the five centuries after Pythagoras lived. However, when authors such as Plutarch and Cicero attributed the theorem to Pythagoras, they did so in a way which suggests that the attribution was widely known and undoubted. "Whether this formula is rightly attributed to Pythagoras personally, [...] one can safely assume that it belongs to the very oldest period of Pythagorean mathematics."

Around 400 BC, according to Proclus, Plato gave a method for finding Pythagorean triples that combined algebra and geometry. Around 300 BC, in Euclid's *Elements*, the oldest extant axiomatic proof of the theorem is presented.

In Popular Culture

Exhibit on the Pythagorean theorem at the Universum museum in Mexico City

The Pythagorean theorem has arisen in popular culture in a variety of ways.

- John Aubrey in his Brief Lives records of Thomas Hobbes that "He was forty years old before he looked on geometry; which happened accidentally. Being in a gentleman's library Euclid's Elements lay open, and 'twas the forty-seventh proposition* in the first book. He read the proposition. 'By G ,' said he, 'this is impossible!' So he reads the demonstration of it, which referred him back to such a proof; which referred him back to another, which he also read. Et sic deinceps, that at last he was demonstratively convinced of that truth. This made him in love with geometry."

- Hans Christian Andersen wrote in 1831 a poem about the Pythagorean theorem: *Formens Evige Magie (Et poetisk Spilfægteri)*.

- A verse of the Major-General's Song in the Gilbert and Sullivan comic opera *The Pirates of Penzance*, "About binomial theorem I'm teeming with a lot o' news, With many cheerful facts about the square of the hypotenuse", makes an oblique reference to the theorem.

- The Scarecrow in the film *The Wizard of Oz* makes a more specific reference to the theorem. Upon receiving his diploma from the Wizard, he immediately exhibits his "knowledge" by reciting a mangled and incorrect version of the theorem: "The sum of the square roots of any two sides of an isosceles triangle is equal to the square root of the remaining side. Oh, joy! Oh, rapture! I've got a brain!"

- In 2000, Uganda released a coin with the shape of an isosceles right triangle. The coin's tail has an image of Pythagoras and the equation $\alpha^2 + \beta^2 = \gamma^2$, accompanied with the mention "PYTHAGORAS MILLENNIUM".

- Greece, Japan, San Marino, Sierra Leone, and Suriname have issued postage stamps depicting Pythagoras and the Pythagorean theorem.

- In Neal Stephenson's speculative fiction *Anathem*, the Pythagorean theorem is referred to as 'the Adrakhonic theorem'. A geometric proof of the theorem is displayed on the side of an alien ship to demonstrate the aliens' understanding of mathematics.

Set Theory

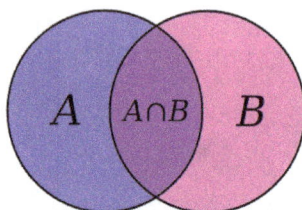

A Venn diagram illustrating the intersection of two sets.

Set theory is the branch of mathematical logic that studies sets, which informally are collections of objects. Although any type of object can be collected into a set, set theory is applied most often to objects that are relevant to mathematics. The language of set theory can be used in the definitions of nearly all mathematical objects.

The modern study of set theory was initiated by Georg Cantor and Richard Dedekind in the 1870s. After the discovery of paradoxes in naive set theory, numerous axiom systems were proposed in the early twentieth century, of which the Zermelo–Fraenkel axioms, with the axiom of choice, are the best-known.

Set theory is commonly employed as a foundational system for mathematics, particularly in the form of Zermelo–Fraenkel set theory with the axiom of choice. Beyond its foundational role, set theory is a branch of mathematics in its own right, with an active research community. Contemporary research into set theory includes a diverse collection of topics, ranging from the structure of the real number line to the study of the consistency of large cardinals.

History

Georg Cantor.

Mathematical topics typically emerge and evolve through interactions among many researchers. Set theory, however, was founded by a single paper in 1874 by Georg Cantor: "On a Property of the Collection of All Real Algebraic Numbers".

Since the 5th century BC, beginning with Greek mathematician Zeno of Elea in the West and early Indian mathematicians in the East, mathematicians had struggled with the concept of infinity. Especially notable is the work of Bernard Bolzano in the first half of the 19th century. Modern understanding of infinity began in 1867–71, with Cantor's work on number theory. An 1872 meeting between Cantor and Richard Dedekind influenced Cantor's thinking and culminated in Cantor's 1874 paper.

Cantor's work initially polarized the mathematicians of his day. While Karl Weierstrass and Dedekind supported Cantor, Leopold Kronecker, now seen as a founder of mathematical constructivism, did not. Cantorian set theory eventually became widespread, due to the utility of Cantorian concepts, such as one-to-one correspondence among sets, his proof that there are more real numbers than integers, and the "infinity of infinities" ("Cantor's paradise") resulting from the power set operation. This utility of set theory led to the article "Mengenlehre" contributed in 1898 by Arthur Schoenflies to Klein's encyclopedia.

The next wave of excitement in set theory came around 1900, when it was discovered that some interpretations of Cantorian set theory gave rise to several contradictions, called antinomies or paradoxes. Bertrand Russell and Ernst Zermelo independently found the simplest and best known paradox, now called Russell's paradox: consider "the set of all sets that are not members of themselves", which leads to a contradiction since it must be a member of itself, and not a member of itself. In 1899 Cantor had himself posed the question "What is the cardinal number of the set of all sets?", and obtained a related paradox. Russell used his paradox as a theme in his 1903 review of continental mathematics in his *The Principles of Mathematics*.

In 1906 English readers gained the book *Theory of Sets of Points* by William Henry Young and his wife Grace Chisholm Young, published by Cambridge University Press.

The momentum of set theory was such that debate on the paradoxes did not lead to its abandonment. The work of Zermelo in 1908 and Abraham Fraenkel in 1922 resulted in the set of axioms ZFC, which became the most commonly used set of axioms for set theory. The work of analysts such as Henri Lebesgue demonstrated the great mathematical utility of set theory, which has since become woven into the fabric of modern mathematics. Set theory is commonly used as a foundational system, although in some areas category theory is thought to be a preferred foundation.

Basic Concepts and Notation

Set theory begins with a fundamental binary relation between an object o and a set A. If o is a member (or element) of A, write $o \in A$. Since sets are objects, the membership relation can relate sets as well.

A derived binary relation between two sets is the subset relation, also called set inclusion. If all the members of set A are also members of set B, then A is a subset of B, denoted $A \subseteq B$. For example, $\{1, 2\}$ is a subset of $\{1, 2, 3\}$, and so is $\{2\}$ but $\{1, 4\}$ is not. From this definition, it is clear that a set is a subset of itself; for cases where one wishes to rule this out, the term proper subset is defined. A

is called a proper subset of B if and only if A is a subset of B, but A is not equal to B. Note also that 1 and 2 and 3 are members (elements) of set $\{1, 2, 3\}$, but are *not* subsets, and the subsets in turn are *not* as such members of the set.

Just as arithmetic features binary operations on numbers, set theory features binary operations on sets. The:

- Union of the sets A and B, denoted $A \cup B$, is the set of all objects that are a member of A, or B, or both. The union of $\{1, 2, 3\}$ and $\{2, 3, 4\}$ is the set $\{1, 2, 3, 4\}$.

- Intersection of the sets A and B, denoted $A \cap B$, is the set of all objects that are members of both A and B. The intersection of $\{1, 2, 3\}$ and $\{2, 3, 4\}$ is the set $\{2, 3\}$.

- Set difference of U and A, denoted $U \setminus A$, is the set of all members of U that are not members of A. The set difference $\{1, 2, 3\} \setminus \{2, 3, 4\}$ is $\{1\}$, while, conversely, the set difference $\{2, 3, 4\} \setminus \{1, 2, 3\}$ is $\{4\}$. When A is a subset of U, the set difference $U \setminus A$ is also called the complement of A in U. In this case, if the choice of U is clear from the context, the notation A^c is sometimes used instead of $U \setminus A$, particularly if U is a universal set as in the study of Venn diagrams.

- Symmetric difference of sets A and B, denoted $A \vartriangle B$ or $A \ominus B$, is the set of all objects that are a member of exactly one of A and B (elements which are in one of the sets, but not in both). For instance, for the sets $\{1, 2, 3\}$ and $\{2, 3, 4\}$, the symmetric difference set is $\{1, 4\}$. It is the set difference of the union and the intersection, $(A \cup B) \setminus (A \cap B)$ or $(A \setminus B) \cup (B \setminus A)$.

- Cartesian product of A and B, denoted $A \times B$, is the set whose members are all possible ordered pairs (a, b) where a is a member of A and b is a member of B. The cartesian product of $\{1, 2\}$ and $\{red, white\}$ is $\{(1, red), (1, white), (2, red), (2, white)\}$.

- Power set of a set A is the set whose members are all possible subsets of A. For example, the power set of $\{1, 2\}$ is $\{ \{\}, \{1\}, \{2\}, \{1, 2\} \}$.

Some basic sets of central importance are the empty set (the unique set containing no elements), the set of natural numbers, and the set of real numbers.

Some Ontology

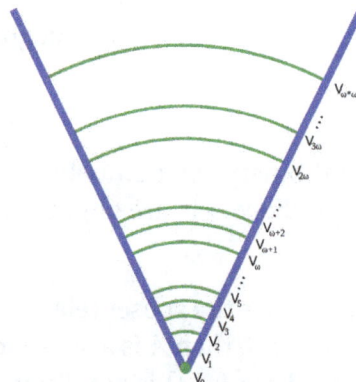

An initial segment of the von Neumann hierarchy.

A set is pure if all of its members are sets, all members of its members are sets, and so on. For example, the set {{}} containing only the empty set is a nonempty pure set. In modern set theory, it is common to restrict attention to the von Neumann universe of pure sets, and many systems of axiomatic set theory are designed to axiomatize the pure sets only. There are many technical advantages to this restriction, and little generality is lost, because essentially all mathematical concepts can be modeled by pure sets. Sets in the von Neumann universe are organized into a cumulative hierarchy, based on how deeply their members, members of members, etc. are nested. Each set in this hierarchy is assigned (by transfinite recursion) an ordinal number α, known as its rank. The rank of a pure set X is defined to be the least upper bound of all successors of ranks of members of X. For example, the empty set is assigned rank 0, while the set {{}} containing only the empty set is assigned rank 1. For each ordinal α, the set V_α is defined to consist of all pure sets with rank less than α. The entire von Neumann universe is denoted V.

Axiomatic Set Theory

Elementary set theory can be studied informally and intuitively, and so can be taught in primary schools using Venn diagrams. The intuitive approach tacitly assumes that a set may be formed from the class of all objects satisfying any particular defining condition. This assumption gives rise to paradoxes, the simplest and best known of which are Russell's paradox and the Burali-Forti paradox. Axiomatic set theory was originally devised to rid set theory of such paradoxes.

The most widely studied systems of axiomatic set theory imply that all sets form a cumulative hierarchy. Such systems come in two flavors, those whose ontology consists of:

- *Sets alone.* This includes the most common axiomatic set theory, Zermelo–Fraenkel set theory (ZFC), which includes the axiom of choice. Fragments of ZFC include:

    - Zermelo set theory, which replaces the axiom schema of replacement with that of separation;

    - General set theory, a small fragment of Zermelo set theory sufficient for the Peano axioms and finite sets;

    - Kripke–Platek set theory, which omits the axioms of infinity, powerset, and choice, and weakens the axiom schemata of separation and replacement.

- *Sets and proper classes.* These include Von Neumann–Bernays–Gödel set theory, which has the same strength as ZFC for theorems about sets alone, and Morse-Kelley set theory and Tarski–Grothendieck set theory, both of which are stronger than ZFC.

The above systems can be modified to allow urelements, objects that can be members of sets but that are not themselves sets and do not have any members.

The systems of New Foundations NFU (allowing urelements) and NF (lacking them) are not based on a cumulative hierarchy. NF and NFU include a "set of everything, " relative to which every set has a complement. In these systems urelements matter, because NF, but not NFU, produces sets for which the axiom of choice does not hold.

Systems of constructive set theory, such as CST, CZF, and IZF, embed their set axioms in intuitionistic instead of classical logic. Yet other systems accept classical logic but feature a nonstandard membership relation. These include rough set theory and fuzzy set theory, in which the value of an atomic formula embodying the membership relation is not simply True or False. The Boolean-valued models of ZFC are a related subject.

An enrichment of ZFC called Internal Set Theory was proposed by Edward Nelson in 1977.

Applications

Many mathematical concepts can be defined precisely using only set theoretic concepts. For example, mathematical structures as diverse as graphs, manifolds, rings, and vector spaces can all be defined as sets satisfying various (axiomatic) properties. Equivalence and order relations are ubiquitous in mathematics, and the theory of mathematical relations can be described in set theory.

Set theory is also a promising foundational system for much of mathematics. Since the publication of the first volume of *Principia Mathematica*, it has been claimed that most or even all mathematical theorems can be derived using an aptly designed set of axioms for set theory, augmented with many definitions, using first or second order logic. For example, properties of the natural and real numbers can be derived within set theory, as each number system can be identified with a set of equivalence classes under a suitable equivalence relation whose field is some infinite set.

Set theory as a foundation for mathematical analysis, topology, abstract algebra, and discrete mathematics is likewise uncontroversial; mathematicians accept that (in principle) theorems in these areas can be derived from the relevant definitions and the axioms of set theory. Few full derivations of complex mathematical theorems from set theory have been formally verified, however, because such formal derivations are often much longer than the natural language proofs mathematicians commonly present. One verification project, Metamath, includes human-written, computer-verified derivations of more than 12,000 theorems starting from ZFC set theory, first order logic and propositional logic.

Areas of Study

Set theory is a major area of research in mathematics, with many interrelated subfields.

Combinatorial Set Theory

Combinatorial set theory concerns extensions of finite combinatorics to infinite sets. This includes the study of cardinal arithmetic and the study of extensions of Ramsey's theorem such as the Erdős–Rado theorem.

Descriptive Set Theory

Descriptive set theory is the study of subsets of the real line and, more generally, subsets of Polish spaces. It begins with the study of pointclasses in the Borel hierarchy and extends to the study of more complex hierarchies such as the projective hierarchy and the Wadge hierarchy. Many properties of Borel sets can be established in ZFC, but proving these properties hold for more complicated sets requires additional axioms related to determinacy and large cardinals.

The field of effective descriptive set theory is between set theory and recursion theory. It includes the study of lightface pointclasses, and is closely related to hyperarithmetical theory. In many cases, results of classical descriptive set theory have effective versions; in some cases, new results are obtained by proving the effective version first and then extending ("relativizing") it to make it more broadly applicable.

A recent area of research concerns Borel equivalence relations and more complicated definable equivalence relations. This has important applications to the study of invariants in many fields of mathematics.

Fuzzy Set Theory

In set theory as Cantor defined and Zermelo and Fraenkel axiomatized, an object is either a member of a set or not. In fuzzy set theory this condition was relaxed by Lotfi A. Zadeh so an object has a *degree of membership* in a set, a number between 0 and 1. For example, the degree of membership of a person in the set of "tall people" is more flexible than a simple yes or no answer and can be a real number such as 0.75.

Inner Model Theory

An inner model of Zermelo–Fraenkel set theory (ZF) is a transitive class that includes all the ordinals and satisfies all the axioms of ZF. The canonical example is the constructible universe L developed by Gödel. One reason that the study of inner models is of interest is that it can be used to prove consistency results. For example, it can be shown that regardless of whether a model V of ZF satisfies the continuum hypothesis or the axiom of choice, the inner model L constructed inside the original model will satisfy both the generalized continuum hypothesis and the axiom of choice. Thus the assumption that ZF is consistent (has at least one model) implies that ZF together with these two principles is consistent.

The study of inner models is common in the study of determinacy and large cardinals, especially when considering axioms such as the axiom of determinacy that contradict the axiom of choice. Even if a fixed model of set theory satisfies the axiom of choice, it is possible for an inner model to fail to satisfy the axiom of choice. For example, the existence of sufficiently large cardinals implies that there is an inner model satisfying the axiom of determinacy (and thus not satisfying the axiom of choice).

Large Cardinals

A large cardinal is a cardinal number with an extra property. Many such properties are studied, including inaccessible cardinals, measurable cardinals, and many more. These properties typically imply the cardinal number must be very large, with the existence of a cardinal with the specified property unprovable in Zermelo-Fraenkel set theory.

Determinacy

Determinacy refers to the fact that, under appropriate assumptions, certain two-player games of perfect information are determined from the start in the sense that one player must have a winning strategy. The existence of these strategies has important consequences in descriptive set theory, as the assumption that a broader class of games is determined often implies that a broader class of sets will have a topological property. The axiom of determinacy (AD) is an important object of study; although incompatible with the axiom of choice, AD implies that all subsets of the real line are well behaved (in particular, measurable and with the perfect set property). AD can be used to prove that the Wadge degrees have an elegant structure.

Forcing

Paul Cohen invented the method of forcing while searching for a model of ZFC in which the continuum hypothesis fails, or a model of ZF in which the axiom of choice fails. Forcing adjoins to some given model of set theory additional sets in order to create a larger model with properties determined (i.e. "forced") by the construction and the original model. For example, Cohen's construction adjoins additional subsets of the natural numbers without changing any of the cardinal numbers of the original model. Forcing is also one of two methods for proving relative consistency by finitistic methods, the other method being Boolean-valued models.

Cardinal Invariants

A cardinal invariant is a property of the real line measured by a cardinal number. For example, a well-studied invariant is the smallest cardinality of a collection of meagre sets of reals whose union is the entire real line. These are invariants in the sense that any two isomorphic models of set theory must give the same cardinal for each invariant. Many cardinal invariants have been studied, and the relationships between them are often complex and related to axioms of set theory.

Set-theoretic Topology

Set-theoretic topology studies questions of general topology that are set-theoretic in nature or that require advanced methods of set theory for their solution. Many of these theorems are independent of ZFC, requiring stronger axioms for their proof. A famous problem is the normal Moore space question, a question in general topology that was the subject of intense research. The answer to the normal Moore space question was eventually proved to be independent of ZFC.

Objections to Set Theory as a Foundation for Mathematics

From set theory's inception, some mathematicians have objected to it as a foundation for mathematics. The most common objection to set theory, one Kronecker voiced in set theory's earliest years, starts from the constructivist view that mathematics is loosely related to computation. If this view is granted, then the treatment of infinite sets, both in naive and in axiomatic set theory, introduces into mathematics methods and objects that are not computable even in principle. The feasibility of constructivism as a substitute foundation for mathematics was greatly increased by Errett Bishop's influential book *Foundations of Constructive Analysis*.

A different objection put forth by Henri Poincaré is that defining sets using the axiom schemas of specification and replacement, as well as the axiom of power set, introduces impredicativity, a type of circularity, into the definitions of mathematical objects. The scope of predicatively founded mathematics, while less than that of the commonly accepted Zermelo-Fraenkel theory, is much greater than that of constructive mathematics, to the point that Solomon Feferman has said that "all of scientifically applicable analysis can be developed [using predicative methods]".

Ludwig Wittgenstein condemned set theory. He wrote that "set theory is wrong", since it builds on the "nonsense" of fictitious symbolism, has "pernicious idioms", and that it is nonsensical to talk about "all numbers". Wittgenstein's views about the foundations of mathematics were later criticised by Georg Kreisel and Paul Bernays, and investigated by Crispin Wright, among others.

Category theorists have proposed topos theory as an alternative to traditional axiomatic set theory. Topos theory can interpret various alternatives to that theory, such as constructivism, finite set theory, and computable set theory. Topoi also give a natural setting for forcing and discussions of the independence of choice from ZF, as well as providing the framework for pointless topology and Stone spaces.

An active area of research is the univalent foundations arising from homotopy type theory. Here, sets may be defined as certain kinds of types, with universal properties of sets arising from higher inductive types. Principles such as the axiom of choice and the law of the excluded middle appear in a spectrum of different forms, some of which can be proven, others which correspond to the classical notions; this allows for a detailed discussion of the effect of these axioms on mathematics.

Model Theory

In mathematics, model theory is the study of classes of mathematical structures (e.g. groups, fields, graphs, universes of set theory) from the perspective of mathematical logic. The objects of study are models of theories in a formal language. We call a set of sentences in a formal language a theory; a model of a theory is a structure (e.g. an interpretation) that satisfies the sentences of that theory.

Model theory recognises and is intimately concerned with a duality: It examines semantical elements (meaning and truth) by means of syntactical elements (formulas and proofs) of a corresponding language. To quote the first page of Chang & Keisler (1990):

universal algebra + logic = model theory.

Model theory developed rapidly during the 1990s, and a more modern definition is provided by Wilfrid Hodges (1997):

model theory = algebraic geometry − fields,

although model theorists are also interested in the study of fields. Other nearby areas of mathematics include combinatorics, number theory, arithmetic dynamics, analytic functions, and non-standard analysis.

In a similar way to proof theory, model theory is situated in an area of interdisciplinarity among mathematics, philosophy, and computer science. The most prominent professional organization in the field of model theory is the Association for Symbolic Logic.

Branches of Model Theory

This article focuses on finitary first order model theory of infinite structures. Finite model theory, which concentrates on finite structures, diverges significantly from the study of infinite structures in both the problems studied and the techniques used. Model theory in higher-order logics or infinitary logics is hampered by the fact that completeness and compactness do not in general hold for these logics. However, a great deal of study has also been done in such logics.

Informally, model theory can be divided into classical model theory, model theory applied to groups and fields, and geometric model theory. A missing subdivision is computable model theory, but this can arguably be viewed as an independent subfield of logic.

Examples of early theorems from classical model theory include Gödel's completeness theorem, the upward and downward Löwenheim–Skolem theorems, Vaught's two-cardinal theorem, Scott's isomorphism theorem, the omitting types theorem, and the Ryll-Nardzewski theorem. Examples of early results from model theory applied to fields are Tarski's elimination of quantifiers for real closed fields, Ax's theorem on pseudo-finite fields, and Robinson's development of non-standard analysis. An important step in the evolution of classical model theory occurred with the birth of stability theory (through Morley's theorem on uncountably categorical theories and Shelah's classification program), which developed a calculus of independence and rank based on syntactical conditions satisfied by theories.

During the last several decades applied model theory has repeatedly merged with the more pure stability theory. The result of this synthesis is called geometric model theory in this article (which is taken to include o-minimality, for example, as well as classical geometric stability theory). An example of a theorem from geometric model theory is Hrushovski's proof of the Mordell–Lang conjecture for function fields. The ambition of geometric model theory is to provide a *geography of mathematics* by embarking on a detailed study of definable sets in various mathematical structures, aided by the substantial tools developed in the study of pure model theory.

Universal Algebra

Fundamental concepts in universal algebra are signatures σ and σ-algebras. Since these concepts are formally defined in the article on structures, the present article can contend itself with an informal introduction which consists in examples of how these terms are used.

The standard signature of rings is $\sigma_{ring} = \{\times,+,-,0,1\}$, where $\times$ and $+$ are binary, $-$ is unary, and 0 and 1 are nullary.

The standard signature of semirings is $\sigma_{smr} = \{\times,+,0,1\}$, where the arities are as above.

The standard signature of groups (with multiplicative notation) is $\sigma_{grp} = \{\times,^{-1},1\}$, where $\times$ is binary, $^{-1}$ is unary and 1 is nullary.

The standard signature of monoids is $\sigma_{mnd} = \{\times,1\}$.

A ring is a σ_{ring}-structure which satisfies the identities $u + (v + w) = (u + v) + w, u + v = v + u, u + 0 = u, u + (-u) = 0, u \times (v \times w) = (u \times v) \times w, u \times 1 = u, 1 \times u = u, u \times (v + w) = (u \times v) + (u \times w)$ and $(v + w) \times u = (v \times u) + (w \times u)$.

A group is a σ_{grp}-structure which satisfies the identities $u \times (v \times w) = (u \times v) \times w$, $u \times 1 = u$, $1 \times u = u$, $u \times u^{-1} = 1$ and $u^{-1} \times u = 1$.

A monoid is a σ_{mnd}-structure which satisfies the identities $u \times (v \times w) = (u \times v) \times w$, $u \times 1 = u$ and $1 \times u = u$.

A semigroup is a $\{\times\}$-structure which satisfies the identity $u \times (v \times w) = (u \times v) \times w$.

A magma is just a $\{\times\}$-structure.

This is a very efficient way to define most classes of algebraic structures, because there is also the concept of σ-homomorphism, which correctly specializes to the usual notions of homomorphism for groups, semigroups, magmas and rings. For this to work, the signature must be chosen well.

Terms such as the σ_{ring}-term $t(u,v,w)$ given by $(u + (v \times w)) + (-1)$ are used to define identities $t = t'$, but also to construct free algebras. An equational class is a class of structures which, like the examples above and many others, is defined as the class of all σ-structures which satisfy a certain set of identities. Birkhoff's theorem states:

> A class of σ-structures is an equational class if and only if it is not empty and closed under subalgebras, homomorphic images, and direct products.

An important non-trivial tool in universal algebra are ultraproducts $\Pi_{i \in I} A_i / U$, where I is an infinite set indexing a system of σ-structures A_i, and U is an ultrafilter on I.

While model theory is generally considered a part of mathematical logic, universal algebra, which grew out of Alfred North Whitehead's (1898) work on abstract algebra, is part of algebra. This is reflected by their respective MSC classifications. Nevertheless, model theory can be seen as an extension of universal algebra.

Finite Model Theory

Finite model theory is the area of model theory which has the closest ties to universal algebra. Like some parts of universal algebra, and in contrast with the other areas of model theory, it is mainly concerned with finite algebras, or more generally, with finite σ-structures for signatures σ which may contain relation symbols as in the following example:

> The standard signature for graphs is $\sigma_{grph}=\{E\}$, where E is a binary relation symbol.

> A graph is a σ_{grph}-structure satisfying the sentences $\forall u \forall v (uEv \rightarrow vEu)$ and $\forall u \neg (uEu)$..

A σ-homomorphism is a map that commutes with the operations and preserves the relations in σ. This definition gives rise to the usual notion of graph homomorphism, which has the interesting property that a bijective homomorphism need not be invertible. Structures are also a part of universal algebra; after all, some algebraic structures such as ordered groups have a binary relation $<$. What distinguishes finite model theory from universal algebra is its use of more general logical sentences (as in the example above) in place of identities. (In a model-theoretic context an identity $t=t'$ is written as a sentence $\forall u_1 u_2 \ldots u_n (t = t')$.)

The logics employed in finite model theory are often substantially more expressive than first-order logic, the standard logic for model theory of infinite structures.

First-order Logic

Whereas universal algebra provides the semantics for a signature, logic provides the syntax. With terms, identities and quasi-identities, even universal algebra has some limited syntactic tools; first-order logic is the result of making quantification explicit and adding negation into the picture.

A first-order formula is built out of atomic formulas such as $R(f(x,y),z)$ or $y = x + 1$ by means of the Boolean connectives $\neg, \wedge, \vee, \rightarrow$ and prefixing of quantifiers $\forall v$ or $\exists v$. A sentence is a formula in which each occurrence of a variable is in the scope of a corresponding quantifier. Examples for formulas are φ (or $\varphi(x)$ to mark the fact that at most x is an unbound variable in φ) and ψ defined as follows:

$$\varphi = \forall u \forall v (\exists w (x \times w = u \times v) \rightarrow (\exists w (x \times w = u) \vee \exists w (x \times w = v))) \wedge x \neq 0 \wedge x \neq 1,$$

$$\psi = \forall u \forall v ((u \times v = x) \rightarrow (u = x) \vee (v = x)) \wedge x \neq 0 \wedge x \neq 1.$$

(Note that the equality symbol has a double meaning here.) It is intuitively clear how to translate such formulas into mathematical meaning. In the σ_{smr}-structure $\mathcal{N}$ of the natural numbers, for example, an element n satisfies the formula φ if and only if n is a prime number. The formula ψ similarly defines irreducibility. Tarski gave a rigorous definition, sometimes called "Tarski's definition of truth", for the satisfaction relation $\vDash$, so that one easily proves:

$$\mathcal{N} \vDash \varphi(n) \leftrightarrow n \text{ is a prime number.}$$

$$\mathcal{N} \vDash \psi(n) \leftrightarrow n \text{ is irreducible.}$$

A set T of sentences is called a (first-order) theory. A theory is satisfiable if it has a model $\mathcal{M} \vDash T$, , i.e. a structure (of the appropriate signature) which satisfies all the sentences in the set T. Consistency of a theory is usually defined in a syntactical way, but in first-order logic by the completeness theorem there is no need to distinguish between satisfiability and consistency. Therefore, model theorists often use "consistent" as a synonym for "satisfiable".

A theory is called categorical if it determines a structure up to isomorphism, but it turns out that this definition is not useful, due to serious restrictions in the expressivity of first-order logic. The Löwenheim–Skolem theorem implies that for every theory T which has an infinite model and for every infinite cardinal number κ, there is a model $\mathcal{M} \vDash T$ such that the number of elements of is exactly κ. Therefore, only finitary structures can be described by a categorical theory.

Lack of expressivity (when compared to higher logics such as second-order logic) has its advantages, though. For model theorists, the Löwenheim–Skolem theorem is an important practical tool rather than the source of Skolem's paradox. In a certain sense made precise by Lindström's theorem, first-order logic is the most expressive logic for which both the Löwenheim–Skolem theorem and the compactness theorem hold.

As a corollary (i.e., its contrapositive), the compactness theorem says that every unsatisfiable first-order theory has a finite unsatisfiable subset. This theorem is of central importance in infinite model theory, where the words "by compactness" are commonplace. One way to prove it is by means of ultraproducts. An alternative proof uses the completeness theorem, which is otherwise reduced to a marginal role in most of modern model theory.

Axiomatizability, Elimination of Quantifiers, and Model-completeness

The first step, often trivial, for applying the methods of model theory to a class of mathematical objects such as groups, or trees in the sense of graph theory, is to choose a signature σ and represent the objects as σ-structures. The next step is to show that the class is an elementary class, i.e. axiomatizable in first-order logic (i.e. there is a theory T such that a σ-structure is in the class if and only if it satisfies T). E.g. this step fails for the trees, since connectedness cannot be expressed in first-order logic. Axiomatizability ensures that model theory can speak about the right objects. Quantifier elimination can be seen as a condition which ensures that model theory does not say too much about the objects.

A theory T has quantifier elimination if every first-order formula $\varphi(x_1,...,x_n)$ over its signature is equivalent modulo T to a first-order formula $\psi(x_1,...,x_n)$ without quantifiers, i.e. $\forall x_1 ... \forall x_n(\phi(x_1,...,x_n) \leftrightarrow \psi(x_1,...,x_n))$ holds in all models of T. For example, the theory of algebraically closed fields in the signature $\sigma_{ring}=(\times,+,-,0,1)$ has quantifier elimination because every formula is equivalent to a Boolean combination of equations between polynomials.

A substructure of a σ-structure is a subset of its domain, closed under all functions in its signature σ, which is regarded as a σ-structure by restricting all functions and relations in σ to the subset. An embedding of a σ-structure $\mathcal{A}$ into another σ-structure $\mathcal{B}$ is a map f: A → B between the domains which can be written as an isomorphism of $\mathcal{A}$ with a substructure of $\mathcal{B}$. Every embedding is an injective homomorphism, but the converse holds only if the signature contains no relation symbols.

If a theory does not have quantifier elimination, one can add additional symbols to its signature so that it does. Early model theory spent much effort on proving axiomatizability and quantifier elimination results for specific theories, especially in algebra. But often instead of quantifier elimination a weaker property suffices:

A theory T is called model-complete if every substructure of a model of T which is itself a model of T is an elementary substructure. There is a useful criterion for testing whether a substructure is an elementary substructure, called the Tarski–Vaught test. It follows from this criterion that a theory T is model-complete if and only if every first-order formula $\varphi(x_1,...,x_n)$ over its signature is equivalent modulo T to an existential first-order formula, i.e. a formula of the following form:

$$\exists v_1 ... \exists v_m \psi(x_1,...,x_n,v_1,...,v_m),$$

where ψ is quantifier free. A theory that is not model-complete may or may not have a model completion, which is a related model-complete theory that is not, in general, an extension of the original theory. A more general notion is that of model companions.

Categoricity

As observed in the section on first-order logic, first-order theories cannot be categorical, i.e. they cannot describe a unique model up to isomorphism, unless that model is finite. But two famous model-theoretic theorems deal with the weaker notion of κ-categoricity for a cardinal κ. A theory T is called κ-categorical if any two models of T that are of cardinality κ are isomorphic. It turns out that the question of κ-categoricity depends critically on whether κ is bigger than the cardinality of the language (i.e. $\aleph_0 + |\sigma|$, where $|\sigma|$ is the cardinality of the signature). For finite or countable signatures this means that there is a fundamental difference between $\aleph_0$-cardinality and κ-cardinality for uncountable κ.

A few characterizations of $\aleph_0$-categoricity include:

For a complete first-order theory T in a finite or countable signature the following conditions are equivalent:

1. *T* is $\aleph_0$-categorical.

2. For every natural number n, the Stone space $S_n(T)$ is finite.

3. For every natural number n, the number of formulas $\varphi(x_1, ..., x_n)$ in n free variables, up to equivalence modulo T, is finite.

This result, due independently to Engeler, Ryll-Nardzewski and Svenonius, is sometimes referred to as the Ryll-Nardzewski theorem.

Further, $\aleph_0$-categorical theories and their countable models have strong ties with oligomorphic groups. They are often constructed as Fraïssé limits.

Michael Morley's highly non-trivial result that (for countable languages) there is only *one* notion of uncountable categoricity was the starting point for modern model theory, and in particular classification theory and stability theory:

> Morley's categoricity theorem
>
> If a first-order theory T in a finite or countable signature is κ-categorical for some uncountable cardinal κ, then T is κ-categorical for all uncountable cardinals κ.

Uncountably categorical (i.e. κ-categorical for all uncountable cardinals κ) theories are from many points of view the most well-behaved theories. A theory that is both $\aleph_0$-categorical and uncountably categorical is called totally categorical.

Model Theory and Set Theory

Set theory (which is expressed in a countable language), if it is consistent, has a countable model; this is known as Skolem's paradox, since there are sentences in set theory which postulate the existence of uncountable sets and yet these sentences are true in our countable model. Particularly the proof of the independence of the continuum hypothesis requires considering sets in models which appear to be uncountable when viewed from *within* the model, but are countable to someone *outside* the model.

The model-theoretic viewpoint has been useful in set theory; for example in Kurt Gödel's work on the constructible universe, which, along with the method of forcing developed by Paul Cohen can be shown to prove the (again philosophically interesting) independence of the axiom of choice and the continuum hypothesis from the other axioms of set theory.

In the other direction, model theory itself can be formalized within ZFC set theory. The development of the fundamentals of model theory (such as the compactness theorem) rely on the axiom of choice, or more exactly the Boolean prime ideal theorem. Other results in model theory depend on set-theoretic axioms beyond the standard ZFC framework. For example, if the Continuum Hypothesis holds then every countable model has an ultrapower which is saturated (in its own cardinality). Similarly, if the Generalized Continuum Hypothesis holds then every model has a saturated elementary extension. Neither of these results are provable in ZFC alone. Finally, some questions arising from model theory (such as compactness for infinitary logics) have been shown to be equivalent to large cardinal axioms.

Other Basic Notions of Model Theory

Reducts and Expansions

A field or a vector space can be regarded as a (commutative) group by simply ignoring some of its structure. The corresponding notion in model theory is that of a reduct of a structure to a subset of the original signature. The opposite relation is called an *expansion* - e.g. the (additive) group of the rational numbers, regarded as a structure in the signature {+,0} can be expanded to a field with the signature {×,+,1,0} or to an ordered group with the signature {+,0,<}.

Similarly, if σ' is a signature that extends another signature σ, then a complete σ'-theory can be restricted to σ by intersecting the set of its sentences with the set of σ-formulas. Conversely, a complete σ-theory can be regarded as a σ'-theory, and one can extend it (in more than one way) to a complete σ'-theory. The terms reduct and expansion are sometimes applied to this relation as well.

Interpretability

Given a mathematical structure, there are very often associated structures which can be constructed as a quotient of part of the original structure via an equivalence relation. An important example is a quotient group of a group.

One might say that to understand the full structure one must understand these quotients. When the equivalence relation is definable, we can give the previous sentence a precise meaning. We say that these structures are interpretable.

A key fact is that one can translate sentences from the language of the interpreted structures to the language of the original structure. Thus one can show that if a structure M interprets another whose theory is undecidable, then M itself is undecidable.

Using the Compactness and Completeness Theorems

Gödel's completeness theorem says that a theory has a model if and only if it is consistent, i.e. no contradiction is proved by the theory. This is the heart of model theory as it lets us answer questions about theories by looking at models and vice versa. One should not confuse the completeness theorem with the notion of a complete theory. A complete theory is a theory that contains every sentence or its negation. Importantly, one can find a complete consistent theory extending any consistent theory. However, as shown by Gödel's incompleteness theorems only in relatively simple cases will it be possible to have a complete consistent theory that is also recursive, i.e. that can be described by a recursively enumerable set of axioms. In particular, the theory of natural numbers has no recursive complete and consistent theory. Non-recursive theories are of little practical use, since it is undecidable if a proposed axiom is indeed an axiom, making proof-checking a supertask.

The compactness theorem states that a set of sentences S is satisfiable if every finite subset of S is satisfiable. In the context of proof theory the analogous statement is trivial, since every proof can have only a finite number of antecedents used in the proof. In the context of model theory, however, this proof is somewhat more difficult. There are two well known proofs, one by Gödel (which goes via proofs) and one by Malcev (which is more direct and allows us to restrict the cardinality of the resulting model).

Model theory is usually concerned with first-order logic, and many important results (such as the completeness and compactness theorems) fail in second-order logic or other alternatives. In first-order logic all infinite cardinals look the same to a language which is countable. This is expressed in the Löwenheim–Skolem theorems, which state that any countable theory with an infinite model $\mathfrak{A}$ has models of all infinite cardinalities (at least that of the language) which agree with $\mathfrak{A}$ on all sentences, i.e. they are 'elementarily equivalent'.

Types

Fix an L-structure M, and a natural number n. The set of definable subsets of M^n over some parameters A is a Boolean algebra. By Stone's representation theorem for Boolean algebras there is a natural dual notion to this. One can consider this to be the topological space consisting of maximal consistent sets of formulae over A. We call this the space of (complete) n-types over A, and write $S_n(A)$.

Now consider an element $m \in M^n$. Then the set of all formulae ϕ with parameters in A in free variables $x_1,\ldots,x_n$ so that $M \vDash \phi(m)$ is consistent and maximal such. It is called the *type* of m over A.

One can show that for any n-type p, there exists some elementary extension N of M and some $a \in N^n$ so that p is the type of a over A.

Many important properties in model theory can be expressed with types. Further many proofs go via constructing models with elements that contain elements with certain types and then using these elements.

Illustrative Example: Suppose M is an algebraically closed field. The theory has quantifier elimination. This allows us to show that a type is determined exactly by the polynomial equations it contains. Thus the space of n-types over a subfield A is bijective with the set of prime ideals of the polynomial ring $A[x_1,\ldots,x_n]$. This is the same set as the spectrum of $A[x_1,\ldots,x_n]$. Note however that the topology considered on the type space is the constructible topology: a set of types is basic open iff it is of the form $\{p : f(x) = 0 \in p\}$ or of the form $\{p : f(x) \neq 0 \in p\}$. This is finer than the Zariski topology.

History

Model theory as a subject has existed since approximately the middle of the 20th century. However some earlier research, especially in mathematical logic, is often regarded as being of a model-theoretical nature in retrospect. The first significant result in what is now model theory was a special case of the downward Löwenheim–Skolem theorem, published by Leopold Löwenheim in 1915.

The compactness theorem was implicit in work by Thoralf Skolem, but it was first published in 1930, as a lemma in Kurt Gödel's proof of his completeness theorem. The Löwenheim–Skolem theorem and the compactness theorem received their respective general forms in 1936 and 1941 from Anatoly Maltsev.

Modern scientific breakthroughs in the development of model theory can be traced to the Polish scientist Alfred Tarski, a member of the Lwów–Warsaw school during the interbellum. Tarski's work included logical consequence, deductive systems, the algebra of logic, the theory of definability, and the semantic definition of truth, among other topics. His semantic methods culminated in the model theory he and a number of his Berkeley students developed in the 1950s and 60s. These modern concepts of model theory profoundly influenced Hilbert's program and modern mathematics.

Mathematical Model

A mathematical model is a description of a system using mathematical concepts and language. The process of developing a mathematical model is termed mathematical modeling. Mathematical models are used in the natural sciences (such as physics, biology, earth science, meteorology) and engineering disciplines (such as computer science, artificial intelligence), as well as in the social sciences (such as economics, psychology, sociology, political science). Physicists, engineers, statisticians, operations research analysts, and economists use mathematical models most extensively. A model may help to explain a system and to study the effects of different components, and to make predictions about behaviour.

Elements of a Mathematical Model

Mathematical models can take many forms, including dynamical systems, statistical models, differential equations, or game theoretic models. These and other types of models can overlap, with a given model involving a variety of abstract structures. In general, mathematical models may include logical models. In many cases, the quality of a scientific field depends on how well the mathematical models developed on the theoretical side agree with results of repeatable experiments. Lack of agreement between theoretical mathematical models and experimental measurements often leads to important advances as better theories are developed.

The traditional mathematical model contains four major elements. These are

1. Governing equations
2. Constitutive equations
3. Constraints
4. Kinematic equations

Classifications

Mathematical models are usually composed of relationships and *variables*. Relationships can be described by *operators*, such as algebraic operators, functions, differential operators, etc. Variables are abstractions of system parameters of interest, that can be quantified. Several classification criteria can be used for mathematical models according to their structure:

- Linear vs. nonlinear: If all the operators in a mathematical model exhibit linearity, the resulting mathematical model is defined as linear. A model is considered to be nonlinear otherwise. The definition of linearity and nonlinearity is dependent on context, and linear models may have nonlinear expressions in them. For example, in a statistical linear model, it is assumed that a relationship is linear in the parameters, but it may be nonlinear in the predictor variables. Similarly, a differential equation is said to be linear if it can be written with linear differential operators, but it can still have non-linear expressions in it. In a mathematical programming model, if the objective functions and constraints are represented entirely by linear equations, then the model is regarded as a linear model. If one or more of the objective functions or constraints are represented with a nonlinear equation, then the model is known as a nonlinear model. Nonlinearity, even in fairly simple systems, is often associated with phenomena such as chaos and irreversibility. Although there are exceptions, nonlinear systems and models tend to be more difficult to study than linear ones. A common approach to nonlinear problems is linearization, but this can be problematic if one is trying to study aspects such as irreversibility, which are strongly tied to nonlinearity.

- Static vs. dynamic: A *dynamic* model accounts for time-dependent changes in the state of the system, while a *static* (or steady-state) model calculates the system in equilibrium, and thus is time-invariant. Dynamic models typically are represented by differential equations.

- Explicit vs. implicit: If all of the input parameters of the overall model are known, and the output parameters can be calculated by a finite series of computations (known as linear programming), the model is said to be *explicit*. But sometimes it is the *output* parameters which are known, and the corresponding inputs must be solved for by an iterative procedure, such as Newton's method (if the model is linear) or Broyden's method (if non-linear). For example, a jet engine's physical properties such as turbine and nozzle throat areas can be explicitly calculated given a design thermodynamic cycle (air and fuel flow rates, pressures, and temperatures) at a specific flight condition and power setting, but the engine's operating cycles at other flight conditions and power settings cannot be explicitly calculated from the constant physical properties.

- Discrete vs. continuous: A discrete model treats objects as discrete, such as the particles in a molecular model or the states in a statistical model; while a continuous model represents the objects in a continuous manner, such as the velocity field of fluid in pipe flows, temperatures and stresses in a solid, and electric field that applies continuously over the entire model due to a point charge.

- Deterministic vs. probabilistic (stochastic): A deterministic model is one in which every set of variable states is uniquely determined by parameters in the model and by sets of previous states of these variables; therefore, a deterministic model always performs the same way for a given set of initial conditions. Conversely, in a stochastic model—usually called a "statistical model"—randomness is present, and variable states are not described by unique values, but rather by probability distributions.

- Deductive, inductive, or floating: A deductive model is a logical structure based on a theory. An inductive model arises from empirical findings and generalization from them. The floating model rests on neither theory nor observation, but is merely the invocation of expected structure. Application of mathematics in social sciences outside of economics has been criticized for unfounded models. Application of catastrophe theory in science has been characterized as a floating model.

Significance in the Natural Sciences

Mathematical models are of great importance in the natural sciences, particularly in physics. Physical theories are almost invariably expressed using mathematical models.

Throughout history, more and more accurate mathematical models have been developed. Newton's laws accurately describe many everyday phenomena, but at certain limits relativity theory and quantum mechanics must be used; even these do not apply to all situations and need further refinement. It is possible to obtain the less accurate models in appropriate limits, for example relativistic mechanics reduces to Newtonian mechanics at speeds much less than the speed of light. Quantum mechanics reduces to classical physics when the quantum numbers are high. For example, the de Broglie wavelength of a tennis ball is insignificantly small, so classical physics is a good approximation to use in this case.

It is common to use idealized models in physics to simplify things. Massless ropes, point particles, ideal gases and the particle in a box are among the many simplified models used in physics. The laws of physics are represented with simple equations such as Newton's laws, Maxwell's equations and the Schrödinger equation. These laws are such as a basis for making mathematical models of real situations. Many real situations are very complex and thus modeled approximate on a computer, a model that is computationally feasible to compute is made from the basic laws or from approximate models made from the basic laws. For example, molecules can be modeled by molecular orbital models that are approximate solutions to the Schrödinger equation. In engineering, physics models are often made by mathematical methods such as finite element analysis.

Different mathematical models use different geometries that are not necessarily accurate descriptions of the geometry of the universe. Euclidean geometry is much used in classical physics, while special relativity and general relativity are examples of theories that use geometries which are not Euclidean.

Some Applications

Since prehistorical times simple models such as maps and diagrams have been used.

Often when engineers analyze a system to be controlled or optimized, they use a mathematical model. In analysis, engineers can build a descriptive model of the system as a hypothesis of how the system could work, or try to estimate how an unforeseeable event could affect the system. Similarly, in control of a system, engineers can try out different control approaches in simulations.

A mathematical model usually describes a system by a set of variables and a set of equations that establish relationships between the variables. Variables may be of many types; real or integer numbers, boolean values or strings, for example. The variables represent some properties of the system, for example, measured system outputs often in the form of signals, timing data, counters, and event occurrence (yes/no). The actual model is the set of functions that describe the relations between the different variables.

Building Blocks

In business and engineering, mathematical models may be used to maximize a certain output. The system under consideration will require certain inputs. The system relating inputs to outputs depends on other variables too: decision variables, state variables, exogenous variables, and random variables.

Decision variables are sometimes known as independent variables. Exogenous variables are sometimes known as parameters or constants. The variables are not independent of each other as the state variables are dependent on the decision, input, random, and exogenous variables. Furthermore, the output variables are dependent on the state of the system (represented by the state variables).

Objectives and constraints of the system and its users can be represented as functions of the output variables or state variables. The objective functions will depend on the perspective of the model's user. Depending on the context, an objective function is also known as an *index of performance*, as it is some measure of interest to the user. Although there is no limit to the number of objective functions and constraints a model can have, using or optimizing the model becomes more involved (computationally) as the number increases.

For example, in economics students often apply linear algebra when using input-output models. Complicated mathematical models that have many variables may be consolidated by use of vectors where one symbol represents several variables.

A Priori Information

Mathematical modeling problems are often classified into black box or white box models, according to how much a priori information on the system is available. A black-box model is a system of which there is no a priori information available. A white-box model (also called glass box or clear box) is a system where all necessary information is available. Practically all systems are somewhere between the black-box and white-box models, so this concept is useful only as an intuitive guide for deciding which approach to take.

Usually it is preferable to use as much a priori information as possible to make the model more accurate. Therefore, the white-box models are usually considered easier, because if you have used the information correctly, then the model will behave correctly. Often the a priori information comes in forms of knowing the type of functions relating different variables. For example, if we make a model of how a medicine works in a human system, we know that usually the amount of medicine in the blood is an exponentially decaying function. But we are still left with several unknown parameters; how rapidly does the medicine amount decay, and what is the initial amount of medicine in blood? This example is therefore not a completely white-box model. These parameters have to be estimated through some means before one can use the model.

In black-box models one tries to estimate both the functional form of relations between variables and the numerical parameters in those functions. Using a priori information we could end up, for example, with a set of functions that probably could describe the system adequately. If there is no a priori information we would try to use functions as general as possible to cover all different models. An often used approach for black-box models are neural networks which usually do not make assumptions about incoming data. Alternatively the NARMAX (Nonlinear AutoRegressive Moving Average model with eXogenous inputs) algorithms which were developed as part of non-linear system identification can be used to select the model terms, determine the model structure, and estimate the unknown parameters in the presence of correlated and nonlinear noise. The advantage of NARMAX models compared to neural networks is that NARMAX produces models that can be written down and related to the underlying process, whereas neural networks produce an approximation that is opaque.

Subjective Information

Sometimes it is useful to incorporate subjective information into a mathematical model. This can be done based on intuition, experience, or expert opinion, or based on convenience of mathematical form. Bayesian statistics provides a theoretical framework for incorporating such subjectivity into a rigorous analysis: we specify a prior probability distribution (which can be subjective), and then update this distribution based on empirical data.

An example of when such approach would be necessary is a situation in which an experimenter bends a coin slightly and tosses it once, recording whether it comes up heads, and is then given the task of predicting the probability that the next flip comes up heads. After bending the coin, the true probability that the coin will come up heads is unknown; so the experimenter would need to make a decision (perhaps by looking at the shape of the coin) about what prior distribution to use. Incorporation of such subjective information might be important to get an accurate estimate of the probability.

Complexity

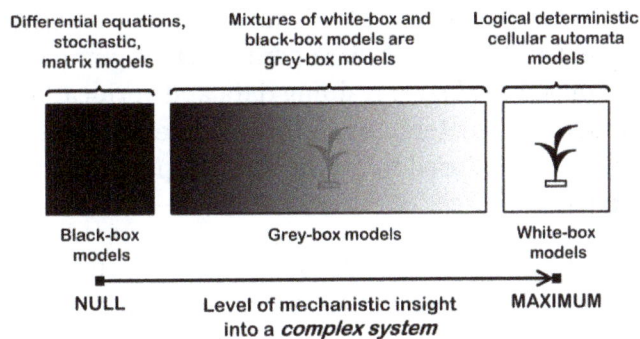

This is a schematic representation of three types of mathematical models of complex systems with the level of their mechanistic understanding.

In general, model complexity involves a trade-off between simplicity and accuracy of the model. Occam's razor is a principle particularly relevant to modeling; the essential idea being that among models with roughly equal predictive power, the simplest one is the most desirable. While added complexity usually improves the realism of a model, it can make the model difficult to understand

and analyze, and can also pose computational problems, including numerical instability. Thomas Kuhn argues that as science progresses, explanations tend to become more complex before a paradigm shift offers radical simplification.

For example, when modeling the flight of an aircraft, we could embed each mechanical part of the aircraft into our model and would thus acquire an almost white-box model of the system. However, the computational cost of adding such a huge amount of detail would effectively inhibit the usage of such a model. Additionally, the uncertainty would increase due to an overly complex system, because each separate part induces some amount of variance into the model. It is therefore usually appropriate to make some approximations to reduce the model to a sensible size. Engineers often can accept some approximations in order to get a more robust and simple model. For example, Newton's classical mechanics is an approximated model of the real world. Still, Newton's model is quite sufficient for most ordinary-life situations, that is, as long as particle speeds are well below the speed of light, and we study macro-particles only.

Training

Any model which is not pure white-box contains some parameters that can be used to fit the model to the system it is intended to describe. If the modeling is done by a neural network, the optimization of parameters is called *training*. In more conventional modeling through explicitly given mathematical functions, parameters are determined by curve fitting.

Model Evaluation

A crucial part of the modeling process is the evaluation of whether or not a given mathematical model describes a system accurately. This question can be difficult to answer as it involves several different types of evaluation.

Fit to Empirical Data

Usually the easiest part of model evaluation is checking whether a model fits experimental measurements or other empirical data. In models with parameters, a common approach to test this fit is to split the data into two disjoint subsets: training data and verification data. The training data are used to estimate the model parameters. An accurate model will closely match the verification data even though these data were not used to set the model's parameters. This practice is referred to as cross-validation in statistics.

Defining a metric to measure distances between observed and predicted data is a useful tool of assessing model fit. In statistics, decision theory, and some economic models, a loss function plays a similar role.

While it is rather straightforward to test the appropriateness of parameters, it can be more difficult to test the validity of the general mathematical form of a model. In general, more mathematical tools have been developed to test the fit of statistical models than models involving differential equations. Tools from non-parametric statistics can sometimes be used to evaluate how well the data fit a known distribution or to come up with a general model that makes only minimal assumptions about the model's mathematical form.

Scope of the Model

Assessing the scope of a model, that is, determining what situations the model is applicable to, can be less straightforward. If the model was constructed based on a set of data, one must determine for which systems or situations the known data is a "typical" set of data.

The question of whether the model describes well the properties of the system between data points is called interpolation, and the same question for events or data points outside the observed data is called extrapolation.

As an example of the typical limitations of the scope of a model, in evaluating Newtonian classical mechanics, we can note that Newton made his measurements without advanced equipment, so he could not measure properties of particles travelling at speeds close to the speed of light. Likewise, he did not measure the movements of molecules and other small particles, but macro particles only. It is then not surprising that his model does not extrapolate well into these domains, even though his model is quite sufficient for ordinary life physics.

Philosophical Considerations

Many types of modeling implicitly involve claims about causality. This is usually (but not always) true of models involving differential equations. As the purpose of modeling is to increase our understanding of the world, the validity of a model rests not only on its fit to empirical observations, but also on its ability to extrapolate to situations or data beyond those originally described in the model. One can think of this as the differentiation between qualitative and quantitative predictions. One can also argue that a model is worthless unless it provides some insight which goes beyond what is already known from direct investigation of the phenomenon being studied.

An example of such criticism is the argument that the mathematical models of Optimal foraging theory do not offer insight that goes beyond the common-sense conclusions of evolution and other basic principles of ecology.

Examples

- One of the popular examples in computer science is the mathematical models of various machines, an example is the deterministic finite automaton which is defined as an abstract mathematical concept, but due to the deterministic nature of a DFA, it is implementable in hardware and software for solving various specific problems. For example, the following is a DFA M with a binary alphabet, which requires that the input contains an even number of 0s.

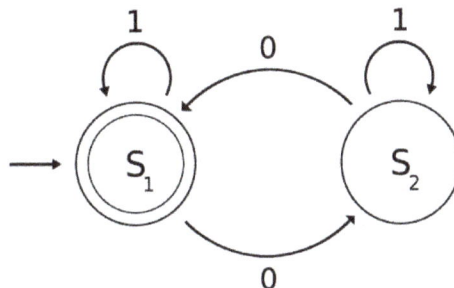

The state diagram for M

$M = (Q, \Sigma, \delta, q_0, F)$ where

- $Q = \{S_1, S_2\}$,

- $\Sigma = \{0, 1\}$,

- $q_0 = S_1$,

- $F = \{S_1\}$, and

- δ is defined by the following state transition table:

	0	1
S_1	S_2	S_1
S_2	S_1	S_2

The state S_1 represents that there has been an even number of 0s in the input so far, while S_2 signifies an odd number. A 1 in the input does not change the state of the automaton. When the input ends, the state will show whether the input contained an even number of 0s or not. If the input did contain an even number of 0s, M will finish in state S_1, an accepting state, so the input string will be accepted.

The language recognized by M is the regular language given by the regular expression 1*(0 (1*) 0 (1*))*, where "*" is the Kleene star, e.g., 1* denotes any non-negative number (possibly zero) of symbols "1".

- Many everyday activities carried out without a thought are uses of mathematical models. A geographical map projection of a region of the earth onto a small, plane surface is a model which can be used for many purposes such as planning travel.

- Another simple activity is predicting the position of a vehicle from its initial position, direction and speed of travel, using the equation that distance traveled is the product of time and speed. This is known as dead reckoning when used more formally. Mathematical modeling in this way does not necessarily require formal mathematics; animals have been shown to use dead reckoning.

- *Population Growth*. A simple (though approximate) model of population growth is the Malthusian growth model. A slightly more realistic and largely used population growth model is the logistic function, and its extensions.

- *Individual-based cellular automata models of population growth*

- *Model of a particle in a potential-field*. In this model we consider a particle as being a point of mass which describes a trajectory in space which is modeled by a function giving its coordinates in space as a function of time. The potential field is given by a function $V: \mathbb{R}^3 \to \mathbb{R}$ and the trajectory, that is a function $\mathbf{r}: \mathbb{R} \to \mathbb{R}^3$, is the solution of the differential equation:

$$-\frac{d^2\mathbf{r}(t)}{dt^2} m = \frac{\partial V[\mathbf{r}(t)]}{\partial x}\hat{\mathbf{x}} + \frac{\partial V[\mathbf{r}(t)]}{\partial y}\hat{\mathbf{y}} + \frac{\partial V[\mathbf{r}(t)]}{\partial z}\hat{\mathbf{z}},$$

that can be written also as:

$$m\frac{\mathrm{d}^2\mathbf{r}(t)}{\mathrm{d}t^2} = -\nabla V[\mathbf{r}(t)].$$

Note this model assumes the particle is a point mass, which is certainly known to be false in many cases in which we use this model; for example, as a model of planetary motion.

- *Model of rational behavior for a consumer.* In this model we assume a consumer faces a choice of n commodities labeled 1,2,...,n each with a market price $p_1, p_2,..., p_n$. The consumer is assumed to have a *cardinal* utility function U (cardinal in the sense that it assigns numerical values to utilities), depending on the amounts of commodities $x_1, x_2,..., x_n$ consumed. The model further assumes that the consumer has a budget M which is used to purchase a vector $x_1, x_2,..., x_n$ in such a way as to maximize $U(x_1, x_2,..., x_n)$. The problem of rational behavior in this model then becomes an optimization problem, that is:

$$\max U(x_1, x_2,\ldots,x_n)$$

subject to:

$$\sum_{i=1}^{n} p_i x_i \leq M.$$

$$x_i \geq 0 \quad \forall i \in \{1, 2,\ldots,n\}$$

This model has been used in general equilibrium theory, particularly to show existence and Pareto efficiency of economic equilibria. However, the fact that this particular formulation assigns *numerical values* to levels of satisfaction is the source of criticism (and even ridicule). However, it is not an essential ingredient of the theory and again this is an idealization.

- *Neighbour-sensing model* explains the mushroom formation from the initially chaotic fungal network.

- *Computer science*: models in Computer Networks, data models, surface model,...

- *Mechanics*: movement of rocket model,...

Modeling requires selecting and identifying relevant aspects of a situation in the real world.

Statistical Model

A statistical model is a class of mathematical model, which embodies a set of assumptions concerning the generation of some sample data, and similar data from a larger population. A statistical model represents, often in considerably idealized form, the data-generating process.

The assumptions embodied by a statistical model describe a set of probability distributions, some of which are assumed to adequately approximate the distribution from which a particular data set is sampled. The probability distributions inherent in statistical models are what distinguishes statistical models from other, non-statistical, mathematical models.

A statistical model is usually specified by mathematical equations that relate one or more random variables and possibly other non-random variables. As such, "a model is a formal representation of a theory" (Herman Adèr quoting Kenneth Bollen).

All statistical hypothesis tests and all statistical estimators are derived from statistical models. More generally, statistical models are part of the foundation of statistical inference.

Formal Definition

In mathematical terms, a statistical model is usually thought of as a pair $(S, \mathcal{P})$, where S is the set of possible observations, i.e. the sample space, and $\mathcal{P}$ is a set of probability distributions on S.

The intuition behind this definition is as follows. It is assumed that there is a "true" probability distribution induced by the process that generates the observed data. We choose $\mathcal{P}$ to represent a set (of distributions) which contains a distribution that adequately approximates the true distribution. Note that we do not require that $\mathcal{P}$ contains the true distribution, and in practice that is rarely the case. Indeed, as Burnham & Anderson state, "A model is a simplification or approximation of reality and hence will not reflect all of reality"—whence the saying "all models are wrong".

The set $\mathcal{P}$ is almost always parameterized: $\mathcal{P} = \{P_\theta : \theta \in \Theta\}$. The set Θ defines the *parameters* of the model. A parameterization is generally required to have distinct parameter values give rise to distinct distributions, i.e. $P_{\theta_1} = P_{\theta_2} \Rightarrow \theta_1 = \theta_2$ must hold (in other words, it must be injective). A parameterization that meets the condition is said to be *identifiable*.

An Example

Height and age are each probabilistically distributed over humans. They are stochastically related: when we know that a person is of age 10, this influences the chance of the person being 5 feet tall. We could formalize that relationship in a linear regression model with the following form: $\text{height}_i = b_0 + b_1\text{age}_i + \varepsilon_i$, where b_0 is the intercept, b_1 is a parameter that age is multiplied by to get a prediction of height, ε is the error term, and i identifies the person. This implies that height is predicted by age, with some error.

An admissible model must be consistent with all the data points. Thus, the straight line ($\text{height}_i = b_0 + b_1\text{age}_i$) is *not* a model of the data. The line cannot be a model, unless it exactly fits all the data points—i.e. all the data points lie perfectly on a straight line. The error term, ε_i, must be included in the model, so that the model is consistent with all the data points.

To do statistical inference, we would first need to assume some probability distributions for the ε_i. For instance, we might assume that the ε_i distributions are i.i.d. Gaussian, with zero mean. In this instance, the model would have 3 parameters: b_0, b_1, and the variance of the Gaussian distribution.

We can formally specify the model in the form $(S, \mathcal{P})$ as follows. The sample space, S, of our model comprises the set of all possible pairs (age, height). Each possible value of $\theta = (b_0, b_1, \sigma^2)$

determines a distribution on S; denote that distribution by P_θ. If Θ is the set of all possible values of θ, then $\mathcal{P} = \{P_\theta : \theta \in \Theta\}$. (The parameterization is identifiable, and this is easy to check.)

In this example, the model is determined by (1) specifying S and (2) making some assumptions relevant to $\mathcal{P}$. There are two assumptions: that height can be approximated by a linear function of age; that errors in the approximation are distributed as i.i.d. Gaussian. The assumptions are sufficient to specify $\mathcal{P}$ —as they are required to do.

General Remarks

A statistical model is a special class of mathematical model. What distinguishes a statistical model from other mathematical models is that a statistical model is non-deterministic. Thus, in a statistical model specified via mathematical equations, some of the variables do not have specific values, but instead have probability distributions; i.e. some of the variables are stochastic. In the example above, ε is a stochastic variable; without that variable, the model would be deterministic.

Statistical models are often used even when the physical process being modeled is deterministic. For instance, coin tossing is, in principle, a deterministic process; yet it is commonly modeled as stochastic (via a Bernoulli process).

There are three purposes for a statistical model, according to Konishi & Kitagawa.

- Predictions

- Extraction of information

- Description of stochastic structures

Dimension of a Model

Suppose that we have a statistical model $(S, \mathcal{P})$ with $\mathcal{P} = \{P_\theta : \theta \in \Theta\}$. The model is said to be parametric if Θ has a finite dimension. In notation, we write that $\Theta \subseteq \mathbb{R}^d$ where d is a positive integer ($\mathbb{R}$ denotes the real numbers; other sets can be used, in principle). Here, d is called the dimension of the model.

As an example, if we assume that data arise from a univariate Gaussian distribution, then we are assuming that

$$\mathcal{P} = \{P_{\mu,\sigma}(x) \equiv \frac{1}{\sqrt{2\pi}\sigma} \exp\left(-\frac{(x-\mu)^2}{2\sigma^2}\right) : \mu \in \mathbb{R}, \sigma > 0\}.$$

In this example, the dimension, d, equals 2.

As another example, suppose that the data consists of points (x, y) that we assume are distributed according to a straight line with i.i.d. Gaussian residuals (with zero mean). Then the dimension of the statistical model is 3: the intercept of the line, the slope of the line, and the variance of the distribution of the residuals. (Note that in geometry, a straight line has dimension 1.)

A statistical model is nonparametric if the parameter set Θ is infinite dimensional. A statistical model is semiparametric if it has both finite-dimensional and infinite-dimensional parameters. Formally, if d is the dimension of $\rightarrow \infty$ and n is the number of samples, both semiparametric and nonparametric models have $d \rightarrow \infty$ as $n \rightarrow \infty$. If $d/n \rightarrow 0$ as $n \rightarrow \infty$, then the model is semiparametric; otherwise, the model is nonparametric.

Parametric models are by far the most commonly used statistical models. Regarding semiparametric and nonparametric models, Sir David Cox has said, "These typically involve fewer assumptions of structure and distributional form but usually contain strong assumptions about independencies".

Nested Models

Two statistical models are nested if the first model can be transformed into the second model by imposing constraints on the parameters of the first model. For example, the set of all Gaussian distributions has, nested within it, the set of zero-mean Gaussian distributions: we constrain the mean in the set of all Gaussian distributions to get the zero-mean distributions.

In that example, the first model has a higher dimension than the second model (the zero-mean model has dimension 1). Such is usually, but not always, the case. As a different example, the set of positive-mean Gaussian distributions, which has dimension 2, is nested within the set of all Gaussian distributions.

Comparing Models

It is assumed that there is a "true" probability distribution underlying the observed data, induced by the process that generated the data. The main goal of model selection is to make statements about which elements of $\mathcal{P}$ are most likely to adequately approximate the true distribution.

Models can be compared to each other by exploratory data analysis or confirmatory data analysis. In exploratory analysis, a variety of models are formulated and an assessment is performed of how well each one describes the data. In confirmatory analysis, a previously formulated model or models are compared to the data. Common criteria for comparing models include R^2, Bayes factor, and the likelihood-ratio test together with its generalization relative likelihood.

Konishi & Kitagawa state: "The majority of the problems in statistical inference can be considered to be problems related to statistical modeling. They are typically formulated as comparisons of several statistical models." Relatedly, Sir David Cox has said, "How [the] translation from subject-matter problem to statistical model is done is often the most critical part of an analysis".

References

- Kolmogoroff (1933). Grundbegriffe der Wahrscheinlichkeitsrechnung. doi:10.1007/978-3-642-49888-6. ISBN 978-3-642-49888-6.

- Olav Kallenberg; Foundations of Modern Probability, 2nd ed. Springer Series in Statistics. (2002). 650 pp. ISBN 0-387-95313-2

- Birkhoff, Garrett (1940). Lattice Theory. 25 (3rd Revised ed.). American Mathematical Society. ISBN 978-0-8218-1025-5.

- Davey, B. A.; Priestley, H. A. (2002). Introduction to Lattices and Order (2nd ed.). Cambridge University Press. ISBN 0-521-78451-4.

- Gierz, G.; Hofmann, K. H.; Keimel, K.; Mislove, M.; Scott, D. S. (2003). Continuous Lattices and Domains. Encyclopedia of Mathematics and its Applications. 93. Cambridge University Press. ISBN 978-0-521-80338-0.

- "The Unreasonable Effectiveness of Number Theory", Stefan Andrus Burr, George E. Andrews, American Mathematical Soc., 1992, ISBN 9780821855010

- "Applications of number theory to numerical analysis", Lo-keng Hua, Luogeng Hua, Yuan Wang, Springer-Verlag, 1981, ISBN 978-3-540-10382-0

- Bell, John L. (1999). The Art of the Intelligible: An Elementary Survey of Mathematics in its Conceptual Development. Kluwer. ISBN 0-7923-5972-0.

- Euclid (1956). Translated by Johan Ludvig Heiberg with an introduction and commentary by Thomas L. Heath, ed. The Elements (3 vols.). Vol. 1 (Books I and II) (Reprint of 1908 ed.). Dover. ISBN 0-486-60088-2.

- Heath, Sir Thomas (1921). "The 'Theorem of Pythagoras'". A History of Greek Mathematics (2 Vols.) (Dover Publications, Inc. (1981) ed.). Clarendon Press, Oxford. p. 144 ff. ISBN 0-486-24073-8.

- Libeskind, Shlomo (2008). Euclidean and transformational geometry: a deductive inquiry. Jones & Bartlett Learning. ISBN 0-7637-4366-6.

- Loomis, Elisha Scott (1968). The Pythagorean proposition (2nd ed.). The National Council of Teachers of Mathematics. ISBN 978-0-87353-036-1.

- Maor, Eli (2007). The Pythagorean Theorem: A 4,000-Year History. Princeton, New Jersey: Princeton University Press. ISBN 978-0-691-12526-8.

- Bolzano, Bernard (1975), Berg, Jan, ed., Einleitung zur Größenlehre und erste Begriffe der allgemeinen Größenlehre, Bernard-Bolzano-Gesamtausgabe, edited by Eduard Winter et al., Vol. II, A, 7, Stuttgart, Bad Cannstatt: Friedrich Frommann Verlag, p. 152, ISBN 3-7728-0466-7

- Jech, Thomas (2003), Set Theory, Springer Monographs in Mathematics (Third Millennium ed.), Berlin, New York: Springer-Verlag, p. 642, ISBN 978-3-540-44085-7, Zbl 1007.03002

- Chang, Chen Chung; Keisler, H. Jerome (1990) [1973]. Model Theory. Studies in Logic and the Foundations of Mathematics (3rd ed.). Elsevier. ISBN 978-0-444-88054-3.

- Billings S.A. (2013), Nonlinear System Identification: NARMAX Methods in the Time, Frequency, and Spatio-Temporal Domains, Wiley.

6

Significant Approaches of Mathematics

These topics depend on mathematical logic and have become disciplines in their own right. Widely applicable, these branches of mathematics contribute greatly to mathematical knowledge. They also help to better understand and learn mathematics. This chapter provides a plethora of topics for better comprehension of mathematics.

Statistics

Statistics is the study of the collection, analysis, interpretation, presentation, and organization of data. In applying statistics to, e.g., a scientific, industrial, or social problem, it is conventional to begin with a statistical population or a statistical model process to be studied. Populations can be diverse topics such as "all people living in a country" or "every atom composing a crystal". Statistics deals with all aspects of data including the planning of data collection in terms of the design of surveys and experiments.

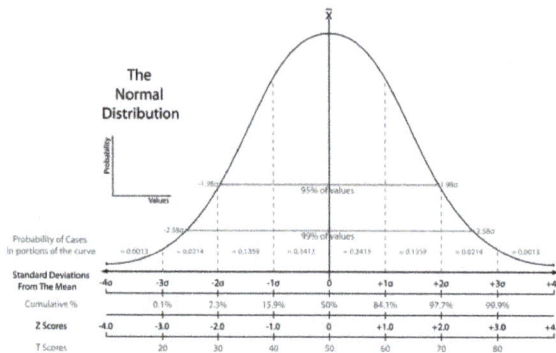

More probability density is found as one gets closer to the expected (mean) value in a normal distribution. Statistics used in standardized testing assessment are shown. The scales include *standard deviations, cumulative percentages, percentile equivalents, Z-scores, T-scores, standard nines,* and *percentages in standard nines.*

Some popular definitions are:

- Merriam-Webster dictionary defines statistics as "classified facts representing the conditions of a people in a state – especially the facts that can be stated in numbers or any other tabular or classified arrangement".

- Statistician Sir Arthur Lyon Bowley defines statistics as "Numerical statements of facts in any department of inquiry placed in relation to each other".

When census data cannot be collected, statisticians collect data by developing specific experiment designs and survey samples. Representative sampling assures that inferences and conclusions can safely extend from the sample to the population as a whole. An experimental study involves taking measurements of the system under study, manipulating the system, and then taking additional measurements using the same procedure to determine if the manipulation has modified the values of the measurements. In contrast, an observational study does not involve experimental manipulation.

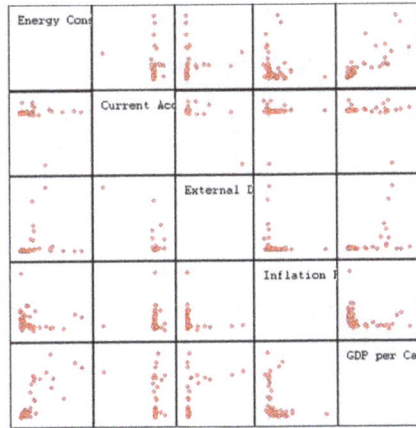

Scatter plots are used in descriptive statistics to show the observed relationships between different variables.

Two main statistical methodologies are used in data analysis: descriptive statistics, which summarizes data from a sample using indexes such as the mean or standard deviation, and inferential statistics, which draws conclusions from data that are subject to random variation (e.g., observational errors, sampling variation). Descriptive statistics are most often concerned with two sets of properties of a *distribution* (sample or population): *central tendency* (or *location*) seeks to characterize the distribution's central or typical value, while *dispersion* (or *variability*) characterizes the extent to which members of the distribution depart from its center and each other. Inferences on mathematical statistics are made under the framework of probability theory, which deals with the analysis of random phenomena.

A standard statistical procedure involves the test of the relationship between two statistical data sets, or a data set and a synthetic data drawn from idealized model. An hypothesis is proposed for the statistical relationship between the two data sets, and this is compared as an alternative to an idealized null hypothesis of no relationship between two data sets. Rejecting or disproving the null hypothesis is done using statistical tests that quantify the sense in which the null can be proven false, given the data that are used in the test. Working from a null hypothesis, two basic forms of error are recognized: Type I errors (null hypothesis is falsely rejected giving a "false positive") and Type II errors (null hypothesis fails to be rejected and an actual difference between populations is missed giving a "false negative"). Multiple problems have come to be associated with this framework: ranging from obtaining a sufficient sample size to specifying an adequate null hypothesis.

Measurement processes that generate statistical data are also subject to error. Many of these errors are classified as random (noise) or systematic (bias), but other types of errors (e.g., blunder, such as when an analyst reports incorrect units) can also be important. The presence of missing data and/or censoring may result in biased estimates and specific techniques have been developed to address these problems.

Statistics can be said to have begun in ancient civilization, going back at least to the 5th century BC, but it was not until the 18th century that it started to draw more heavily from calculus and probability theory. Statistics continues to be an area of active research, for example on the problem of how to analyze Big data.

Scope

Statistics is a mathematical body of science that pertains to the collection, analysis, interpretation or explanation, and presentation of data, or as a branch of mathematics. Some consider statistics to be a distinct mathematical science rather than a branch of mathematics. While many scientific investigations make use of data, statistics is concerned with the use of data in the context of uncertainty and decision making in the face of uncertainty.

Mathematical Statistics

Mathematical statistics is the application of mathematics to statistics, which was originally conceived as the science of the state — the collection and analysis of facts about a country: its economy, land, military, population, and so forth. Mathematical techniques used for this include mathematical analysis, linear algebra, stochastic analysis, differential equations, and measure-theoretic probability theory.

Overview

In applying statistics to a problem, it is common practice to start with a population or process to be studied. Populations can be diverse topics such as "all persons living in a country" or "every atom composing a crystal".

Ideally, statisticians compile data about the entire population (an operation called census). This may be organized by governmental statistical institutes. Descriptive statistics can be used to summarize the population data. Numerical descriptors include mean and standard deviation for continuous data types (like income), while frequency and percentage are more useful in terms of describing categorical data (like race).

When a census is not feasible, a chosen subset of the population called a sample is studied. Once a sample that is representative of the population is determined, data is collected for the sample members in an observational or experimental setting. Again, descriptive statistics can be used to summarize the sample data. However, the drawing of the sample has been subject to an element of randomness, hence the established numerical descriptors from the sample are also due to uncertainty. To still draw meaningful conclusions about the entire population, inferential statistics is needed. It uses patterns in the sample data to draw inferences about the population represented, accounting for randomness. These inferences may take the form of: answering yes/no questions about the data (hypothesis testing), estimating numerical characteristics of the data (estimation), describing associations within the data (correlation) and modeling relationships within the data (for example, using regression analysis). Inference can extend to forecasting, prediction and estimation of unobserved values either in or associated with the population being studied; it can include extrapolation and interpolation of time series or spatial data, and can also include data mining.

Data Collection

Sampling

When full census data cannot be collected, statisticians collect sample data by developing specific experiment designs and survey samples. Statistics itself also provides tools for prediction and forecasting the use of data through statistical models. To use a sample as a guide to an entire population, it is important that it truly represents the overall population. Representative sampling assures that inferences and conclusions can safely extend from the sample to the population as a whole. A major problem lies in determining the extent that the sample chosen is actually representative. Statistics offers methods to estimate and correct for any bias within the sample and data collection procedures. There are also methods of experimental design for experiments that can lessen these issues at the outset of a study, strengthening its capability to discern truths about the population.

Sampling theory is part of the mathematical discipline of probability theory. Probability is used in mathematical statistics to study the sampling distributions of sample statistics and, more generally, the properties of statistical procedures. The use of any statistical method is valid when the system or population under consideration satisfies the assumptions of the method. The difference in point of view between classic probability theory and sampling theory is, roughly, that probability theory starts from the given parameters of a total population to deduce probabilities that pertain to samples. Statistical inference, however, moves in the opposite direction—inductively inferring from samples to the parameters of a larger or total population.

Experimental and Observational Studies

A common goal for a statistical research project is to investigate causality, and in particular to draw a conclusion on the effect of changes in the values of predictors or independent variables on dependent variables. There are two major types of causal statistical studies: experimental studies and observational studies. In both types of studies, the effect of differences of an independent variable (or variables) on the behavior of the dependent variable are observed. The difference between the two types lies in how the study is actually conducted. Each can be very effective. An experimental study involves taking measurements of the system under study, manipulating the system, and then taking additional measurements using the same procedure to determine if the manipulation has modified the values of the measurements. In contrast, an observational study does not involve experimental manipulation. Instead, data are gathered and correlations between predictors and response are investigated. While the tools of data analysis work best on data from randomized studies, they are also applied to other kinds of data – like natural experiments and observational studies – for which a statistician would use a modified, more structured estimation method (e.g., Difference in differences estimation and instrumental variables, among many others) that produce consistent estimators.

Experiments

The basic steps of a statistical experiment are:

1. Planning the research, including finding the number of replicates of the study, using the

following information: preliminary estimates regarding the size of treatment effects, alternative hypotheses, and the estimated experimental variability. Consideration of the selection of experimental subjects and the ethics of research is necessary. Statisticians recommend that experiments compare (at least) one new treatment with a standard treatment or control, to allow an unbiased estimate of the difference in treatment effects.

2. Design of experiments, using blocking to reduce the influence of confounding variables, and randomized assignment of treatments to subjects to allow unbiased estimates of treatment effects and experimental error. At this stage, the experimenters and statisticians write the *experimental protocol* that will guide the performance of the experiment and which specifies the *primary analysis* of the experimental data.

3. Performing the experiment following the experimental protocol and analyzing the data following the experimental protocol.

4. Further examining the data set in secondary analyses, to suggest new hypotheses for future study.

5. Documenting and presenting the results of the study.

Experiments on human behavior have special concerns. The famous Hawthorne study examined changes to the working environment at the Hawthorne plant of the Western Electric Company. The researchers were interested in determining whether increased illumination would increase the productivity of the assembly line workers. The researchers first measured the productivity in the plant, then modified the illumination in an area of the plant and checked if the changes in illumination affected productivity. It turned out that productivity indeed improved (under the experimental conditions). However, the study is heavily criticized today for errors in experimental procedures, specifically for the lack of a control group and blindness. The Hawthorne effect refers to finding that an outcome (in this case, worker productivity) changed due to observation itself. Those in the Hawthorne study became more productive not because the lighting was changed but because they were being observed.

Observational Study

An example of an observational study is one that explores the association between smoking and lung cancer. This type of study typically uses a survey to collect observations about the area of interest and then performs statistical analysis. In this case, the researchers would collect observations of both smokers and non-smokers, perhaps through a case-control study, and then look for the number of cases of lung cancer in each group.

Types of Data

Various attempts have been made to produce a taxonomy of levels of measurement. The psychophysicist Stanley Smith Stevens defined nominal, ordinal, interval, and ratio scales. Nominal measurements do not have meaningful rank order among values, and permit any one-to-one transformation. Ordinal measurements have imprecise differences between consecutive values, but have a meaningful order to those values, and permit any order-preserving transformation. Interval measurements have meaningful distances between measurements defined, but the zero value is arbitrary (as in the case with longitude and temperature measurements in Celsius or Fahrenheit), and permit any linear transformation. Ratio measurements have both a meaningful zero value and the distances between different measurements defined, and permit any rescaling transformation.

Because variables conforming only to nominal or ordinal measurements cannot be reasonably measured numerically, sometimes they are grouped together as categorical variables, whereas ratio and interval measurements are grouped together as quantitative variables, which can be either discrete or continuous, due to their numerical nature. Such distinctions can often be loosely correlated with data type in computer science, in that dichotomous categorical variables may be represented with the Boolean data type, polytomous categorical variables with arbitrarily assigned integers in the integral data type, and continuous variables with the real data type involving floating point computation. But the mapping of computer science data types to statistical data types depends on which categorization of the latter is being implemented.

Other categorizations have been proposed. For example, Mosteller and Tukey (1977) distinguished grades, ranks, counted fractions, counts, amounts, and balances. Nelder (1990) described continuous counts, continuous ratios, count ratios, and categorical modes of data.

The issue of whether or not it is appropriate to apply different kinds of statistical methods to data obtained from different kinds of measurement procedures is complicated by issues concerning the transformation of variables and the precise interpretation of research questions. "The relationship between the data and what they describe merely reflects the fact that certain kinds of statistical statements may have truth values which are not invariant under some transformations. Whether or not a transformation is sensible to contemplate depends on the question one is trying to answer" (Hand, 2004, p. 82).

Terminology and Theory of Inferential Statistics

Statistics, Estimators and Pivotal Quantities

Consider independent identically distributed (IID) random variables with a given probability distribution: standard statistical inference and estimation theory defines a random sample as the random vector given by the column vector of these IID variables. The population being examined is described by a probability distribution that may have unknown parameters.

A statistic is a random variable that is a function of the random sample, but *not a function of unknown parameters*. The probability distribution of the statistic, though, may have unknown parameters.

Consider now a function of the unknown parameter: an estimator is a statistic used to estimate such function. Commonly used estimators include sample mean, unbiased sample variance and sample covariance.

A random variable that is a function of the random sample and of the unknown parameter, but whose probability distribution *does not depend on the unknown parameter* is called a pivotal quantity or pivot. Widely used pivots include the z-score, the chi square statistic and Student's t-value.

Between two estimators of a given parameter, the one with lower mean squared error is said to be more efficient. Furthermore, an estimator is said to be unbiased if its expected value is equal to the true value of the unknown parameter being estimated, and asymptotically unbiased if its expected value converges at the limit to the true value of such parameter.

Other desirable properties for estimators include: UMVUE estimators that have the lowest variance for all possible values of the parameter to be estimated (this is usually an easier property to verify than efficiency) and consistent estimators which converges in probability to the true value of such parameter.

This still leaves the question of how to obtain estimators in a given situation and carry the computation, several methods have been proposed: the method of moments, the maximum likelihood method, the least squares method and the more recent method of estimating equations.

Null Hypothesis and Alternative Hypothesis

Interpretation of statistical information can often involve the development of a null hypothesis which is usually (but not necessarily) that no relationship exists among variables or that no change occurred over time.

The best illustration for a novice is the predicament encountered by a criminal trial. The null hypothesis, H_0, asserts that the defendant is innocent, whereas the alternative hypothesis, H_1, asserts that the defendant is guilty. The indictment comes because of suspicion of the guilt. The H_0 (status quo) stands in opposition to H_1 and is maintained unless H_1 is supported by evidence "beyond a reasonable doubt". However, "failure to reject H_0" in this case does not imply innocence, but merely that the evidence was insufficient to convict. So the jury does not necessarily *accept* H_0 but *fails to reject* H_0. While one can not "prove" a null hypothesis, one can test how close it is to being true with a power test, which tests for type II errors.

What statisticians call an alternative hypothesis is simply an hypothesis that contradicts the null hypothesis.

Error

Working from a null hypothesis, two basic forms of error are recognized:

- Type I errors where the null hypothesis is falsely rejected giving a "false positive".
- Type II errors where the null hypothesis fails to be rejected and an actual difference between populations is missed giving a "false negative".

Standard deviation refers to the extent to which individual observations in a sample differ from a central value, such as the sample or population mean, while Standard error refers to an estimate of difference between sample mean and population mean.

A statistical error is the amount by which an observation differs from its expected value, a residual is the amount an observation differs from the value the estimator of the expected value assumes on a given sample.

Mean squared error is used for obtaining efficient estimators, a widely used class of estimators. Root mean square error is simply the square root of mean squared error.

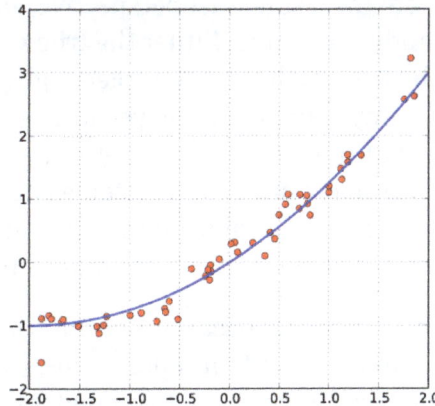

A least squares fit: in red the points to be fitted, in blue the fitted line.

Many statistical methods seek to minimize the residual sum of squares, and these are called "methods of least squares" in contrast to Least absolute deviations. The latter gives equal weight to small and big errors, while the former gives more weight to large errors. Residual sum of squares is also differentiable, which provides a handy property for doing regression. Least squares applied to linear regression is called ordinary least squares method and least squares applied to nonlinear regression is called non-linear least squares. Also in a linear regression model the non deterministic part of the model is called error term, disturbance or more simply noise. Both linear regression and non-linear regression are addressed in polynomial least squares, which also describes the variance in a prediction of the dependent variable (y axis) as a function of the independent variable (x axis) and the deviations (errors, noise, disturbances) from the estimated (fitted) curve.

Measurement processes that generate statistical data are also subject to error. Many of these errors are classified as random (noise) or systematic (bias), but other types of errors (e.g., blunder, such as when an analyst reports incorrect units) can also be important. The presence of missing data and/or censoring may result in biased estimates and specific techniques have been developed to address these problems.

Interval Estimation

Confidence intervals: the red line is true value for the mean in this example, the blue lines are random confidence intervals for 100 realizations.

Most studies only sample part of a population, so results don't fully represent the whole population. Any estimates obtained from the sample only approximate the population value. Confidence intervals allow statisticians to express how closely the sample estimate matches the true value in the whole population. Often they are expressed as 95% confidence intervals. Formally, a 95% confidence interval for a value is a range where, if the sampling and analysis were repeated under the same conditions (yielding a different dataset), the interval would include the true (population) value in 95% of all possible cases. This does *not* imply that the probability that the true value is in the confidence interval is 95%. From the frequentist perspective, such a claim does not even make sense, as the true value is not a random variable. Either the true value is or is not within the given interval. However, it is true that, before any data are sampled and given a plan for how to construct the confidence interval, the probability is 95% that the yet-to-be-calculated interval will cover the true value: at this point, the limits of the interval are yet-to-be-observed random variables. One approach that does yield an interval that can be interpreted as having a given probability of containing the true value is to use a credible interval from Bayesian statistics: this approach depends on a different way of interpreting what is meant by "probability", that is as a Bayesian probability.

In principle confidence intervals can be symmetrical or asymmetrical. An interval can be asymmetrical because it works as lower or upper bound for a parameter (left-sided interval or right sided interval), but it can also be asymmetrical because the two sided interval is built violating symmetry around the estimate. Sometimes the bounds for a confidence interval are reached asymptotically and these are used to approximate the true bounds.

Significance

Statistics rarely give a simple Yes/No type answer to the question under analysis. Interpretation often comes down to the level of statistical significance applied to the numbers and often refers to the probability of a value accurately rejecting the null hypothesis (sometimes referred to as the p-value).

Important:

Pr (observation | hypothesis) ≠ Pr (hypothesis | observation)

The probability of observing a result given that some hypothesis is true is *not equivalent* to the probability that a hypothesis is true given that some result has been observed.

Using the p-value as a "score" is committing an egregious logical error: **the transposed conditional fallacy.**

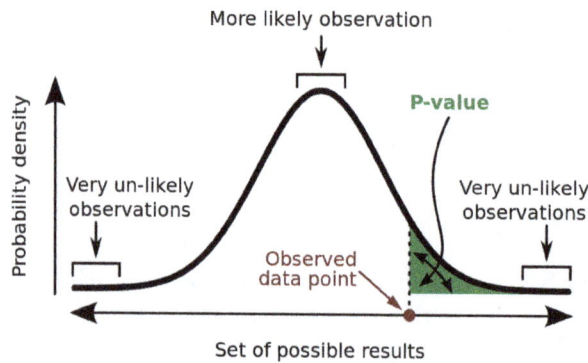

A **p-value** (shaded green area) is the probability of an observed (or more extreme) result assuming that the null hypothesis is true.

In this graph the black line is probability distribution for the test statistic, the critical region is the set of values to the right of the observed data point (observed value of the test statistic) and the p-value is represented by the green area.

The standard approach is to test a null hypothesis against an alternative hypothesis. A critical region is the set of values of the estimator that leads to refuting the null hypothesis. The probability of type I error is therefore the probability that the estimator belongs to the critical region given that null hypothesis is true (statistical significance) and the probability of type II error is the probability that the estimator doesn't belong to the critical region given that the alternative hypothesis is true. The statistical power of a test is the probability that it correctly rejects the null hypothesis when the null hypothesis is false.

Referring to statistical significance does not necessarily mean that the overall result is significant in real world terms. For example, in a large study of a drug it may be shown that the drug has a statistically significant but very small beneficial effect, such that the drug is unlikely to help the patient noticeably.

While in principle the acceptable level of statistical significance may be subject to debate, the p-value is the smallest significance level that allows the test to reject the null hypothesis. This is logically equivalent to saying that the p-value is the probability, assuming the null hypothesis is true, of observing a result at least as extreme as the test statistic. Therefore, the smaller the p-value, the lower the probability of committing type I error.

Some problems are usually associated with this framework:

- A difference that is highly statistically significant can still be of no practical significance, but it is possible to properly formulate tests to account for this. One response involves going beyond reporting only the significance level to include the p-value when reporting whether a hypothesis is rejected or accepted. The p-value, however, does not indicate the size or importance of the observed effect and can also seem to exaggerate the importance of minor differences in large studies. A better and increasingly common approach is to report confidence intervals. Although these are produced from the same calculations as those of hypothesis tests or p-values, they describe both the size of the effect and the uncertainty surrounding it.

- Fallacy of the transposed conditional, aka prosecutor's fallacy: criticisms arise because the hypothesis testing approach forces one hypothesis (the null hypothesis) to be favored, since what is being evaluated is probability of the observed result given the null hypothesis and not probability of the null hypothesis given the observed result. An alternative to this approach is offered by Bayesian inference, although it requires establishing a prior probability.

- Rejecting the null hypothesis does not automatically prove the alternative hypothesis.

- As everything in inferential statistics it relies on sample size, and therefore under fat tails p-values may be seriously mis-computed.

Examples

Some well-known statistical tests and procedures are:

- Analysis of variance (ANOVA)
- Chi-squared test

- Correlation

- Factor analysis

- Mann–Whitney U

- Mean square weighted deviation (MSWD)

- Pearson product-moment correlation coefficient

- Regression analysis

- Spearman's rank correlation coefficient

- Student's t-test

- Time series analysis

- Conjoint Analysis

Misuse of Statistics

Misuse of statistics can produce subtle, but serious errors in description and interpretation—subtle in the sense that even experienced professionals make such errors, and serious in the sense that they can lead to devastating decision errors. For instance, social policy, medical practice, and the reliability of structures like bridges all rely on the proper use of statistics.

Even when statistical techniques are correctly applied, the results can be difficult to interpret for those lacking expertise. The statistical significance of a trend in the data—which measures the extent to which a trend could be caused by random variation in the sample—may or may not agree with an intuitive sense of its significance. The set of basic statistical skills (and skepticism) that people need to deal with information in their everyday lives properly is referred to as statistical literacy.

There is a general perception that statistical knowledge is all-too-frequently intentionally misused by finding ways to interpret only the data that are favorable to the presenter. A mistrust and misunderstanding of statistics is associated with the quotation, "There are three kinds of lies: lies, damned lies, and statistics". Misuse of statistics can be both inadvertent and intentional, and the book *How to Lie with Statistics* outlines a range of considerations. In an attempt to shed light on the use and misuse of statistics, reviews of statistical techniques used in particular fields are conducted (e.g. Warne, Lazo, Ramos, and Ritter (2012)).

Ways to avoid misuse of statistics include using proper diagrams and avoiding bias. Misuse can occur when conclusions are overgeneralized and claimed to be representative of more than they really are, often by either deliberately or unconsciously overlooking sampling bias. Bar graphs are arguably the easiest diagrams to use and understand, and they can be made either by hand or with simple computer programs. Unfortunately, most people do not look for bias or errors, so they are not noticed. Thus, people may often believe that something is true even if it is not well represented. To make data gathered from statistics believable and accurate, the sample taken must be representative of the whole. According to Huff, "The dependability of a sample can be destroyed by [bias]... allow yourself some degree of skepticism."

To assist in the understanding of statistics Huff proposed a series of questions to be asked in each case:

- Who says so? (Does he/she have an axe to grind?)

- How does he/she know? (Does he/she have the resources to know the facts?)

- What's missing? (Does he/she give us a complete picture?)

- Did someone change the subject? (Does he/she offer us the right answer to the wrong problem?)

- Does it make sense? (Is his/her conclusion logical and consistent with what we already know?)

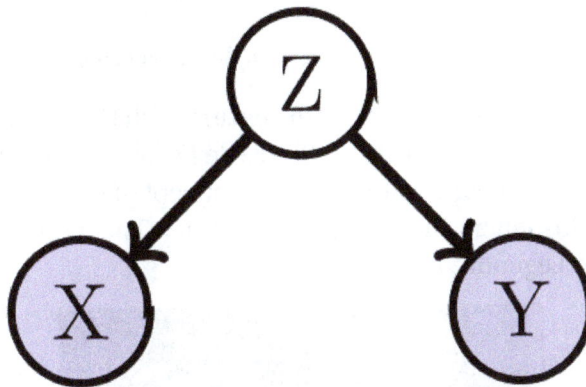

The confounding variable problem: X and Y may be correlated, not because there is causal relationship between them, but because both depend on a third variable Z. Z is called a confounding factor.

Misinterpretation: Correlation

The concept of correlation is particularly noteworthy for the potential confusion it can cause. Statistical analysis of a data set often reveals that two variables (properties) of the population under consideration tend to vary together, as if they were connected. For example, a study of annual income that also looks at age of death might find that poor people tend to have shorter lives than affluent people. The two variables are said to be correlated; however, they may or may not be the cause of one another. The correlation phenomena could be caused by a third, previously unconsidered phenomenon, called a lurking variable or confounding variable. For this reason, there is no way to immediately infer the existence of a causal relationship between the two variables.

History of Statistical Science

Statistical methods date back at least to the 5th century BC.

Some scholars pinpoint the origin of statistics to 1663, with the publication of *Natural and Political Observations upon the Bills of Mortality* by John Graunt. Early applications of statistical thinking revolved around the needs of states to base policy on demographic and economic data, hence its *stat-* etymology. The scope of the discipline of statistics broadened in the early 19th century to include the collection and analysis of data in general. Today, statistics is widely employed in government, business, and natural and social sciences.

Gerolamo Cardano, the earliest pioneer on the mathematics of probability.

Its mathematical foundations were laid in the 17th century with the development of the probability theory by Gerolamo Cardano, Blaise Pascal and Pierre de Fermat. Mathematical probability theory arose from the study of games of chance, although the concept of probability was already examined in medieval law and by philosophers such as Juan Caramuel. The method of least squares was first described by Adrien-Marie Legendre in 1805.

Karl Pearson, a founder of mathematical statistics.

The modern field of statistics emerged in the late 19th and early 20th century in three stages. The first wave, at the turn of the century, was led by the work of Francis Galton and Karl Pearson, who transformed statistics into a rigorous mathematical discipline used for analysis, not just in science, but in industry and politics as well. Galton's contributions included introducing the concepts of standard deviation, correlation, regression analysis and the application of these methods to the study of the variety of human characteristics – height, weight, eyelash length among others. Pearson developed the Pearson product-moment correlation coefficient, defined as a product-moment, the method of moments for the fitting of distributions to samples and the Pearson distribution, among many other things. Galton and Pearson founded *Biometrika* as the first journal of mathematical statistics and biostatistics (then called biometry), and the latter founded the world's first university statistics department at University College London.

Ronald Fisher coined the term null hypothesis during the Lady tasting tea experiment, which "is never proved or established, but is possibly disproved, in the course of experimentation".

The second wave of the 1910s and 20s was initiated by William Gosset, and reached its culmination in the insights of Ronald Fisher, who wrote the textbooks that were to define the academic discipline in universities around the world. Fisher's most important publications were his 1918 seminal paper *The Correlation between Relatives on the Supposition of Mendelian Inheritance*, which was the first to use the statistical term, variance, his classic 1925 work *Statistical Methods for Research Workers* and his 1935 *The Design of Experiments*, where he developed rigorous design of experiments models. He originated the concepts of sufficiency, ancillary statistics, Fisher's linear discriminator and Fisher information. In his 1930 book *The Genetical Theory of Natural Selection* he applied statistics to various biological concepts such as Fisher's principle). Nevertheless, A. W. F. Edwards has remarked that it is "probably the most celebrated argument in evolutionary biology". (about the sex ratio), the Fisherian runaway, a concept in sexual selection about a positive feedback runaway affect found in evolution.

The final wave, which mainly saw the refinement and expansion of earlier developments, emerged from the collaborative work between Egon Pearson and Jerzy Neyman in the 1930s. They introduced the concepts of "Type II" error, power of a test and confidence intervals. Jerzy Neyman in 1934 showed that stratified random sampling was in general a better method of estimation than purposive (quota) sampling.

Today, statistical methods are applied in all fields that involve decision making, for making accurate inferences from a collated body of data and for making decisions in the face of uncertainty based on statistical methodology. The use of modern computers has expedited large-scale statistical computations, and has also made possible new methods that are impractical to perform manually. Statistics continues to be an area of active research, for example on the problem of how to analyze Big data.

Applications

Applied Statistics, Theoretical Statistics and Mathematical Statistics

"Applied statistics" comprises descriptive statistics and the application of inferential statistics. *Theoretical statistics* concerns both the logical arguments underlying justification of approaches to statistical inference, as well encompassing *mathematical statistics*. Mathematical statistics includes not only the manipulation of probability distributions necessary for deriving results related to methods of estimation and inference, but also various aspects of computational statistics and the design of experiments.

Machine Learning and Data Mining

There are two applications for machine learning and data mining: data management and data analysis. Statistics tools are necessary for the data analysis.

Statistics in Society

Statistics is applicable to a wide variety of academic disciplines, including natural and social sciences, government, and business. Statistical consultants can help organizations and companies that don't have in-house expertise relevant to their particular questions.

Statistical Computing

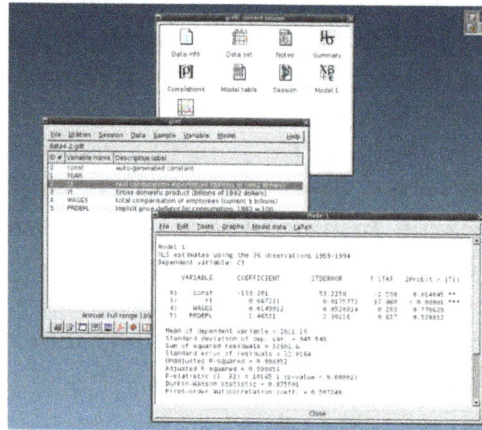

gretl, an example of an open source statistical package

The rapid and sustained increases in computing power starting from the second half of the 20th century have had a substantial impact on the practice of statistical science. Early statistical models were almost always from the class of linear models, but powerful computers, coupled with suitable numerical algorithms, caused an increased interest in nonlinear models (such as neural networks) as well as the creation of new types, such as generalized linear models and multilevel models.

Increased computing power has also led to the growing popularity of computationally intensive methods based on resampling, such as permutation tests and the bootstrap, while techniques such as Gibbs sampling have made use of Bayesian models more feasible. The computer revolution has implications for the future of statistics with new emphasis on "experimental" and "empirical" statistics. A large number of both general and special purpose statistical software are now available.

Statistics Applied to Mathematics or the Arts

Traditionally, statistics was concerned with drawing inferences using a semi-standardized methodology that was "required learning" in most sciences. This has changed with use of statistics in non-inferential contexts. What was once considered a dry subject, taken in many fields as a degree-requirement, is now viewed enthusiastically. Initially derided by some mathematical purists, it is now considered essential methodology in certain areas.

- In number theory, scatter plots of data generated by a distribution function may be transformed with familiar tools used in statistics to reveal underlying patterns, which may then lead to hypotheses.

- Methods of statistics including predictive methods in forecasting are combined with chaos theory and fractal geometry to create video works that are considered to have great beauty.

- The process art of Jackson Pollock relied on artistic experiments whereby underlying distributions in nature were artistically revealed. With the advent of computers, statistical methods were applied to formalize such distribution-driven natural processes to make and analyze moving video art.

- Methods of statistics may be used predicatively in performance art, as in a card trick based on a Markov process that only works some of the time, the occasion of which can be predicted using statistical methodology.

- Statistics can be used to predicatively create art, as in the statistical or stochastic music invented by Iannis Xenakis, where the music is performance-specific. Though this type of artistry does not always come out as expected, it does behave in ways that are predictable and tunable using statistics.

Specialized Disciplines

Statistical techniques are used in a wide range of types of scientific and social research, including: biostatistics, computational biology, computational sociology, network biology, social science, sociology and social research. Some fields of inquiry use applied statistics so extensively that they have specialized terminology. These disciplines include:

- Actuarial science (assesses risk in the insurance and finance industries)

- Applied information economics

- Astrostatistics (statistical evaluation of astronomical data)

- Biostatistics

- Business statistics

- Chemometrics (for analysis of data from chemistry)

- Data mining (applying statistics and pattern recognition to discover knowledge from data)

- Data science

- Demography

- Econometrics (statistical analysis of economic data)

- Energy statistics

- Engineering statistics

- Epidemiology (statistical analysis of disease)

- Geography and Geographic Information Systems, specifically in Spatial analysis

- Image processing

- Medical Statistics

- Psychological statistics

- Reliability engineering

- Social statistics

- Statistical Mechanics

In addition, there are particular types of statistical analysis that have also developed their own specialised terminology and methodology:

- Bootstrap / Jackknife resampling

- Multivariate statistics

- Statistical classification

- Structured data analysis (statistics)

- Structural equation modelling

- Survey methodology

- Survival analysis

- Statistics in various sports, particularly baseball - known as Sabermetrics - and cricket

Statistics form a key basis tool in business and manufacturing as well. It is used to understand measurement systems variability, control processes (as in statistical process control or SPC), for summarizing data, and to make data-driven decisions. In these roles, it is a key tool, and perhaps the only reliable tool.

Probability

Probability is the measure of the likelihood that an event will occur. Probability is quantified as a number between 0 and 1 (where 0 indicates impossibility and 1 indicates certainty). The higher the probability of an event, the more certain that the event will occur. A simple example is the tossing of a fair (unbiased) coin. Since the coin is unbiased, the two outcomes ("head" and "tail") are both equally probable; the probability of "head" equals the probability of "tail." Since no other outcomes are possible, the probability is 1/2 (or 50%), of either "head" or "tail". In other words, the probability of "head" is 1 out of 2 outcomes and the probability of "tail" is also 1 out of 2 outcomes, expressed as 0.5 when converted to decimal, with the above mentioned quantification system.

These concepts have been given an axiomatic mathematical formalization in probability theory, which is used widely in such areas of study as mathematics, statistics, finance, gambling, science (in particular physics), artificial intelligence/machine learning, computer science, game theory, and philosophy to, for example, draw inferences about the expected frequency of events. Probability theory is also used to describe the underlying mechanics and regularities of complex systems.

Interpretations

When dealing with experiments that are random and well-defined in a purely theoretical setting (like tossing a fair coin), probabilities can be numerically described by the number of desired outcomes divided by the total number of all outcomes. For example, tossing a fair coin twice will yield "head-head", "head-tail", "tail-head", and "tail-tail" outcomes. The probability of getting an outcome of "head-head" is 1 out of 4 outcomes or 1 divided by four or 1/4 (or 25%). When it comes to practical application however, there are two major competing categories of probability interpretations, whose adherents possess different views about the fundamental nature of probability:

1. Objectivists assign numbers to describe some objective or physical state of affairs. The most popular version of objective probability is frequentist probability, which claims that the probability of a random event denotes the *relative frequency of occurrence* of an experiment's outcome, when repeating the experiment. This interpretation considers probability to be the relative frequency "in the long run" of outcomes. A modification of this is propensity probability, which interprets probability as the tendency of some experiment to yield a certain outcome, even if it is performed only once.

2. Subjectivists assign numbers per subjective probability, i.e., as a degree of belief. The degree of belief has been interpreted as, "the price at which you would buy or sell a bet that pays 1 unit of utility if E, 0 if not E." The most popular version of subjective probability is Bayesian probability, which includes expert knowledge as well as experimental data to produce probabilities. The expert knowledge is represented by some (subjective) prior probability distribution. The data is incorporated in a likelihood function. The product of the prior and the likelihood, normalized, results in a posterior probability distribution that incorporates all the information known to date. Starting from arbitrary, subjective probabilities for a group of agents, some Bayesians claim that all agents will eventually have sufficiently similar assessments of probabilities, given enough evidence.

Etymology

The word *probability* derives from the Latin *probabilitas*, which can also mean "probity", a measure of the authority of a witness in a legal case in Europe, and often correlated with the witness's nobility. In a sense, this differs much from the modern meaning of *probability*, which, in contrast, is a measure of the weight of empirical evidence, and is arrived at from inductive reasoning and statistical inference.

History

scientific study of probability is a modern development of mathematics. Gambling shows that there has been an interest in quantifying the ideas of probability for millennia, but exact mathematical descriptions arose much later. There are reasons of course, for the slow development of the mathematics of probability. Whereas games of chance provided the impetus for the mathematical study of probability, fundamental issues are still obscured by the superstitions of gamblers.

Christiaan Huygens probably published the first book on probability

According to Richard Jeffrey, "Before the middle of the seventeenth century, the term 'probable' (Latin *probabilis*) meant *approvable*, and was applied in that sense, univocally, to opinion and to action. A probable action or opinion was one such as sensible people would undertake or hold, in the circumstances." However, in legal contexts especially, 'probable' could also apply to propositions for which there was good evidence.

Gerolamo Cardano

The sixteenth century Italian polymath Gerolamo Cardano demonstrated the efficacy of defining odds as the ratio of favourable to unfavourable outcomes (which implies that the probability of an event is given by the ratio of favourable outcomes to the total number of possible outcomes). Aside from the elementary work by Cardano, the doctrine of probabilities dates to the correspondence of Pierre de Fermat and Blaise Pascal (1654). Christiaan Huygens (1657) gave the earliest known scientific treatment of the subject. Jakob Bernoulli's *Ars Conjectandi* (posthumous, 1713) and Abraham de Moivre's *Doctrine of Chances* (1718) treated the subject as a branch of mathematics. See Ian Hacking's *The Emergence of Probability* and James Franklin's *The Science of Conjecture* for histories of the early development of the very concept of mathematical probability.

The theory of errors may be traced back to Roger Cotes's *Opera Miscellanea* (posthumous, 1722), but a memoir prepared by Thomas Simpson in 1755 (printed 1756) first applied the theory to the discussion of errors of observation. The reprint (1757) of this memoir lays down the axioms that positive and negative errors are equally probable, and that certain assignable limits define the range of all errors. Simpson also discusses continuous errors and describes a probability curve.

The first two laws of error that were proposed both originated with Pierre-Simon Laplace. The first law was published in 1774 and stated that the frequency of an error could be expressed as an exponential function of the numerical magnitude of the error, disregarding sign. The second law of error was proposed in 1778 by Laplace and stated that the frequency of the error is an exponential function of the square of the error. The second law of error is called the normal distribution or the Gauss law. "It is difficult historically to attribute that law to Gauss, who in spite of his well-known precocity had probably not made this discovery before he was two years old."

Daniel Bernoulli (1778) introduced the principle of the maximum product of the probabilities of a system of concurrent errors.

Carl Friedrich Gauss

Adrien-Marie Legendre (1805) developed the method of least squares, and introduced it in his *Nouvelles méthodes pour la détermination des orbites des comètes* (*New Methods for Determining the Orbits of Comets*). In ignorance of Legendre's contribution, an Irish-American writer, Robert Adrain, editor of "The Analyst" (1808), first deduced the law of facility of error,

$$\phi(x) = ce^{-h^2x^2},$$

where h is a constant depending on precision of observation, and c is a scale factor ensuring that the area under the curve equals 1. He gave two proofs, the second being essentially the same as John Herschel's (1850). Gauss gave the first proof that seems to have been known in Europe (the third after Adrain's) in 1809. Further proofs were given by Laplace (1810, 1812), Gauss (1823), James Ivory (1825, 1826), Hagen (1837), Friedrich Bessel (1838), W. F. Donkin (1844, 1856), and Morgan Crofton (1870). Other contributors were Ellis (1844), De Morgan (1864), Glaisher (1872), and Giovanni Schiaparelli (1875). Peters's (1856) formula for r, the probable error of a single observation, is well known.

In the nineteenth century authors on the general theory included Laplace, Sylvestre Lacroix (1816), Littrow (1833), Adolphe Quetelet (1853), Richard Dedekind (1860), Helmert (1872), Hermann Laurent (1873), Liagre, Didion, and Karl Pearson. Augustus De Morgan and George Boole improved the exposition of the theory.

Andrey Markov introduced the notion of Markov chains (1906), which played an important role in stochastic processes theory and its applications. The modern theory of probability based on the measure theory was developed by Andrey Kolmogorov (1931).

On the geometric side contributors to *The Educational Times* were influential (Miller, Crofton, McColl, Wolstenholme, Watson, and Artemas Martin).

Theory

Like other theories, the theory of probability is a representation of probabilistic concepts in formal terms—that is, in terms that can be considered separately from their meaning. These formal terms are manipulated by the rules of mathematics and logic, and any results are interpreted or translated back into the problem domain.

There have been at least two successful attempts to formalize probability, namely the Kolmogorov formulation and the Cox formulation. In Kolmogorov's formulation, sets are interpreted as events and probability itself as a measure on a class of sets. In Cox's theorem, probability is taken as a primitive (that is, not further analyzed) and the emphasis is on constructing a consistent assignment of probability values to propositions. In both cases, the laws of probability are the same, except for technical details.

There are other methods for quantifying uncertainty, such as the Dempster–Shafer theory or possibility theory, but those are essentially different and not compatible with the laws of probability as usually understood.

Applications

Probability theory is applied in everyday life in risk assessment and modeling. The insurance industry and markets use actuarial science to determine pricing and make trading decisions. Governments apply probabilistic methods in environmental regulation, entitlement analysis (Reliability theory of aging and longevity), and financial regulation.

A good example of the use of probibility theory in equity trading is the effect of the perceived probability of any widespread Middle East conflict on oil prices, which have ripple effects in the economy as a whole. An assessment by a commodity trader that a war is more likely can send that commodity's prices up or down, and signals other traders of that opinion. Accordingly, the probabilities are neither assessed independently nor necessarily very rationally. The theory of behavioral finance emerged to describe the effect of such groupthink on pricing, on policy, and on peace and conflict.

In addition to financial assessment, probability can be used to analyze trends in biology (e.g. disease spread) as well as ecology (e.g. biological Punnett squares). As with finance, risk assessment can be used as a statistical tool to calculate the likelihood of undesirable events occurring and can assist with implementing protocols to avoid encountering such circumstances. Probability is used to design games of chance so that casinos can make a guaranteed profit, yet provide payouts to players that are frequent enough to encourage continued play.

The discovery of rigorous methods to assess and combine probability assessments has changed society. It is important for most citizens to understand how probability assessments are made, and how they contribute to decisions.

Another significant application of probability theory in everyday life is reliability. Many consumer products, such as automobiles and consumer electronics, use reliability theory in product design to reduce the probability of failure. Failure probability may influence a manufacturer's decisions on a product's warranty.

The cache language model and other statistical language models that are used in natural language processing are also examples of applications of probability theory.

Mathematical Treatment

Consider an experiment that can produce a number of results. The collection of all possible results is called the sample space of the experiment. The power set of the sample space is formed by considering all different collections of possible results. For example, rolling a dice can produce six possible results. One collection of possible results gives an odd number on the dice. Thus, the subset {1,3,5} is an element of the power set of the sample space of dice rolls. These collections are called "events." In this case, {1,3,5} is the event that the dice falls on some odd number. If the results that actually occur fall in a given event, the event is said to have occurred.

A probability is a way of assigning every event a value between zero and one, with the requirement that the event made up of all possible results (in our example, the event {1,2,3,4,5,6}) is assigned a value of one. To qualify as a probability, the assignment of values must satisfy the requirement that if you look at a collection of mutually exclusive events (events with no common results, e.g., the events {1,6}, {3}, and {2,4} are all mutually exclusive), the probability that at least one of the events will occur is given by the sum of the probabilities of all the individual events.

The probability of an event A is written as $P(A)$, $p(A)$, or $\Pr(A)$. This mathematical definition of probability can extend to infinite sample spaces, and even uncountable sample spaces, using the concept of a measure.

The *opposite* or *complement* of an event A is the event [not A] (that is, the event of A not occurring), often denoted as $\bar{A}, A^{C}, \neg A$, or $\sim A$; its probability is given by $P(\text{not } A) = 1 - P(A)$. As an example, the chance of not rolling a six on a six-sided die is $1 - (\text{chance of rolling a six}) = 1 - \dfrac{1}{6} = \dfrac{5}{6}$.

If two events A and B occur on a single performance of an experiment, this is called the intersection or joint probability of A and B, denoted as $P(A \cap B)$.

Independent Events

If two events, A and B are independent then the joint probability is

$$P(A \text{ and } B) = P(A \cap B) = P(A)P(B),$$

for example, if two coins are flipped the chance of both being heads is $\frac{1}{2} \times \frac{1}{2} = \frac{1}{4}$.

Mutually Exclusive Events

If either event A or event B occurs on a single performance of an experiment this is called the union of the events A and B denoted as $P(A \cup B)$. If two events are mutually exclusive then the probability of either occurring is

$$P(A \text{ or } B) = P(A \cup B) = P(A) + P(B).$$

For example, the chance of rolling a 1 or 2 on a six-sided die is $P(1 \text{ or } 2) = P(1) + P(2) = \frac{1}{6} + \frac{1}{6} = \frac{1}{3}$.

Not Mutually Exclusive Events

If the events are not mutually exclusive then

$$P\left(A \text{ or } B\right) = P\left(A\right) + P\left(B\right) - P\left(A \text{ and } B\right).$$

For example, when drawing a single card at random from a regular deck of cards, the chance of getting a heart or a face card (J,Q,K) (or one that is both) is $\frac{13}{52} + \frac{12}{52} - \frac{3}{52} = \frac{11}{26}$, because of the 52 cards of a deck 13 are hearts, 12 are face cards, and 3 are both: here the possibilities included in the "3 that are both" are included in each of the "13 hearts" and the "12 face cards" but should only be counted once.

Conditional Probability

Conditional probability is the probability of some event A, given the occurrence of some other event B. Conditional probability is written $P(A|B)$, and is read "the probability of A, given B". It is defined by

$$P(A|B) = \frac{P(A \cap B)}{P(B)}.$$

If $P(B) = 0$ then $P(A|B)$ is formally undefined by this expression. However, it is possible to define a conditional probability for some zero-probability events using a σ-algebra of such events (such as those arising from a continuous random variable).

For example, in a bag of 2 red balls and 2 blue balls (4 balls in total), the probability of taking a red ball is $1/2$; however, when taking a second ball, the probability of it being either a red ball or a blue ball depends on the ball previously taken, such as, if a red ball was taken, the probability of picking a red ball again would be $1/3$ since only 1 red and 2 blue balls would have been remaining.

Inverse Probability

In probability theory and applications, Bayes' rule relates the odds of event A_1 to event A_2, before (prior to) and after (posterior to) conditioning on another event B. The odds on A_1 to event A_2 is simply the ratio of the probabilities of the two events. When arbitrarily many events A are of interest, not just two, the rule can be rephrased as posterior is proportional to prior times likelihood, $P(A|B) \propto P(A)P(B|A)$ where the proportionality symbol means that the left hand side is proportional to (i.e., equals a constant times) the right hand side as A varies, for fixed or given B (Lee, 2012; Bertsch McGrayne, 2012). In this form it goes back to Laplace (1774) and to Cournot (1843).

Summary of Probabilities

Summary of probabilities	
Event	**Probability**
A	$P(A) \in [0,1]$
not A	$P(A^c) = 1 - P(A)$

A or B	$P(A \cup B) = P(A) + P(B) - P(A \cap B)$
	$P(A \cup B) = P(A) + P(B)$ if A and B are mutually exclusive
A and B	$P(A \cap B) = P(A \mid B)P(B) = P(B \mid A)P(A)$
	$P(A \cap B) = P(A)P(B)$ if A and B are independent
A given B	$P(A \mid B) = \dfrac{P(A \cap B)}{P(B)} = \dfrac{P(B \mid A)P(A)}{P(B)}$

Relation to Randomness

In a deterministic universe, based on Newtonian concepts, there would be no probability if all conditions were known (Laplace's demon), (but there are situations in which sensitivity to initial conditions exceeds our ability to measure them, i.e. know them). In the case of a roulette wheel, if the force of the hand and the period of that force are known, the number on which the ball will stop would be a certainty (though as a practical matter, this would likely be true only of a roulette wheel that had not been exactly levelled — as Thomas A. Bass' Newtonian Casino revealed). Of course, this also assumes knowledge of inertia and friction of the wheel, weight, smoothness and roundness of the ball, variations in hand speed during the turning and so forth. A probabilistic description can thus be more useful than Newtonian mechanics for analyzing the pattern of outcomes of repeated rolls of a roulette wheel. Physicists face the same situation in kinetic theory of gases, where the system, while deterministic *in principle*, is so complex (with the number of molecules typically the order of magnitude of Avogadro constant $6.02 \cdot 10^{23}$) that only a statistical description of its properties is feasible.

Probability theory is required to describe quantum phenomena. A revolutionary discovery of early 20th century physics was the random character of all physical processes that occur at sub-atomic scales and are governed by the laws of quantum mechanics. The objective wave function evolves deterministically but, according to the Copenhagen interpretation, it deals with probabilities of observing, the outcome being explained by a wave function collapse when an observation is made. However, the loss of determinism for the sake of instrumentalism did not meet with universal approval. Albert Einstein famously remarked in a letter to Max Born: "I am convinced that God does not play dice". Like Einstein, Erwin Schrödinger, who discovered the wave function, believed quantum mechanics is a statistical approximation of an underlying deterministic reality. In some modern interpretations of the statistical mechanics of measurement, quantum decoherence is invoked to account for the appearance of subjectively probabilistic experimental outcomes.

Discrete Mathematics

Discrete mathematics is the study of mathematical structures that are fundamentally discrete rather than continuous. In contrast to real numbers that have the property of varying "smoothly", the objects studied in discrete mathematics – such as integers, graphs, and statements in logic – do not vary smoothly in this way, but have distinct, separated values. Discrete mathematics therefore excludes topics in "continuous mathematics" such as calculus and analysis. Discrete objects can often be enumerated by integers. More formally, discrete mathematics has been characterized as the branch of mathematics dealing with countable sets (sets that have the same cardinality as subsets of the natural numbers, including rational numbers but not real numbers). However, there is no exact definition of the term "discrete mathematics." Indeed, discrete mathematics is described less by what is included than by what is excluded: continuously varying quantities and related notions.

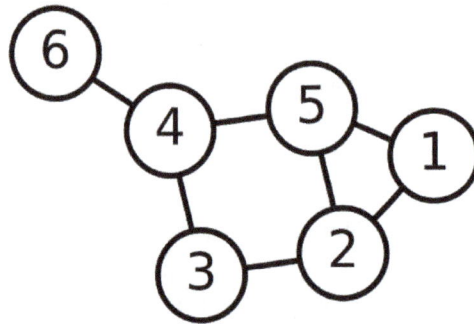

Graphs like this are among the objects studied by discrete mathematics, for their interesting mathematical properties, their usefulness as models of real-world problems, and their importance in developing computer algorithms.

The set of objects studied in discrete mathematics can be finite or infinite. The term finite mathematics is sometimes applied to parts of the field of discrete mathematics that deals with finite sets, particularly those areas relevant to business.

Research in discrete mathematics increased in the latter half of the twentieth century partly due to the development of digital computers which operate in discrete steps and store data in discrete bits. Concepts and notations from discrete mathematics are useful in studying and describing objects and problems in branches of computer science, such as computer algorithms, programming languages, cryptography, automated theorem proving, and software development. Conversely, computer implementations are significant in applying ideas from discrete mathematics to real-world problems, such as in operations research.

Although the main objects of study in discrete mathematics are discrete objects, analytic methods from continuous mathematics are often employed as well.

In university curricula, "Discrete Mathematics" appeared in the 1980s, initially as a computer science support course; its contents were somewhat haphazard at the time. The curriculum has thereafter developed in conjunction with efforts by ACM and MAA into a course that is basically intended to develop mathematical maturity in freshmen; as such it is nowadays a prerequisite for mathematics majors in some universities as well. Some high-school-level discrete mathematics textbooks have appeared as well. At this level, discrete mathematics is sometimes seen as a preparatory course, not unlike precalculus in this respect.

The Fulkerson Prize is awarded for outstanding papers in discrete mathematics.

Grand Challenges, Past and Present

The history of discrete mathematics has involved a number of challenging problems which have focused attention within areas of the field. In graph theory, much research was motivated by attempts to prove the four color theorem, first stated in 1852, but not proved until 1976 (by Kenneth Appel and Wolfgang Haken, using substantial computer assistance).

Much research in graph theory was motivated by attempts to prove that all maps, like this one, could be colored using only four colors so that no areas of the same color touched. Kenneth Appel and Wolfgang Haken proved this in 1976.

In logic, the second problem on David Hilbert's list of open problems presented in 1900 was to prove that the axioms of arithmetic are consistent. Gödel's second incompleteness theorem, proved in 1931, showed that this was not possible – at least not within arithmetic itself. Hilbert's tenth problem was to determine whether a given polynomial Diophantine equation with integer coefficients has an integer solution. In 1970, Yuri Matiyasevich proved that this could not be done.

The need to break German codes in World War II led to advances in cryptography and theoretical computer science, with the first programmable digital electronic computer being developed at England's Bletchley Park with the guidance of Alan Turing and his seminal work, On Computable Numbers. At the same time, military requirements motivated advances in operations research. The Cold War meant that cryptography remained important, with fundamental advances such as public-key cryptography being developed in the following decades. Operations research remained important as a tool in business and project management, with the critical path method being developed in the 1950s. The telecommunication industry has also motivated advances in discrete mathematics, particularly in graph theory and information theory. Formal verification of statements in logic has been necessary for software development of safety-critical systems, and advances in automated theorem proving have been driven by this need.

Computational geometry has been an important part of the computer graphics incorporated into modern video games and computer-aided design tools.

Several fields of discrete mathematics, particularly theoretical computer science, graph theory, and combinatorics, are important in addressing the challenging bioinformatics problems associated with understanding the tree of life.

Currently, one of the most famous open problems in theoretical computer science is the P = NP problem, which involves the relationship between the complexity classes P and NP. The Clay Mathematics Institute has offered a $1 million USD prize for the first correct proof, along with prizes for six other mathematical problems.

Topics in Discrete Mathematics

Theoretical Computer Science

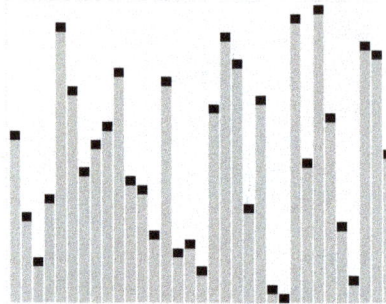

Complexity studies the time taken by algorithms, such as this sorting routine.

Theoretical computer science includes areas of discrete mathematics relevant to computing. It draws heavily on graph theory and mathematical logic. Included within theoretical computer science is the study of algorithms for computing mathematical results. Computability studies what can be computed in principle, and has close ties to logic, while complexity studies the time taken by computations. Automata theory and formal language theory are closely related to computability. Petri nets and process algebras are used to model computer systems, and methods from discrete mathematics are used in analyzing VLSI electronic circuits. Computational geometry applies algorithms to geometrical problems, while computer image analysis applies them to representations of images. Theoretical computer science also includes the study of various continuous computational topics.

Information Theory

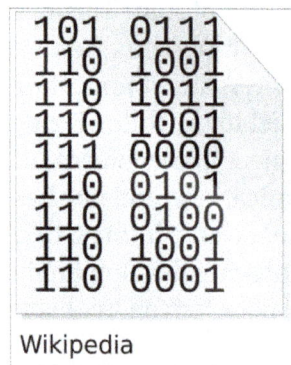

The ASCII codes for the word "Wikipedia", given here in binary, provide a way of representing the word in information theory, as well as for information-processing algorithms.

Information theory involves the quantification of information. Closely related is coding theory which is used to design efficient and reliable data transmission and storage methods. Information theory also includes continuous topics such as: analog signals, analog coding, analog encryption.

Logic

Logic is the study of the principles of valid reasoning and inference, as well as of consistency, soundness, and completeness. For example, in most systems of logic (but not in intuitionistic logic) Peirce's law $(((P{\rightarrow}Q){\rightarrow}P){\rightarrow}P)$ is a theorem. For classical logic, it can be easily verified with a truth table. The study of mathematical proof is particularly important in logic, and has applications to automated theorem proving and formal verification of software.

Logical formulas are discrete structures, as are proofs, which form finite trees or, more generally, directed acyclic graph structures (with each inference step combining one or more premise branches to give a single conclusion). The truth values of logical formulas usually form a finite set, generally restricted to two values: *true* and *false*, but logic can also be continuous-valued, e.g., fuzzy logic. Concepts such as infinite proof trees or infinite derivation trees have also been studied, e.g. infinitary logic.

Set Theory

Set theory is the branch of mathematics that studies sets, which are collections of objects, such as {blue, white, red} or the (infinite) set of all prime numbers. Partially ordered sets and sets with other relations have applications in several areas.

In discrete mathematics, countable sets (including finite sets) are the main focus. The beginning of set theory as a branch of mathematics is usually marked by Georg Cantor's work distinguishing between different kinds of infinite set, motivated by the study of trigonometric series, and further development of the theory of infinite sets is outside the scope of discrete mathematics. Indeed, contemporary work in descriptive set theory makes extensive use of traditional continuous mathematics.

Combinatorics

Combinatorics studies the way in which discrete structures can be combined or arranged. Enumerative combinatorics concentrates on counting the number of certain combinatorial objects - e.g. the twelvefold way provides a unified framework for counting permutations, combinations and partitions. Analytic combinatorics concerns the enumeration (i.e., determining the number) of combinatorial structures using tools from complex analysis and probability theory. In contrast with enumerative combinatorics which uses explicit combinatorial formulae and generating functions to describe the results, analytic combinatorics aims at obtaining asymptotic formulae. Design theory is a study of combinatorial designs, which are collections of subsets with certain intersection properties. Partition theory studies various enumeration and asymptotic problems related to integer partitions, and is closely related to q-series, special functions and orthogonal polynomials. Originally a part of number theory and analysis, partition theory is now considered a part of combinatorics or an independent field. Order theory is the study of partially ordered sets, both finite and infinite.

Graph Theory

Graph theory, the study of graphs and networks, is often considered part of combinatorics, but has grown large enough and distinct enough, with its own kind of problems, to be regarded as a subject in its own right. Graphs are one of the prime objects of study in discrete mathematics. They are among the most ubiquitous models of both natural and human-made structures. They can model many types of relations and process dynamics in physical, biological and social systems. In computer science, they can represent networks of communication, data organization, computational devices, the flow of computation, etc. In mathematics, they are useful in geometry and certain parts of topology, e.g. knot theory. Algebraic graph theory has close links with group theory. There are also continuous graphs, however for the most part research in graph theory falls within the domain of discrete mathematics.

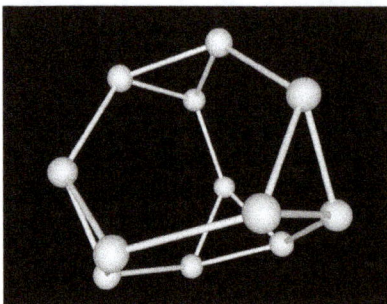

Graph theory has close links to group theory. This truncated tetrahedron graph is related to the alternating group A_4.

Probability

Discrete probability theory deals with events that occur in countable sample spaces. For example, count observations such as the numbers of birds in flocks comprise only natural number values $\{0, 1, 2, ...\}$. On the other hand, continuous observations such as the weights of birds comprise real number values and would typically be modeled by a continuous probability distribution such as the normal. Discrete probability distributions can be used to approximate continuous ones and vice versa. For highly constrained situations such as throwing dice or experiments with decks of cards, calculating the probability of events is basically enumerative combinatorics.

Number Theory

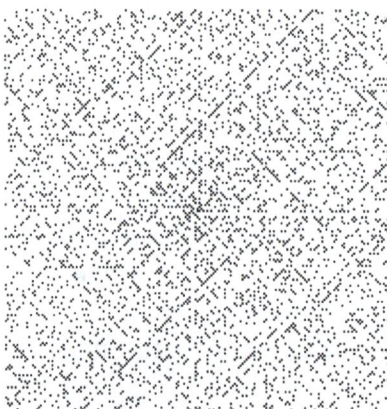

The Ulam spiral of numbers, with black pixels showing prime numbers. This diagram hints at patterns in the distribution of prime numbers.

Number theory is concerned with the properties of numbers in general, particularly integers. It has applications to cryptography, cryptanalysis, and cryptology, particularly with regard to modular arithmetic, diophantine equations, linear and quadratic congruences, prime numbers and primality testing. Other discrete aspects of number theory include geometry of numbers. In analytic number theory, techniques from continuous mathematics are also used. Topics that go beyond discrete objects include transcendental numbers, diophantine approximation, p-adic analysis and function fields.

Algebra

Algebraic structures occur as both discrete examples and continuous examples. Discrete algebras include: boolean algebra used in logic gates and programming; relational algebra used in databases; discrete and finite versions of groups, rings and fields are important in algebraic coding theory; discrete semigroups and monoids appear in the theory of formal languages.

Calculus of Finite Differences, Discrete Calculus or Discrete Analysis

A function defined on an interval of the integers is usually called a sequence. A sequence could be a finite sequence from a data source or an infinite sequence from a discrete dynamical system. Such a discrete function could be defined explicitly by a list (if its domain is finite), or by a formula for its general term, or it could be given implicitly by a recurrence relation or difference equation. Difference equations are similar to a differential equations, but replace differentiation by taking the difference between adjacent terms; they can be used to approximate differential equations or (more often) studied in their own right. Many questions and methods concerning differential equations have counterparts for difference equations. For instance where there are integral transforms in harmonic analysis for studying continuous functions or analog signals, there are discrete transforms for discrete functions or digital signals. As well as the discrete metric there are more general discrete or finite metric spaces and finite topological spaces.

Geometry

Computational geometry applies computer algorithms to representations of geometrical objects.

Discrete geometry and combinatorial geometry are about combinatorial properties of *discrete collections* of geometrical objects. A long-standing topic in discrete geometry is tiling of the plane. Computational geometry applies algorithms to geometrical problems.

Topology

Although topology is the field of mathematics that formalizes and generalizes the intuitive notion of "continuous deformation" of objects, it gives rise to many discrete topics; this can be attributed in part to the focus on topological invariants, which themselves usually take discrete values. See combinatorial topology, topological graph theory, topological combinatorics, computational topology, discrete topological space, finite topological space, topology (chemistry).

Operations Research

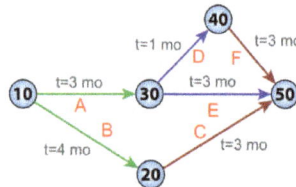

PERT charts like this provide a business management technique based on graph theory.

Operations research provides techniques for solving practical problems in business and other fields — problems such as allocating resources to maximize profit, or scheduling project activities to minimize risk. Operations research techniques include linear programming and other areas of optimization, queuing theory, scheduling theory, network theory. Operations research also includes continuous topics such as continuous-time Markov process, continuous-time martingales, process optimization, and continuous and hybrid control theory.

Game Theory, Decision Theory, Utility Theory, Social Choice Theory

	Cooperate	Defect
Cooperate	-1, -1	-10, 0
Defect	0, -10	-5, -5

Payoff matrix for the Prisoner's dilemma, a common example in game theory. One player chooses a row, the other a column; the resulting pair gives their payoffs

Decision theory is concerned with identifying the values, uncertainties and other issues relevant in a given decision, its rationality, and the resulting optimal decision.

Utility theory is about measures of the relative economic satisfaction from, or desirability of, consumption of various goods and services.

Social choice theory is about voting. A more puzzle-based approach to voting is ballot theory.

Game theory deals with situations where success depends on the choices of others, which makes choosing the best course of action more complex. There are even continuous games. Topics include auction theory and fair division.

Discretization

Discretization concerns the process of transferring continuous models and equations into discrete counterparts, often for the purposes of making calculations easier by using approximations. Numerical analysis provides an important example.

Discrete Analogues of Continuous Mathematics

There are many concepts in continuous mathematics which have discrete versions, such as discrete calculus, discrete probability distributions, discrete Fourier transforms, discrete geometry, discrete logarithms, discrete differential geometry, discrete exterior calculus, discrete Morse theory, difference equations, discrete dynamical systems, and discrete vector measures.

In applied mathematics, discrete modelling is the discrete analogue of continuous modelling. In discrete modelling, discrete formulae are fit to data. A common method in this form of modelling is to use recurrence relation.

In algebraic geometry, the concept of a curve can be extended to discrete geometries by taking the spectra of polynomial rings over finite fields to be models of the affine spaces over that field, and letting subvarieties or spectra of other rings provide the curves that lie in that space. Although the space in which the curves appear has a finite number of points, the curves are not so much sets of points as analogues of curves in continuous settings. For example, every point of the form $V(x-c) \subset Spec K[x] = \mathbb{A}^1$ for K a field can be studied either as $\operatorname{Spec} K[x]/(x-c) \cong \operatorname{Spec} K$, a point, or as the spectrum $Spec K[x]_{(x-c)}$ of the local ring at (x-c), a point together with a neighborhood around it. Algebraic varieties also have a well-defined notion of tangent space called the Zariski tangent space, making many features of calculus applicable even in finite settings.

Hybrid Discrete and Continuous Mathematics

The time scale calculus is a unification of the theory of difference equations with that of differential equations, which has applications to fields requiring simultaneous modelling of discrete and continuous data. Another way of modeling such a situation is the notion of hybrid dynamical system.

Mathematical Logic

Mathematical logic is a subfield of mathematics exploring the applications of formal logic to mathematics. It bears close connections to metamathematics, the foundations of mathematics, and theoretical computer science. The unifying themes in mathematical logic include the study of the expressive power of formal systems and the deductive power of formal proof systems.

Mathematical logic is often divided into the fields of set theory, model theory, recursion theory, and proof theory. These areas share basic results on logic, particularly first-order logic, and definability. In computer science (particularly in the ACM Classification) mathematical logic encompasses additional topics not detailed in this article.

Since its inception, mathematical logic has both contributed to, and has been motivated by, the study of foundations of mathematics. This study began in the late 19th century with the development of axiomatic frameworks for geometry, arithmetic, and analysis. In the early 20th century it was shaped by David Hilbert's program to prove the consistency of foundational theories. Results of Kurt Gödel, Gerhard Gentzen, and others provided partial resolution to the program, and clarified the issues involved in proving consistency. Work in set theory showed that almost all ordinary mathematics can be formalized in terms of sets, although there are some theorems that cannot be proven in common axiom systems for set theory. Contemporary work in the foundations of mathematics often focuses on establishing which parts of mathematics can be formalized in particular formal systems (as in reverse mathematics) rather than trying to find theories in which all of mathematics can be developed.

Subfields and scope

The *Handbook of Mathematical Logic* (Barwise 1989) makes a rough division of contemporary mathematical logic into four areas:

1.   set theory

2.   model theory

3.   recursion theory, and

4.   proof theory and constructive mathematics (considered as parts of a single area).

Each area has a distinct focus, although many techniques and results are shared among multiple areas. The borderlines amongst these fields, and the lines separating mathematical logic and other fields of mathematics, are not always sharp. Gödel's incompleteness theorem marks not only a milestone in recursion theory and proof theory, but has also led to Löb's theorem in modal logic. The method of forcing is employed in set theory, model theory, and recursion theory, as well as in the study of intuitionistic mathematics.

The mathematical field of category theory uses many formal axiomatic methods, and includes the study of categorical logic, but category theory is not ordinarily considered a subfield of mathematical logic. Because of its applicability in diverse fields of mathematics, mathematicians including Saunders Mac Lane have proposed category theory as a foundational system for mathematics, independent of set theory. These foundations use toposes, which resemble generalized models of set theory that may employ classical or nonclassical logic.

History

Mathematical logic emerged in the mid-19th century as a subfield of mathematics independent of the traditional study of logic (Ferreirós 2001, p. 443). Before this emergence, logic was studied with rhetoric, through the syllogism, and with philosophy. The first half of the 20th century saw an explosion of fundamental results, accompanied by vigorous debate over the foundations of mathematics.

Early History

Theories of logic were developed in many cultures in history, including China, India, Greece and the Islamic world. In 18th-century Europe, attempts to treat the operations of formal logic in a symbolic or algebraic way had been made by philosophical mathematicians including Leibniz and Lambert, but their labors remained isolated and little known.

19th Century

In the middle of the nineteenth century, George Boole and then Augustus De Morgan presented systematic mathematical treatments of logic. Their work, building on work by algebraists such as George Peacock, extended the traditional Aristotelian doctrine of logic into a sufficient framework for the study of foundations of mathematics (Katz 1998, p. 686).

Charles Sanders Peirce built upon the work of Boole to develop a logical system for relations and quantifiers, which he published in several papers from 1870 to 1885. Gottlob Frege presented an independent development of logic with quantifiers in his *Begriffsschrift*, published in 1879, a work generally considered as marking a turning point in the history of logic. Frege's work remained obscure, however, until Bertrand Russell began to promote it near the turn of the century. The two-dimensional notation Frege developed was never widely adopted and is unused in contemporary texts.

From 1890 to 1905, Ernst Schröder published *Vorlesungen über die Algebra der Logik* in three volumes. This work summarized and extended the work of Boole, De Morgan, and Peirce, and was a comprehensive reference to symbolic logic as it was understood at the end of the 19th century.

Foundational Theories

Concerns that mathematics had not been built on a proper foundation led to the development of axiomatic systems for fundamental areas of mathematics such as arithmetic, analysis, and geometry.

In logic, the term *arithmetic* refers to the theory of the natural numbers. Giuseppe Peano (1889) published a set of axioms for arithmetic that came to bear his name (Peano axioms), using a variation of the logical system of Boole and Schröder but adding quantifiers. Peano was unaware of Frege's work at the time. Around the same time Richard Dedekind showed that the natural numbers are uniquely characterized by their induction properties. Dedekind (1888) proposed a different characterization, which lacked the formal logical character of Peano's axioms. Dedekind's work, however, proved theorems inaccessible in Peano's system, including the uniqueness of the set of natural numbers (up to isomorphism) and the recursive definitions of addition and multiplication from the successor function and mathematical induction.

In the mid-19th century, flaws in Euclid's axioms for geometry became known (Katz 1998, p. 774). In addition to the independence of the parallel postulate, established by Nikolai Lobachevsky in 1826 (Lobachevsky 1840), mathematicians discovered that certain theorems taken for granted by Euclid were not in fact provable from his axioms. Among these is the theorem that a line contains at least two points, or that circles of the same radius whose centers are separated by that radius must intersect. Hilbert (1899) developed a complete set of axioms for geometry, building on previous work by Pasch (1882). The success in axiomatizing geometry motivated Hilbert to seek complete axiomatizations of other areas of mathematics, such as the natural numbers and the real line. This would prove to be a major area of research in the first half of the 20th century.

The 19th century saw great advances in the theory of real analysis, including theories of convergence of functions and Fourier series. Mathematicians such as Karl Weierstrass began to construct functions that stretched intuition, such as nowhere-differentiable continuous functions. Previous conceptions of a function as a rule for computation, or a smooth graph, were no longer adequate. Weierstrass began to advocate the arithmetization of analysis, which sought to axiomatize analysis using properties of the natural numbers. The modern (ε, δ)-definition of limit and continuous functions was already developed by Bolzano in 1817 (Felscher 2000), but remained relatively unknown. Cauchy in 1821 defined continuity in terms of infinitesimals. In 1858, Dedekind proposed a definition of the real numbers in terms of Dedekind cuts of rational numbers (Dedekind 1872), a definition still employed in contemporary texts.

Georg Cantor developed the fundamental concepts of infinite set theory. His early results developed the theory of cardinality and proved that the reals and the natural numbers have different cardinalities (Cantor 1874). Over the next twenty years, Cantor developed a theory of transfinite numbers in a series of publications. In 1891, he published a new proof of the uncountability of the real numbers that introduced the diagonal argument, and used this method to prove Cantor's theorem that no set can have the same cardinality as its powerset. Cantor believed that every set could be well-ordered, but was unable to produce a proof for this result, leaving it as an open problem in 1895 (Katz 1998, p. 807).

20th Century

In the early decades of the 20th century, the main areas of study were set theory and formal logic. The discovery of paradoxes in informal set theory caused some to wonder whether mathematics itself is inconsistent, and to look for proofs of consistency.

In 1900, Hilbert posed a famous list of 23 problems for the next century. The first two of these were to resolve the continuum hypothesis and prove the consistency of elementary arithmetic, respectively; the tenth was to produce a method that could decide whether a multivariate polynomial equation over the integers has a solution. Subsequent work to resolve these problems shaped the direction of mathematical logic, as did the effort to resolve Hilbert's *Entscheidungsproblem*, posed in 1928. This problem asked for a procedure that would decide, given a formalized mathematical statement, whether the statement is true or false.

Set Theory and Paradoxes

Ernst Zermelo (1904) gave a proof that every set could be well-ordered, a result Georg Cantor had been unable to obtain. To achieve the proof, Zermelo introduced the axiom of choice, which drew heated debate and research among mathematicians and the pioneers of set theory. The immediate criticism of the method led Zermelo to publish a second exposition of his result, directly addressing criticisms of his proof (Zermelo 1908a). This paper led to the general acceptance of the axiom of choice in the mathematics community.

Skepticism about the axiom of choice was reinforced by recently discovered paradoxes in naive set theory. Cesare Burali-Forti (1897) was the first to state a paradox: the Burali-Forti paradox shows that the collection of all ordinal numbers cannot form a set. Very soon thereafter, Bertrand Russell discovered Russell's paradox in 1901, and Jules Richard (1905) discovered Richard's paradox.

Zermelo (1908b) provided the first set of axioms for set theory. These axioms, together with the additional axiom of replacement proposed by Abraham Fraenkel, are now called Zermelo–Fraenkel set theory (ZF). Zermelo's axioms incorporated the principle of limitation of size to avoid Russell's paradox.

In 1910, the first volume of *Principia Mathematica* by Russell and Alfred North Whitehead was published. This seminal work developed the theory of functions and cardinality in a completely formal framework of type theory, which Russell and Whitehead developed in an effort to avoid the paradoxes. *Principia Mathematica* is considered one of the most influential works of the 20th century, although the framework of type theory did not prove popular as a foundational theory for mathematics (Ferreirós 2001, p. 445).

Fraenkel (1922) proved that the axiom of choice cannot be proved from the remaining axioms of Zermelo's set theory with urelements. Later work by Paul Cohen (1966) showed that the addition of urelements is not needed, and the axiom of choice is unprovable in ZF. Cohen's proof developed the method of forcing, which is now an important tool for establishing independence results in set theory.

Symbolic Logic

Leopold Löwenheim (1915) and Thoralf Skolem (1920) obtained the Löwenheim–Skolem theorem, which says that first-order logic cannot control the cardinalities of infinite structures. Skolem realized that this theorem would apply to first-order formalizations of set theory, and that it implies any such formalization has a countable model. This counterintuitive fact became known as Skolem's paradox.

In his doctoral thesis, Kurt Gödel (1929) proved the completeness theorem, which establishes a correspondence between syntax and semantics in first-order logic. Gödel used the completeness theorem to prove the compactness theorem, demonstrating the finitary nature of first-order logical consequence. These results helped establish first-order logic as the dominant logic used by mathematicians.

In 1931, Gödel published *On Formally Undecidable Propositions of Principia Mathematica and Related Systems*, which proved the incompleteness (in a different meaning of the word) of all sufficiently strong, effective first-order theories. This result, known as Gödel's incompleteness theorem, establishes severe limitations on axiomatic foundations for mathematics, striking a strong blow to Hilbert's program. It showed the impossibility of providing a consistency proof of arithmetic within any formal theory of arithmetic. Hilbert, however, did not acknowledge the importance of the incompleteness theorem for some time.

Gödel's theorem shows that a consistency proof of any sufficiently strong, effective axiom system cannot be obtained in the system itself, if the system is consistent, nor in any weaker system. This leaves open the possibility of consistency proofs that cannot be formalized within the system they consider. Gentzen (1936) proved the consistency of arithmetic using a finitistic system together with a principle of transfinite induction. Gentzen's result introduced the ideas of cut elimination and proof-theoretic ordinals, which became key tools in proof theory. Gödel (1958) gave a different consistency proof, which reduces the consistency of classical arithmetic to that of intutitionistic arithmetic in higher types.

Beginnings of the Other Branches

Alfred Tarski developed the basics of model theory.

Beginning in 1935, a group of prominent mathematicians collaborated under the pseudonym Nicolas Bourbaki to publish a series of encyclopedic mathematics texts. These texts, written in an austere and axiomatic style, emphasized rigorous presentation and set-theoretic foundations. Terminology coined by these texts, such as the words *bijection*, *injection*, and *surjection*, and the set-theoretic foundations the texts employed, were widely adopted throughout mathematics.

The study of computability came to be known as recursion theory, because early formalizations by Gödel and Kleene relied on recursive definitions of functions. When these definitions were shown equivalent to Turing's formalization involving Turing machines, it became clear that a new concept – the computable function – had been discovered, and that this definition was robust enough to admit numerous independent characterizations. In his work on the incompleteness theorems in 1931, Gödel lacked a rigorous concept of an effective formal system; he immediately realized that the new definitions of computability could be used for this purpose, allowing him to state the incompleteness theorems in generality that could only be implied in the original paper.

Numerous results in recursion theory were obtained in the 1940s by Stephen Cole Kleene and Emil Leon Post. Kleene (1943) introduced the concepts of relative computability, foreshadowed by Turing (1939), and the arithmetical hierarchy. Kleene later generalized recursion theory to higher-order functionals. Kleene and Kreisel studied formal versions of intuitionistic mathematics, particularly in the context of proof theory.

Formal Logical Systems

At its core, mathematical logic deals with mathematical concepts expressed using formal logical systems. These systems, though they differ in many details, share the common property of considering only expressions in a fixed formal language. The systems of propositional logic and first-order logic are the most widely studied today, because of their applicability to foundations of mathematics and because of their desirable proof-theoretic properties. Stronger classical logics such as second-order logic or infinitary logic are also studied, along with nonclassical logics such as intuitionistic logic.

First-order Logic

First-order logic is a particular formal system of logic. Its syntax involves only finite expressions as well-formed formulas, while its semantics are characterized by the limitation of all quantifiers to a fixed domain of discourse.

Early results from formal logic established limitations of first-order logic. The Löwenheim–Skolem theorem (1919) showed that if a set of sentences in a countable first-order language has an infinite model then it has at least one model of each infinite cardinality. This shows that it is impossible for a set of first-order axioms to characterize the natural numbers, the real numbers, or any other infinite structure up to isomorphism. As the goal of early foundational studies was to produce axiomatic theories for all parts of mathematics, this limitation was particularly stark.

Gödel's completeness theorem (Gödel 1929) established the equivalence between semantic and syntactic definitions of logical consequence in first-order logic. It shows that if a particular sentence is true in every model that satisfies a particular set of axioms, then there must be a finite deduction of the sentence from the axioms. The compactness theorem first appeared as a lemma in Gödel's proof of the completeness theorem, and it took many years before logicians grasped its significance and began to apply it routinely. It says that a set of sentences has a model if and only if every finite subset has a model, or in other words that an inconsistent set of formulas must have a finite inconsistent subset. The completeness and compactness theorems allow for sophisticated analysis of logical consequence in first-order logic and the development of model theory, and they are a key reason for the prominence of first-order logic in mathematics.

Gödel's incompleteness theorems (Gödel 1931) establish additional limits on first-order axiomatizations. The first incompleteness theorem states that for any consistent, effectively given (defined below) logical system that is capable of interpreting arithmetic (that is, of expressing the Peano axioms) there exists a statement (the Gödel sentence) which is true (in the sense that it holds for the natural numbers) but not provable within that logical system (and which indeed may fail in some non-standard models of arithmetic which may be consistent with the logical system.) Here a logical system is said to be effectively given if it is possible to decide, given any formula in the language of the system, whether the formula is an axiom, and one which can express the Peano axioms is called "sufficiently strong." When applied to first-order logic, the first incompleteness theorem implies that any sufficiently strong, consistent, effective first-order theory has models that are not elementarily equivalent, a stronger limitation than the one established by the Löwenheim–Skolem theorem. The second incompleteness theorem states that no sufficiently strong, consistent, effective axiom system for arithmetic can prove its own consistency, which has been interpreted to show that Hilbert's program cannot be completed.

Other Classical Logics

Many logics besides first-order logic are studied. These include infinitary logics, which allow for formulas to provide an infinite amount of information, and higher-order logics, which include a portion of set theory directly in their semantics.

The most well studied infinitary logic is $L_{\omega_1,\omega}$. In this logic, quantifiers may only be nested to finite depths, as in first-order logic, but formulas may have finite or countably infinite conjunctions and disjunctions within them. Thus, for example, it is possible to say that an object is a whole number using a formula of $L_{\omega_1,\omega}$ such as

$$(x = 0) \vee (x = 1) \vee (x = 2) \vee \cdots.$$

Higher-order logics allow for quantification not only of elements of the domain of discourse, but subsets of the domain of discourse, sets of such subsets, and other objects of higher type. The semantics are defined so that, rather than having a separate domain for each higher-type quantifier to range over, the quantifiers instead range over all objects of the appropriate type. The logics studied before the development of first-order logic, for example Frege's logic, had similar set-theoretic aspects. Although higher-order logics are more expressive, allowing complete axiomatizations of structures such as the natural numbers, they do not satisfy analogues of the completeness and compactness theorems from first-order logic, and are thus less amenable to proof-theoretic analysis.

Another type of logics are fixed-point logics that allow inductive definitions, like one writes for primitive recursive functions.

One can formally define an extension of first-order logic — a notion which encompasses all logics in this section because they behave like first-order logic in certain fundamental ways, but does not encompass all logics in general, e.g. it does not encompass intuitionistic, modal or fuzzy logic. Lindström's theorem implies that the only extension of first-order logic satisfying both the compactness theorem and the Downward Löwenheim–Skolem theorem is first-order logic.

Nonclassical and Modal Logic

Modal logics include additional modal operators, such as an operator which states that a particular formula is not only true, but necessarily true. Although modal logic is not often used to axiomatize mathematics, it has been used to study the properties of first-order provability (Solovay 1976) and set-theoretic forcing (Hamkins and Löwe 2007).

Intuitionistic logic was developed by Heyting to study Brouwer's program of intuitionism, in which Brouwer himself avoided formalization. Intuitionistic logic specifically does not include the law of the excluded middle, which states that each sentence is either true or its negation is true. Kleene's work with the proof theory of intuitionistic logic showed that constructive information can be recovered from intuitionistic proofs. For example, any provably total function in intuitionistic arithmetic is computable; this is not true in classical theories of arithmetic such as Peano arithmetic.

Algebraic Logic

Algebraic logic uses the methods of abstract algebra to study the semantics of formal logics. A fundamental example is the use of Boolean algebras to represent truth values in classical propositional logic, and the use of Heyting algebras to represent truth values in intuitionistic propositional logic. Stronger logics, such as first-order logic and higher-order logic, are studied using more complicated algebraic structures such as cylindric algebras.

Set Theory

Set theory is the study of sets, which are abstract collections of objects. Many of the basic notions, such as ordinal and cardinal numbers, were developed informally by Cantor before formal axiomatizations of set theory were developed. The first such axiomatization, due to Zermelo (1908b), was extended slightly to become Zermelo–Fraenkel set theory (ZF), which is now the most widely used foundational theory for mathematics.

Other formalizations of set theory have been proposed, including von Neumann–Bernays–Gödel set theory (NBG), Morse–Kelley set theory (MK), and New Foundations (NF). Of these, ZF, NBG, and MK are similar in describing a cumulative hierarchy of sets. New Foundations takes a different approach; it allows objects such as the set of all sets at the cost of restrictions on its set-existence axioms. The system of Kripke–Platek set theory is closely related to generalized recursion theory.

Two famous statements in set theory are the axiom of choice and the continuum hypothesis. The axiom of choice, first stated by Zermelo (1904), was proved independent of ZF by Fraenkel (1922), but has come to be widely accepted by mathematicians. It states that given a collection of nonempty sets there is a single set C that contains exactly one element from each set in the collection. The set C is said to "choose" one element from each set in the collection. While the ability to make such a choice is considered obvious by some, since each set in the collection is nonempty, the lack of a general, concrete rule by which the choice can be made renders the axiom nonconstructive. Stefan Banach and Alfred Tarski (1924) showed that the axiom of choice can be used to decompose a solid ball into a finite number of pieces which can then be rearranged, with no scaling, to make two solid balls of the original size. This theorem, known as the Banach–Tarski paradox, is one of many counterintuitive results of the axiom of choice.

The continuum hypothesis, first proposed as a conjecture by Cantor, was listed by David Hilbert as one of his 23 problems in 1900. Gödel showed that the continuum hypothesis cannot be disproven from the axioms of Zermelo–Fraenkel set theory (with or without the axiom of choice), by developing the constructible universe of set theory in which the continuum hypothesis must hold. In 1963, Paul Cohen showed that the continuum hypothesis cannot be proven from the axioms of Zermelo–Fraenkel set theory (Cohen 1966). This independence result did not completely settle Hilbert's question, however, as it is possible that new axioms for set theory could resolve the hypothesis. Recent work along these lines has been conducted by W. Hugh Woodin, although its importance is not yet clear (Woodin 2001).

Contemporary research in set theory includes the study of large cardinals and determinacy. Large cardinals are cardinal numbers with particular properties so strong that the existence of such cardinals cannot be proved in ZFC. The existence of the smallest large cardinal typically studied, an inaccessible cardinal, already implies the consistency of ZFC. Despite the fact that large cardinals have extremely high cardinality, their existence has many ramifications for the structure of the real line. *Determinacy* refers to the possible existence of winning strategies for certain two-player games (the games are said to be *determined*). The existence of these strategies implies structural properties of the real line and other Polish spaces.

Model Theory

Model theory studies the models of various formal theories. Here a theory is a set of formulas in a particular formal logic and signature, while a model is a structure that gives a concrete interpretation of the theory. Model theory is closely related to universal algebra and algebraic geometry, although the methods of model theory focus more on logical considerations than those fields.

The set of all models of a particular theory is called an elementary class; classical model theory seeks to determine the properties of models in a particular elementary class, or determine whether certain classes of structures form elementary classes.

The method of quantifier elimination can be used to show that definable sets in particular theories cannot be too complicated. Tarski (1948) established quantifier elimination for real-closed fields, a result which also shows the theory of the field of real numbers is decidable. (He also noted that his methods were equally applicable to algebraically closed fields of arbitrary characteristic.) A modern subfield developing from this is concerned with o-minimal structures.

Morley's categoricity theorem, proved by Michael D. Morley (1965), states that if a first-order theory in a countable language is categorical in some uncountable cardinality, i.e. all models of this cardinality are isomorphic, then it is categorical in all uncountable cardinalities.

A trivial consequence of the continuum hypothesis is that a complete theory with less than continuum many nonisomorphic countable models can have only countably many. Vaught's conjecture, named after Robert Lawson Vaught, says that this is true even independently of the continuum hypothesis. Many special cases of this conjecture have been established.

Recursion Theory

Recursion theory, also called computability theory, studies the properties of computable functions and the Turing degrees, which divide the uncomputable functions into sets that have the same level of uncomputability. Recursion theory also includes the study of generalized computability and definability. Recursion theory grew from the work of Alonzo Church and Alan Turing in the 1930s, which was greatly extended by Kleene and Post in the 1940s.

Classical recursion theory focuses on the computability of functions from the natural numbers to the natural numbers. The fundamental results establish a robust, canonical class of computable functions with numerous independent, equivalent characterizations using Turing machines, λ calculus, and other systems. More advanced results concern the structure of the Turing degrees and the lattice of recursively enumerable sets.

Generalized recursion theory extends the ideas of recursion theory to computations that are no longer necessarily finite. It includes the study of computability in higher types as well as areas such as hyperarithmetical theory and α-recursion theory.

Contemporary research in recursion theory includes the study of applications such as algorithmic randomness, computable model theory, and reverse mathematics, as well as new results in pure recursion theory.

Algorithmically Unsolvable Problems

An important subfield of recursion theory studies algorithmic unsolvability; a decision problem or function problem is algorithmically unsolvable if there is no possible computable algorithm that returns the correct answer for all legal inputs to the problem. The first results about unsolvability, obtained independently by Church and Turing in 1936, showed that the Entscheidungsproblem is algorithmically unsolvable. Turing proved this by establishing the unsolvability of the halting problem, a result with far-ranging implications in both recursion theory and computer science.

There are many known examples of undecidable problems from ordinary mathematics. The word problem for groups was proved algorithmically unsolvable by Pyotr Novikov in 1955 and independently by W. Boone in 1959. The busy beaver problem, developed by Tibor Radó in 1962, is another well-known example.

Hilbert's tenth problem asked for an algorithm to determine whether a multivariate polynomial equation with integer coefficients has a solution in the integers. Partial progress was made by Julia Robinson, Martin Davis and Hilary Putnam. The algorithmic unsolvability of the problem was proved by Yuri Matiyasevich in 1970 (Davis 1973).

Proof Theory and Constructive Mathematics

Proof theory is the study of formal proofs in various logical deduction systems. These proofs are represented as formal mathematical objects, facilitating their analysis by mathematical techniques. Several deduction systems are commonly considered, including Hilbert-style deduction systems, systems of natural deduction, and the sequent calculus developed by Gentzen.

The study of constructive mathematics, in the context of mathematical logic, includes the study of systems in non-classical logic such as intuitionistic logic, as well as the study of predicative systems. An early proponent of predicativism was Hermann Weyl, who showed it is possible to develop a large part of real analysis using only predicative methods (Weyl 1918).

Because proofs are entirely finitary, whereas truth in a structure is not, it is common for work in constructive mathematics to emphasize provability. The relationship between provability in classical (or nonconstructive) systems and provability in intuitionistic (or constructive, respectively) systems is of particular interest. Results such as the Gödel–Gentzen negative translation show that it is possible to embed (or *translate*) classical logic into intuitionistic logic, allowing some properties about intuitionistic proofs to be transferred back to classical proofs.

Recent developments in proof theory include the study of proof mining by Ulrich Kohlenbach and the study of proof-theoretic ordinals by Michael Rathjen.

Connections with Computer Science

The study of computability theory in computer science is closely related to the study of computability in mathematical logic. There is a difference of emphasis, however. Computer scientists often focus on concrete programming languages and feasible computability, while researchers in mathematical logic often focus on computability as a theoretical concept and on noncomputability.

The theory of semantics of programming languages is related to model theory, as is program verification (in particular, model checking). The Curry–Howard isomorphism between proofs and programs relates to proof theory, especially intuitionistic logic. Formal calculi such as the lambda calculus and combinatory logic are now studied as idealized programming languages.

Computer science also contributes to mathematics by developing techniques for the automatic checking or even finding of proofs, such as automated theorem proving and logic programming.

Descriptive complexity theory relates logics to computational complexity. The first significant result in this area, Fagin's theorem (1974) established that NP is precisely the set of languages expressible by sentences of existential second-order logic.

Foundations of Mathematics

In the 19th century, mathematicians became aware of logical gaps and inconsistencies in their field. It was shown that Euclid's axioms for geometry, which had been taught for centuries as an example of the axiomatic method, were incomplete. The use of infinitesimals, and the very definition of function, came into question in analysis, as pathological examples such as Weierstrass' nowhere-differentiable continuous function were discovered.

Cantor's study of arbitrary infinite sets also drew criticism. Leopold Kronecker famously stated "God made the integers; all else is the work of man," endorsing a return to the study of finite, concrete objects in mathematics. Although Kronecker's argument was carried forward by constructivists in the 20th century, the mathematical community as a whole rejected them. David Hilbert argued in favor of the study of the infinite, saying "No one shall expel us from the Paradise that Cantor has created."

Mathematicians began to search for axiom systems that could be used to formalize large parts of mathematics. In addition to removing ambiguity from previously naive terms such as function, it was hoped that this axiomatization would allow for consistency proofs. In the 19th century, the main method of proving the consistency of a set of axioms was to provide a model for it. Thus, for example, non-Euclidean geometry can be proved consistent by defining *point* to mean a point on a fixed sphere and *line* to mean a great circle on the sphere. The resulting structure, a model of elliptic geometry, satisfies the axioms of plane geometry except the parallel postulate.

With the development of formal logic, Hilbert asked whether it would be possible to prove that an axiom system is consistent by analyzing the structure of possible proofs in the system, and showing through this analysis that it is impossible to prove a contradiction. This idea led to the study of proof theory. Moreover, Hilbert proposed that the analysis should be entirely concrete, using the term *finitary* to refer to the methods he would allow but not precisely defining them. This project, known as Hilbert's program, was seriously affected by Gödel's incompleteness theorems, which show that the consistency of formal theories of arithmetic cannot be established using methods formalizable in those theories. Gentzen showed that it is possible to produce a proof of the consistency of arithmetic in a finitary system augmented with axioms of transfinite induction, and the techniques he developed to do so were seminal in proof theory.

A second thread in the history of foundations of mathematics involves nonclassical logics and constructive mathematics. The study of constructive mathematics includes many different programs with various definitions of *constructive*. At the most accommodating end, proofs in ZF set theory that do not use the axiom of choice are called constructive by many mathematicians. More limited versions of constructivism limit themselves to natural numbers, number-theoretic functions, and sets of natural numbers (which can be used to represent real numbers, facilitating the study of mathematical analysis). A common idea is that a concrete means of computing the values of the function must be known before the function itself can be said to exist.

In the early 20th century, Luitzen Egbertus Jan Brouwer founded intuitionism as a philosophy of mathematics. This philosophy, poorly understood at first, stated that in order for a mathematical statement to be true to a mathematician, that person must be able to *intuit* the statement, to not only believe its truth but understand the reason for its truth. A consequence of this definition of truth was the rejection of the law of the excluded middle, for there are statements that, according to Brouwer, could not be claimed to be true while their negations also could not be claimed true. Brouwer's philosophy was influential, and the cause of bitter disputes among prominent mathematicians. Later, Kleene and Kreisel would study formalized versions of intuitionistic logic (Brouwer rejected formalization, and presented his work in unformalized natural language). With the advent of the BHK interpretation and Kripke models, intuitionism became easier to reconcile with classical mathematics.

References

- Boolos, George; Burgess, John; Jeffrey, Richard (2002), Computability and Logic (4th ed.), Cambridge: Cambridge University Press, ISBN 978-0-521-00758-0 .

- Crossley, J.N.; Ash, C.J.; Brickhill, C.J.; Stillwell, J.C.; Williams, N.H. (1972), What is mathematical logic?, London-Oxford-New York: Oxford University Press, ISBN 0-19-888087-1, Zbl 0251.02001

- Enderton, Herbert (2001), A mathematical introduction to logic (2nd ed.), Boston, MA: Academic Press, ISBN 978-0-12-238452-3 .

- Ebbinghaus, H.-D.; Flum, J.; Thomas, W. (1994), Mathematical Logic (2nd ed.), New York: Springer, ISBN 0-387-94258-0 .

- Rautenberg, Wolfgang (2010), A Concise Introduction to Mathematical Logic (3rd ed.), New York: Springer Science+Business Media, doi:10.1007/978-1-4419-1221-3, ISBN 978-1-4419-1220-6 .

- Shawn Hedman, A first course in logic: an introduction to model theory, proof theory, computability, and complexity, Oxford University Press, 2004, ISBN 0-19-852981-3.

- Schwichtenberg, Helmut (2003–2004), Mathematical Logic (PDF), Munich, Germany: Mathematisches Institut der Universität München, retrieved 2016-02-24 .

- Biggs, Norman L. (2002), Discrete mathematics, Oxford Science Publications (2nd ed.), New York: The Clarendon Press Oxford University Press, p. 89, ISBN 9780198507178.

- Albert Geoffrey Howson, ed. (1988). Mathematics as a Service Subject. Cambridge University Press. pp. 77–78. ISBN 978-0-521-35395-3.

- Trevor R. Hodkinson; John A. N. Parnell (2007). Reconstruction the Tree of Life: Taxonomy And Systematics of Large And Species Rich Taxa. CRC PressINC. p. 97. ISBN 978-0-8493-9579-6.

- A. S. Troelstra; H. Schwichtenberg (2000-07-27). Basic Proof Theory. Cambridge University Press. p. 186. ISBN 978-0-521-77911-1.

- Franz Baader; Gerhard Brewka; Thomas Eiter (2001-10-16). KI 2001: Advances in Artificial Intelligence: Joint German/Austrian Conference on AI, Vienna, Austria, September 19-21, 2001. Proceedings. Springer. p. 325. ISBN 978-3-540-42612-7.

- Moore, David (1992). "Teaching Statistics as a Respectable Subject". In F. Gordon and S. Gordon. Statistics for the Twenty-First Century. Washington, DC: The Mathematical Association of America. pp. 14–25. ISBN 978-0-88385-078-7.

- Chance, Beth L.; Rossman, Allan J. (2005). "Preface". Investigating Statistical Concepts, Applications, and Methods (PDF). Duxbury Press. ISBN 978-0-495-05064-3.

- Lakshmikantham,, ed. by D. Kannan,... V. (2002). Handbook of stochastic analysis and applications. New York: M. Dekker. ISBN 0824706609.

- Everitt, Brian (1998). The Cambridge Dictionary of Statistics. Cambridge, UK New York: Cambridge University Press. ISBN 0521593468.

- Drennan, Robert D. (2008). "Statistics in archaeology". In Pearsall, Deborah M. Encyclopedia of Archaeology. Elsevier Inc. pp. 2093–2100. ISBN 978-0-12-373962-9.

- Anderson, D.R.; Sweeney, D.J.; Williams, T.A. (1994) Introduction to Statistics: Concepts and Applications, pp. 5–9. West Group. ISBN 978-0-314-03309-3

- "What Is the Difference Between Type I and Type II Hypothesis Testing Errors?". About.com Education. Retrieved 2015-11-27.

7

Applications of Mathemathics

Mathematics is widely applicable. The disciplines listed here borrow from mathematical understanding in as much as theorems developed from pure mathematics are applied in these topics. Some of the topics listed in this chapter are mathematical physics and mathematical economics. The subjects discussed in the chapter are of great importance to broaden the existing knowledge on this field.

Mathematical Physics

Mathematical physics refers to development of mathematical methods for application to problems in physics. The *Journal of Mathematical Physics* defines the field as "the application of mathematics to problems in physics and the development of mathematical methods suitable for such applications and for the formulation of physical theories". It is a branch of applied mathematics, but deals with physical problems.

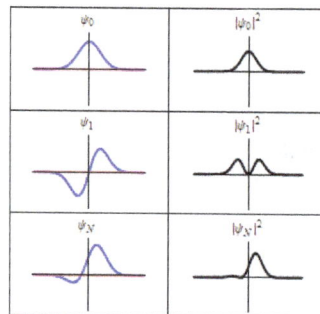

An example of mathematical physics: solutions of Schrödinger's equation for quantum harmonic oscillators (left) with their amplitudes (right).

Scope

There are several distinct branches of mathematical physics, and these roughly correspond to particular historical periods.

Classical Mechanics

The rigorous, abstract and advanced re-formulation of Newtonian mechanics adopting the Lagrangian mechanics and the Hamiltonian mechanics even in the presence of constraints. Both formulations are embodied in the so-called analytical mechanics. It leads, for instance, to discover the deep interplay of the notion of symmetry and that of conserved quantities during the dynamical evolution, stated within the most elementary formulation of Noether's theorem. These approaches and ideas can be and, in fact, have been extended to other areas of physics as statistical mechanics, continuum mechanics, classical field theory and quantum field theory. Moreover, they have provided several examples and basic ideas in differential geometry (e.g. the theory of vector bundles and several notions in symplectic geometry).

Partial Differential Equations

The theory of partial differential equations (and the related areas of variational calculus, Fourier analysis, potential theory, and vector analysis) are perhaps most closely associated with mathematical physics. These were developed intensively from the second half of the eighteenth century (by, for example, D'Alembert, Euler, and Lagrange) until the 1930s. Physical applications of these developments include hydrodynamics, celestial mechanics, continuum mechanics, elasticity theory, acoustics, thermodynamics, electricity, magnetism, and aerodynamics.

Quantum Theory

The theory of atomic spectra (and, later, quantum mechanics) developed almost concurrently with the mathematical fields of linear algebra, the spectral theory of operators, operator algebras and more broadly, functional analysis. Nonrelativistic quantum mechanics includes Schrödinger operators, and it has connections to atomic and molecular physics. Quantum information theory is another subspecialty.

Relativity and Quantum Relativistic Theories

The special and general theories of relativity require a rather different type of mathematics. This was group theory, which played an important role in both quantum field theory and differential geometry. This was, however, gradually supplemented by topology and functional analysis in the mathematical description of cosmological as well as quantum field theory phenomena. In this area both homological algebra and category theory are important nowadays.

Statistical Mechanics

Statistical mechanics forms a separate field, which includes the theory of phase transitions. It relies upon the Hamiltonian mechanics (or its quantum version) and it is closely related with the more mathematical ergodic theory and some parts of probability theory. There are increasing interactions between combinatorics and physics, in particular statistical physics.

Usage

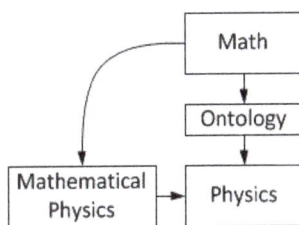

The usage of the term "mathematical physics" is sometimes idiosyncratic. Certain parts of mathematics that initially arose from the development of physics are not, in fact, considered parts of mathematical physics, while other closely related fields are. For example, ordinary differential equations and symplectic geometry are generally viewed as purely mathematical disciplines, whereas dynamical systems and Hamiltonian mechanics belong to mathematical physics.

Mathematical Vs. Theoretical Physics

The term "mathematical physics" is sometimes used to denote research aimed at studying and solving problems inspired by physics or thought experiments within a mathematically rigorous framework. In this sense, mathematical physics covers a very broad academic realm distinguished only by the blending of pure mathematics and physics. Although related to theoretical physics, mathematical physics in this sense emphasizes the mathematical rigour of the same type as found in mathematics.

On the other hand, theoretical physics emphasizes the links to observations and experimental physics, which often requires theoretical physicists (and mathematical physicists in the more general sense) to use heuristic, intuitive, and approximate arguments. Such arguments are not considered rigorous by mathematicians. Arguably, rigorous mathematical physics is closer to mathematics, and theoretical physics is closer to physics. This is reflected institutionally: mathematical physicists are often members of the mathematics department.

Such mathematical physicists primarily expand and elucidate physical theories. Because of the required level of mathematical rigour, these researchers often deal with questions that theoretical physicists have considered to already be solved. However, they can sometimes show (but neither commonly nor easily) that the previous solution was incomplete, incorrect, or simply, too naive. Issues about attempts to infer the second law of thermodynamics from statistical mechanics are examples. Other examples concern the subtleties involved with synchronisation procedures in special and general relativity (Sagnac effect and Einstein synchronisation).

The effort to put physical theories on a mathematically rigorous footing has inspired many mathematical developments. For example, the development of quantum mechanics and some aspects of functional analysis parallel each other in many ways. The mathematical study of quantum mechanics, quantum field theory and quantum statistical mechanics has motivated results in operator algebras. The attempt to construct a rigorous quantum field theory has also brought about progress in fields such as representation theory. Use of geometry and topology plays an important role in string theory.

Prominent Mathematical Physicists

Before Newton

The roots of mathematical physics can be traced back to the likes of Archimedes in Greece, Ptolemy in Egypt, Alhazen in Iraq, and Al-Biruni in Persia.

In the first decade of the 16th century, amateur astronomer Nicolaus Copernicus proposed heliocentrism, and published a treatise on it in 1543. Not quite radical, Copernicus merely sought to simplify astronomy and achieve orbits of more perfect circles, stated by Aristotelian physics to be the intrinsic motion of Aristotle's fifth element—the quintessence or universal essence known in Greek as *aither* for the English *pure air*—that was the pure substance beyond the sublunary sphere, and thus was celestial entities' pure composition. The German Johannes Kepler [1571–1630], Tycho Brahe's assistant, modified Copernican orbits to *ellipses*, however, formalized in the equations of Kepler's laws of planetary motion.

An enthusiastic atomist, Galileo Galilei in his 1623 book *The Assayer* asserted that the "book of nature" is written in mathematics. His 1632 book, upon his telescopic observations, supported heliocentrism. Having introduced experimentation, Galileo then refuted geocentric cosmology by refuting Aristotelian physics itself. Galilei's 1638 book *Discourse on Two New Sciences* established law of equal free fall as well as the principles of inertial motion, founding the central concepts of what would become today's classical mechanics. By the Galilean law of inertia as well as the principle Galilean invariance, also called Galilean relativity, for any object experiencing inertia, there is empirical justification of knowing only its being at *relative* rest or *relative* motion—rest or motion with respect to another object.

René Descartes adopted Galilean principles and developed a complete system of heliocentric cosmology, anchored on the principle of vortex motion, Cartesian physics, whose widespread acceptance brought demise of Aristotelian physics. Descartes sought to formalize mathematical reasoning in science, and developed Cartesian coordinates for geometrically plotting locations in 3D space and marking their progressions along the flow of time.

Newtonian and Post Newtonian

Isaac Newton [1642–1727] developed new mathematics, including calculus and several numerical methods such as Newton's method to solve problems in physics. Newton's theory of motion, published in 1687, modeled three Galilean laws of motion along with Newton's law of universal gravitation on a framework of absolute space—hypothesized by Newton as a physically real entity of Euclidean geometric structure extending infinitely in all directions—while presuming absolute time, supposedly justifying knowledge of absolute motion, the object's motion with respect to absolute space. The principle Galilean invariance/relativity was merely implicit in Newton's theory of motion. Having ostensibly reduced Keplerian celestial laws of motion as well as Galilean terrestrial laws of motion to a unifying force, Newton achieved great mathematic rigor if theoretical laxity.

In the 18th century, the Swiss Daniel Bernoulli [1700–1782] made contributions to fluid dynamics, and vibrating strings. The Swiss Leonhard Euler [1707–1783] did special work in variational calculus, dynamics, fluid dynamics, and other areas. Also notable was the Italian-born Frenchman, Joseph-Louis Lagrange [1736–1813] for work in analytical mechanics (he formulated the so-called Lagrangian mechanics) and variational methods. A major contribution to the formulation of Analytical Dynamics called Hamiltonian Dynamics was also made by the Irish physicist, astronomer and mathematician, William Rowan Hamilton [1805-1865]. Hamiltonian Dynamics had played an important role in the formulation of modern theories in physics including field theory and quantum mechanics. The French mathematical physicist Joseph Fourier [1768 – 1830] introduced the notion of Fourier series to solve the heat equation giving rise to a new approach to handle partial differential equations by means of integral transforms.

Into the early 19th century, the French Pierre-Simon Laplace [1749–1827] made paramount contributions to mathematical astronomy, potential theory, and probability theory. Siméon Denis Poisson [1781–1840] worked in analytical mechanics and potential theory. In Germany, Carl Friedrich Gauss [1777–1855] made key contributions to the theoretical foundations of electricity, magnetism, mechanics, and fluid dynamics.

A couple of decades ahead of Newton's publication of a particle theory of light, the Dutch Christiaan Huygens [1629–1695] developed the wave theory of light, published in 1690. By 1804, Thomas Young's double-slit experiment revealed an interference pattern as though light were a wave, and thus Huygens's wave theory of light, as well as Huygens's inference that that light waves were vibrations of the luminiferous aether was accepted. Jean-Augustin Fresnel modeled hypothetical behavior of the aether. Michael Faraday introduced the theoretical concept of a field—not action at a distance. Mid-19th century, the Scottish James Clerk Maxwell [1831–1879] reduced electricity and magnetism to Maxwell's electromagnetic field theory, whittled down by others to the four Maxwell's equations. Initially, optics was found consequent of Maxwell's field. Later, radiation and then today's known electromagnetic spectrum were found also consequent of this electromagnetic field.

The English physicist Lord Rayleigh [1842–1919] worked on sound. The Irishmen William Rowan Hamilton [1805–1865], George Gabriel Stokes [1819–1903] and Lord Kelvin [1824–1907] produced several major works: Stokes was a leader in optics and fluid dynamics; Kelvin made substantial discoveries in thermodynamics; Hamilton did notable work on analytical mechanics finding out a new and powerful approach nowadays known as Hamiltonian mechanics. Very relevant contributions to this approach are due to his German colleague Carl Gustav Jacobi [1804–1851] in particular referring to the so-called canonical transformations. The German Hermann von Helmholtz [1821–1894] is greatly contributed to electromagnetism, waves, fluids, and sound. In the United States, the pioneering work of Josiah Willard Gibbs [1839–1903] became the basis for statistical mechanics. Fundamental theoretical results in this area were achieved by the German Ludwig Boltzmann [1844-1906]. Together, these individuals laid the foundations of electromagnetic theory, fluid dynamics, and statistical mechanics.

Relativistic

By the 1880s, prominent was the paradox that an observer within Maxwell's electromagnetic field measured it at approximately constant speed regardless of the observer's speed relative to other objects within the electromagnetic field. Thus, although the observer's speed was continually lost relative to the electromagnetic field, it was preserved relative to other objects *in* the electromagnetic field. And yet no violation of Galilean invariance within physical interactions among objects was detected. As Maxwell's electromagnetic field was modeled as oscillations of the aether, physicists inferred that motion within the aether resulted in aether drift, shifting the electromagnetic field, explaining the observer's missing speed relative to it. Physicists' mathematical process to translate the positions in one reference frame to predictions of positions in another reference frame, all plotted on Cartesian coordinates, had been the Galilean transformation, which was newly replaced with Lorentz transformation, modeled by the Dutch Hendrik Lorentz [1853–1928].

In 1887, experimentalists Michelson and Morley failed to detect aether drift, however. It was hypothesized that motion *into* the aether prompted aether's shortening, too, as modeled in the Lorentz contraction. Hypotheses at the aether thus kept Maxwell's electromagnetic field aligned with the principle Galilean invariance across all inertial frames of reference, while Newton's theory of motion was spared.

In the 19th century, Gauss's contributions to non-Euclidean geometry, or geometry on curved surfaces, laid the groundwork for the subsequent development of Riemannian geometry by Bernhard Riemann [1826–1866]. Austrian theoretical physicist and philosopher Ernst Mach criticized Newton's postulated absolute space. Mathematician Jules-Henri Poincaré [1854–1912] questioned even absolute time. In 1905, Pierre Duhem published a devastating criticism of the foundation of Newton's theory of motion. Also in 1905, Albert Einstein [1879–1955] published special theory of relativity, newly explaining both the electromagnetic field's invariance and Galilean invariance by discarding all hypotheses at aether, including aether itself. Refuting the framework of Newton's theory—absolute space and absolute time—special relativity states *relative space* and *relative time*, whereby *length* contracts and *time* dilates along the travel pathway of an object experiencing kinetic energy.

In 1908, Einstein's former professor Hermann Minkowski modeled 3D space together with the 1D axis of time by treating the temporal axis like a fourth spatial dimension—altogether 4D spacetime—and declared the imminent demise of the separation of space and time. Einstein initially called this "superfluous learnedness", but later used Minkowski spacetime to great elegance in general theory of relativity, extending invariance to all reference frames—whether perceived as inertial or as accelerated—and thanked Minkowski, by then deceased. General relativity replaces Cartesian coordinates with Gaussian coordinates, and replaces Newton's claimed empty yet Euclidean space traversed instantly by Newton's vector of hypothetical gravitational force—an instant action at a distance—with a gravitational *field*. The gravitational field is Minkowski spacetime itself, the 4D topology of Einstein aether modeled on a Lorentzian manifold that "curves" geometrically, according to the Riemann curvature tensor, in the vicinity of either mass or energy. (By special relativity—a special case of general relativity—even massless energy exerts gravitational effect by its mass equivalence locally "curving" the geometry of the four, unified dimensions of space and time.)

Quantum

Another revolutionary development of the twentieth century has been quantum theory, which emerged from the seminal contributions of Max Planck [1856–1947] (on black body radiation) and Einstein's work on the photoelectric effect. This was, at first, followed by a heuristic framework devised by Arnold Sommerfeld [1868–1951] and Niels Bohr [1885–1962], but this was soon replaced by the quantum mechanics developed by Max Born [1882–1970], Werner Heisenberg [1901–1976], Paul Dirac [1902–1984], Erwin Schrödinger [1887–1961], Satyendra Nath Bose [1894 –1974], and Wolfgang Pauli [1900–1958]. This revolutionary theoretical framework is based on a probabilistic interpretation of states, and evolution and measurements in terms of self-adjoint operators on an infinite dimensional vector space. That is the so-called Hilbert space, introduced in its elementary form by David Hilbert [1862–1943] and Frigyes Riesz [1880-1956], and rigorously defined within the axiomatic modern version by John von Neumann in his celebrated book on mathematical foundations of quantum mechanics, where he built up a relevant part of modern functional analysis on Hilbert spaces, the spectral theory in particular. Paul Dirac used algebraic constructions to produce a relativistic model for the electron, predicting its magnetic moment and the existence of its antiparticle, the positron.

List of Important Mathematical Physicists in the 20th Century

Prominent contributors to the 20th century's mathematical physics (although the list contains some typically theoretical, not mathematical, physicists and leaves many contributors out) include (ordered by birth date) Jules Henri Poincaré [1854-1912] , David Hilbert [1862–1943], Arnold Sommerfeld [1868–1951], Constantin Caratheodory [1873-1950], Albert Einstein [1879–1955], Max Born [1882–1970], George David Birkhoff [1884-1944], Niels Bohr [1885–1962], Hermann Weyl [1885–1955], Satyendra Nath Bose [1894–1974], Wolfgang Pauli [1900–1958], Werner Heisenberg [1901–1976], Paul Dirac [1902–1984], Eugene Wigner [1902–1995], Lars Onsager [1903-1976], John von Neumann [1903–1957], Sin-Itiro Tomonaga [1906–1979], Hideki Yukawa [1907–1981], Lev Davidovich Landau [1908-1968], Nikolay Bogolyubov [1909–1992], Subrahmanyan Chandrasekhar [1910-1995], Mark Kac [1914–1984], Julian Schwinger [1918–1994], Richard Feynman [1918–1988], Irving Ezra Segal [1918-1998], Arthur Strong Wightman [1922–2013], Chen-Ning Yang [1922–], Rudolf Haag [1922–], Freeman Dyson [1923–], Martin Gutzwiller [1925–2014], Abdus Salam [1926–1996], Jürgen Moser [1928–1999], Michael Francis Atiyah [1929–], Joel Louis Lebowitz [1930–], Roger Penrose [1931–], Elliott H. Lieb [1932–], Sheldon Lee Glashow [1932–], Steven Weinberg [1933–], Ludvig D. Faddeev [1934–], David Ruelle [1935–], Yakov G. Sinai [1935–], Vladimir Igorevich Arnold [1937–2010], Arthur Jaffe [1937–], Roman Jackiw [1939–], Leonard Susskind [1940–], Rodney J. Baxter [1940–], Michael Victor Berry [1941-] Giovanni Gallavotti [1941-], Stephen William Hawking [1942–], Jerrold Eldon Marsden [1942–2010], Alexander M. Polyakov [1945–], Barry Simon [1946–], Gerardus 't Hooft [1946-], John L. Cardy [1947–], Edward Witten [1951–], Herbert Spohn [1951?-], and Juan M. Maldacena [1968–].

Mathematical Economics

Mathematical economics is the application of mathematical methods to represent theories and analyze problems in economics. By convention, the applied methods refer to those beyond simple geometry, such as differential and integral calculus, difference and differential equations, matrix algebra, mathematical programming, and other computational methods. An advantage claimed for the approach is its allowing formulation of theoretical relationships with rigor, generality, and simplicity.

Mathematics allows economists to form meaningful, testable propositions about wide-ranging and complex subjects which could less easily be expressed informally. Further, the language of mathematics allows economists to make specific, positive claims about controversial or contentious subjects that would be impossible without mathematics. Much of economic theory is currently presented in terms of mathematical economic models, a set of stylized and simplified mathematical relationships asserted to clarify assumptions and implications.

Broad applications include:

- optimization problems as to goal equilibrium, whether of a household, business firm, or policy maker

- static (or equilibrium) analysis in which the economic unit (such as a household) or economic system (such as a market or the economy) is modeled as not changing

- comparative statics as to a change from one equilibrium to another induced by a change in one or more factors

- dynamic analysis, tracing changes in an economic system over time, for example from economic growth.

Formal economic modeling began in the 19th century with the use of differential calculus to represent and explain economic behavior, such as utility maximization, an early economic application of mathematical optimization. Economics became more mathematical as a discipline throughout the first half of the 20th century, but introduction of new and generalized techniques in the period around the Second World War, as in game theory, would greatly broaden the use of mathematical formulations in economics.

This rapid systematizing of economics alarmed critics of the discipline as well as some noted economists. John Maynard Keynes, Robert Heilbroner, Friedrich Hayek and others have criticized the broad use of mathematical models for human behavior, arguing that some human choices are irreducible to mathematics.

History

The use of mathematics in the service of social and economic analysis dates back to the 17th century. Then, mainly in German universities, a style of instruction emerged which dealt specifically with detailed presentation of data as it related to public administration. Gottfried Achenwall lectured in this fashion, coining the term statistics. At the same time, a small group of professors in England established a method of "reasoning by figures upon things relating to government" and referred to this practice as *Political Arithmetick*. Sir William Petty wrote at length on issues that would later concern economists, such as taxation, Velocity of money and national income, but while his analysis was numerical, he rejected abstract mathematical methodology. Petty's use of detailed numerical data (along with John Graunt) would influence statisticians and economists for some time, even though Petty's works were largely ignored by English scholars.

The mathematization of economics began in earnest in the 19th century. Most of the economic analysis of the time was what would later be called classical economics. Subjects were discussed and dispensed with through algebraic means, but calculus was not used. More importantly, until Johann Heinrich von Thünen's *The Isolated State* in 1826, economists did not develop explicit and abstract models for behavior in order to apply the tools of mathematics. Thünen's model of farmland use represents the first example of marginal analysis. Thünen's work was largely theoretical, but he also mined empirical data in order to attempt to support his generalizations. In comparison to his contemporaries, Thünen built economic models and tools, rather than applying previous tools to new problems.

Meanwhile, a new cohort of scholars trained in the mathematical methods of the physical sciences gravitated to economics, advocating and applying those methods to their subject, and described today as moving from geometry to mechanics. These included W.S. Jevons who presented paper on a "general mathematical theory of political economy" in 1862, providing an outline for use of the theory of marginal utility in political economy. In 1871, he published *The Principles of Political Economy*, declaring that the subject as science "must be mathematical simply because it deals with quantities." Jevons expected the only collection of statistics for price and quantities would permit the subject as presented to become an exact science. Others preceded and followed in expanding mathematical representations of economic problems.

Marginalists and the Roots of Neoclassical Economics

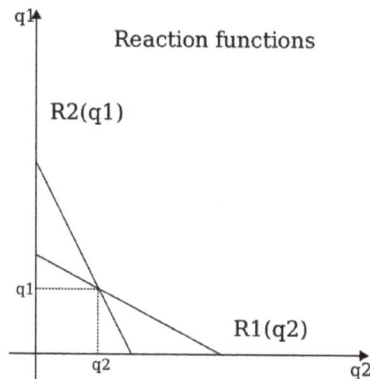

Equilibrium quantities as a solution to two reaction functions in Cournot duopoly. Each reaction function is expressed as a linear equation dependent upon quantity demanded.

Augustin Cournot and Léon Walras built the tools of the discipline axiomatically around utility, arguing that individuals sought to maximize their utility across choices in a way that could be described mathematically. At the time, it was thought that utility was quantifiable, in units known as utils. Cournot, Walras and Francis Ysidro Edgeworth are considered the precursors to modern mathematical economics.

Augustin Cournot

Cournot, a professor of mathematics, developed a mathematical treatment in 1838 for duopoly—a market condition defined by competition between two sellers. This treatment of competition, first published in *Researches into the Mathematical Principles of Wealth*, is referred to as Cournot duopoly. It is assumed that both sellers had equal access to the market and could produce their goods without cost. Further, it assumed that both goods were homogeneous. Each seller would vary her output based on the output of the other and the market price would be determined by the total quantity supplied. The profit for each firm would be determined by multiplying their output and the per unit Market price. Differentiating the profit function with respect to quantity supplied for each firm left a system of linear equations, the simultaneous solution of which gave the equilibrium quantity, price and profits. Cournot's contributions to the mathematization of economics would be neglected for decades, but eventually influenced many of the marginalists. Cournot's models of duopoly and Oligopoly also represent one of the first formulations of non-cooperative games. Today the solution can be given as a Nash equilibrium but Cournot's work preceded modern game theory by over 100 years.

Léon Walras

While Cournot provided a solution for what would later be called partial equilibrium, Léon Walras attempted to formalize discussion of the economy as a whole through a theory of general competitive equilibrium. The behavior of every economic actor would be considered on both the production and consumption side. Walras originally presented four separate models of exchange, each recursively included in the next. The solution of the resulting system of equations (both linear and non-linear) is the general equilibrium. At the time, no general solution could be expressed for a system of arbitrarily many equations, but Walras's attempts produced two famous results in economics. The first is Walras' law and the second is the principle of tâtonnement. Walras' method was considered highly mathematical for the time and Edgeworth commented at length about this fact in his review of *Éléments d›économie politique pure* (Elements of Pure Economics).

Walras' law was introduced as a theoretical answer to the problem of determining the solutions in general equilibrium. His notation is different from modern notation but can be constructed using more modern summation notation. Walras assumed that in equilibrium, all money would be spent on all goods: every good would be sold at the market price for that good and every buyer would expend their last dollar on a basket of goods. Starting from this assumption, Walras could then show that if there were n markets and n-1 markets cleared (reached equilibrium conditions) that the nth market would clear as well. This is easiest to visualize with two markets (considered in most texts as a market for goods and a market for money). If one of two markets has reached an equilibrium state, no additional goods (or conversely, money) can enter or exit the second market, so it must be in a state of equilibrium as well. Walras used this statement to move toward a proof of existence of solutions to general equilibrium but it is commonly used today to illustrate market clearing in money markets at the undergraduate level.

Tâtonnement (roughly, French for *groping toward*) was meant to serve as the practical expression of Walrasian general equilibrium. Walras abstracted the marketplace as an auction of goods where the auctioneer would call out prices and market participants would wait until they could each satisfy their personal reservation prices for the quantity desired (remembering here that this is an auction on *all* goods, so everyone has a reservation price for their desired basket of goods).

Only when all buyers are satisfied with the given market price would transactions occur. The market would "clear" at that price—no surplus or shortage would exist. The word *tâtonnement* is used to describe the directions the market takes in *groping toward* equilibrium, settling high or low prices on different goods until a price is agreed upon for all goods. While the process appears dynamic, Walras only presented a static model, as no transactions would occur until all markets were in equilibrium. In practice very few markets operate in this manner.

Francis Ysidro Edgeworth

Edgeworth introduced mathematical elements to Economics explicitly in *Mathematical Psychics: An Essay on the Application of Mathematics to the Moral Sciences*, published in 1881. He adopted Jeremy Bentham's felicific calculus to economic behavior, allowing the outcome of each decision to be converted into a change in utility. Using this assumption, Edgeworth built a model of exchange on three assumptions: individuals are self-interested, individuals act to maximize utility, and individuals are "free to recontract with another independently of...any third party."

Given two individuals, the set of solutions where the both individuals can maximize utility is described by the *contract curve* on what is now known as an Edgeworth Box. Technically, the construction of the two-person solution to Edgeworth's problem was not developed graphically until 1924 by Arthur Lyon Bowley. The contract curve of the Edgeworth box (or more generally on any set of solutions to Edgeworth's problem for more actors) is referred to as the core of an economy.

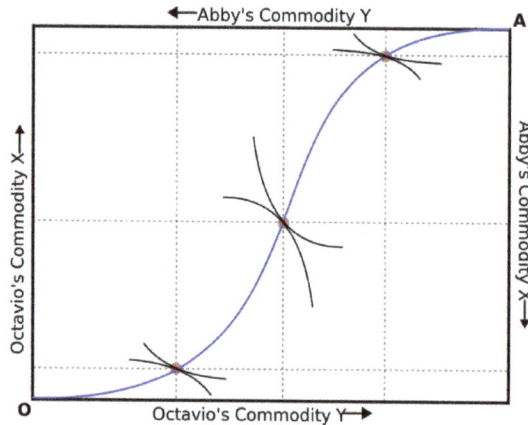

An Edgeworth box displaying the contract curve on an economy with two participants. Referred to as the "core" of the economy in modern parlance, there are infinitely many solutions along the curve for economies with two participants

Edgeworth devoted considerable effort to insisting that mathematical proofs were appropriate for all schools of thought in economics. While at the helm of *The Economic Journal*, he published several articles criticizing the mathematical rigor of rival researchers, including Edwin Robert Anderson Seligman, a noted skeptic of mathematical economics. The articles focused on a back and forth over tax incidence and responses by producers. Edgeworth noticed that a monopoly producing a good that had jointness of supply but not jointness of demand (such as first class and economy on an airplane, if the plane flies, both sets of seats fly with it) might actually lower the price seen by the consumer for one of the two commodities if a tax were applied. Common sense and more traditional, numerical analysis seemed to indicate that this was preposterous. Seligman insisted that the results Edgeworth achieved were a quirk of his mathematical formulation. He suggested that the assumption of a continuous demand function and an infinitesimal change in the tax resulted in the paradoxical predictions. Harold Hotelling later showed that Edgeworth was correct and that the same result (a "diminution of price as a result of the tax") could occur with a discontinuous demand function and large changes in the tax rate.

Modern Mathematical Economics

From the later-1930s, an array of new mathematical tools from the differential calculus and differential equations, convex sets, and graph theory were deployed to advance economic theory in a way similar to new mathematical methods earlier applied to physics. The process was later described as moving from mechanics to axiomatics.

Differential Calculus

Vilfredo Pareto analyzed microeconomics by treating decisions by economic actors as attempts to change a given allotment of goods to another, more preferred allotment. Sets of allocations could then be treated as Pareto efficient (Pareto optimal is an equivalent term) when no exchanges could occur between actors that could make at least one individual better off without making any other individual worse off. Pareto's proof is commonly conflated with Walrassian equilibrium or informally ascribed to Adam Smith's Invisible hand hypothesis. Rather, Pareto's statement was the first formal assertion of what would be known as the first fundamental theorem of welfare economics. These models lacked the inequalities of the next generation of mathematical economics.

In the landmark treatise *Foundations of Economic Analysis* (1947), Paul Samuelson identified a common paradigm and mathematical structure across multiple fields in the subject, building on previous work by Alfred Marshall. *Foundations* took mathematical concepts from physics and applied them to economic problems. This broad view (for example, comparing Le Chatelier's principle to tâtonnement) drives the fundamental premise of mathematical economics: systems of economic actors may be modeled and their behavior described much like any other system. This extension followed on the work of the marginalists in the previous century and extended it significantly. Samuelson approached the problems of applying individual utility maximization over aggregate groups with comparative statics, which compares two different equilibrium states after an exogenous change in a variable. This and other methods in the book provided the foundation for mathematical economics in the 20th century.

Linear Models

Restricted models of general equilibrium were formulated by John von Neumann in 1937. Unlike earlier versions, the models of von Neumann had inequality constraints. For his model of an expanding economy, von Neumann proved the existence and uniqueness of an equilibrium using his generalization of Brouwer's fixed point theorem. Von Neumann's model of an expanding economy considered the matrix pencil $A - \lambda B$ with nonnegative matrices A and B; von Neumann sought probability vectors p and q and a positive number λ that would solve the complementarity equation

$$p^T (A - \lambda B) q = 0,$$

along with two inequality systems expressing economic efficiency. In this model, the (transposed) probability vector p represents the prices of the goods while the probability vector q represents the "intensity" at which the production process would run. The unique solution λ represents the rate of growth of the economy, which equals the interest rate. Proving the existence of a positive growth rate and proving that the growth rate equals the interest rate were remarkable achievements, even for von Neumann. Von Neumann's results have been viewed as a special case of linear programming, where von Neumann's model uses only nonnegative matrices. The study of von Neumann's model of an expanding economy continues to interest mathematical economists with interests in computational economics.

Input-output Economics

In 1936, the Russian–born economist Wassily Leontief built his model of input-output analysis from the 'material balance' tables constructed by Soviet economists, which themselves followed earlier work by the physiocrats. With his model, which described a system of production and demand processes, Leontief described how changes in demand in one economic sector would influence production in another. In practice, Leontief estimated the coefficients of his simple models, to address economically interesting questions. In production economics, "Leontief technologies" produce outputs using constant proportions of inputs, regardless of the price of inputs, reducing the value of Leontief models for understanding economies but allowing their parameters to be estimated relatively easily. In contrast, the von Neumann model of an expanding economy allows for choice of techniques, but the coefficients must be estimated for each technology.

Mathematical Optimization

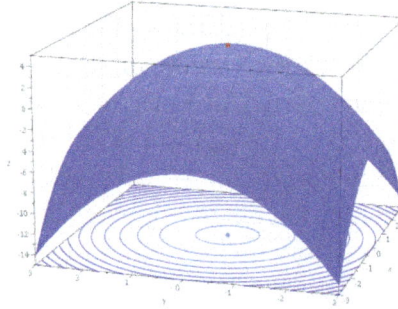

Red dot in z direction as maximum for paraboloid function of (x, y) inputs

In mathematics, mathematical optimization (or optimization or mathematical programming) refers to the selection of a best element from some set of available alternatives. In the simplest case, an optimization problem involves maximizing or minimizing a real function by selecting input values of the function and computing the corresponding values of the function. The solution process includes satisfying general necessary and sufficient conditions for optimality. For optimization problems, specialized notation may be used as to the function and its input(s). More generally, optimization includes finding the best available element of some function given a defined domain and may use a variety of different computational optimization techniques.

Economics is closely enough linked to optimization by agents in an economy that an influential definition relatedly describes economics *qua* science as the "study of human behavior as a relationship between ends and scarce means" with alternative uses. Optimization problems run through modern economics, many with explicit economic or technical constraints. In microeconomics, the utility maximization problem and its dual problem, the expenditure minimization problem for a given level of utility, are economic optimization problems. Theory posits that consumers maximize their utility, subject to their budget constraints and that firms maximize their profits, subject to their production functions, input costs, and market demand.

Economic equilibrium is studied in optimization theory as a key ingredient of economic theorems that in principle could be tested against empirical data. Newer developments have occurred in dynamic programming and modeling optimization with risk and uncertainty, including applications to portfolio theory, the economics of information, and search theory.

Optimality properties for an entire market system may be stated in mathematical terms, as in formulation of the two fundamental theorems of welfare economics and in the Arrow–Debreu model of general equilibrium (also discussed below). More concretely, many problems are amenable to analytical (formulaic) solution. Many others may be sufficiently complex to require numerical methods of solution, aided by software. Still others are complex but tractable enough to allow computable methods of solution, in particular computable general equilibrium models for the entire economy.

Linear and nonlinear programming have profoundly affected microeconomics, which had earlier considered only equality constraints. Many of the mathematical economists who received Nobel Prizes in Economics had conducted notable research using linear programming: Leonid Kantorovich, Leonid Hurwicz, Tjalling Koopmans, Kenneth J. Arrow, and Robert Dorfman, Paul Samuelson, and Robert Solow. Both Kantorovich and Koopmans acknowledged that George B. Dantzig deserved to share their Nobel Prize for linear programming. Economists who conducted research in nonlinear programming also have won the Nobel prize, notably Ragnar Frisch in addition to Kantorovich, Hurwicz, Koopmans, Arrow, and Samuelson.

Linear Optimization

Linear programming was developed to aid the allocation of resources in firms and in industries during the 1930s in Russia and during the 1940s in the United States. During the Berlin airlift (1948), linear programming was used to plan the shipment of supplies to prevent Berlin from starving after the Soviet blockade.

Nonlinear Programming

Extensions to nonlinear optimization with inequality constraints were achieved in 1951 by Albert W. Tucker and Harold Kuhn, who considered the nonlinear optimization problem:

Minimize $f(x)$ subject to $g_i(x) \leq 0$ and $h_j(x) = 0$ where

$f(\cdot)$ is the function to be minimized

$g_i(\cdot)$ ($j = 1, ..., m$) are the functions of the m inequality constraints

$h_j(\cdot)$ ($j = 1,..., l$) are the functions of the l equality constraints.

In allowing inequality constraints, the Kuhn–Tucker approach generalized the classic method of Lagrange multipliers, which (until then) had allowed only equality constraints. The Kuhn–Tucker approach inspired further research on Lagrangian duality, including the treatment of inequality constraints. The duality theory of nonlinear programming is particularly satisfactory when applied to convex minimization problems, which enjoy the convex-analytic duality theory of Fenchel and Rockafellar; this convex duality is particularly strong for polyhedral convex functions, such as those arising in linear programming. Lagrangian duality and convex analysis are used daily in operations research, in the scheduling of power plants, the planning of production schedules for factories, and the routing of airlines (routes, flights, planes, crews).

Variational Calculus and Optimal Control

Economic dynamics allows for changes in economic variables over time, including in dynamic systems. The problem of finding optimal functions for such changes is studied in variational calculus and in optimal control theory. Before the Second World War, Frank Ramsey and Harold Hotelling used the calculus of variations to that end.

Following Richard Bellman's work on dynamic programming and the 1962 English translation of L. Pontryagin *et al.*'s earlier work, optimal control theory was used more extensively in economics in addressing dynamic problems, especially as to economic growth equilibrium and stability of economic systems, of which a textbook example is optimal consumption and saving. A crucial distinction is between deterministic and stochastic control models. Other applications of optimal control theory include those in finance, inventories, and production for example.

Functional Analysis

It was in the course of proving of the existence of an optimal equilibrium in his 1937 model of economic growth that John von Neumann introduced functional analytic methods to include topology in economic theory, in particular, fixed-point theory through his generalization of Brouwer's fixed-point theorem. Following von Neumann's program, Kenneth Arrow and Gérard Debreu formulated abstract models of economic equilibria using convex sets and fixed–point theory. In introducing the Arrow–Debreu model in 1954, they proved the existence (but not the uniqueness) of an equilibrium and also proved that every Walras equilibrium is Pareto efficient; in general, equilibria need not be unique. In their models, the ("primal") vector space represented *quantitites* while the "dual" vector space represented *prices*.

In Russia, the mathematician Leonid Kantorovich developed economic models in partially ordered vector spaces, that emphasized the duality between quantities and prices. Kantorovich renamed *prices* as "objectively determined valuations" which were abbreviated in Russian as "o. o. o.", alluding to the difficulty of discussing prices in the Soviet Union.

Even in finite dimensions, the concepts of functional analysis have illuminated economic theory, particularly in clarifying the role of prices as normal vectors to a hyperplane supporting a convex set, representing production or consumption possibilities. However, problems of describing optimization over time or under uncertainty require the use of infinite–dimensional function spaces, because agents are choosing among functions or stochastic processes.

Differential Decline and Rise

John von Neumann's work on functional analysis and topology in broke new ground in mathematics and economic theory. It also left advanced mathematical economics with fewer applications of differential calculus. In particular, general equilibrium theorists used general topology, convex geometry, and optimization theory more than differential calculus, because the approach of differential calculus had failed to establish the existence of an equilibrium.

However, the decline of differential calculus should not be exaggerated, because differential calculus has always been used in graduate training and in applications. Moreover, differential calculus has returned to the highest levels of mathematical economics, general equilibrium theory (GET), as practiced by the "GET-set" (the humorous designation due to Jacques H. Drèze). In the 1960s and 1970s, however, Gérard Debreu and Stephen Smale led a revival of the use of differential calculus in mathematical economics. In particular, they were able to prove the existence of a general equilibrium, where earlier writers had failed, because of their novel mathematics: Baire category from general topology and Sard's lemma from differential topology. Other economists associated with the use of differential analysis include Egbert Dierker, Andreu Mas-Colell, and Yves Balasko. These advances have changed the traditional narrative of the history of mathematical economics, following von Neumann, which celebrated the abandonment of differential calculus.

Game Theory

John von Neumann, working with Oskar Morgenstern on the theory of games, broke new mathematical ground in 1944 by extending functional analytic methods related to convex sets and topological fixed-point theory to economic analysis. Their work thereby avoided the traditional differential calculus, for which the maximum–operator did not apply to non-differentiable functions. Continuing von Neumann's work in cooperative game theory, game theorists Lloyd S. Shapley, Martin Shubik, Hervé Moulin, Nimrod Megiddo, Bezalel Peleg influenced economic research in politics and economics. For example, research on the fair prices in cooperative games and fair values for voting games led to changed rules for voting in legislatures and for accounting for the costs in public–works projects. For example, cooperative game theory was used in designing the water distribution system of Southern Sweden and for setting rates for dedicated telephone lines in the USA.

Earlier neoclassical theory had bounded only the *range* of bargaining outcomes and in special cases, for example bilateral monopoly or along the contract curve of the Edgeworth box. Von Neumann and Morgenstern's results were similarly weak. Following von Neumann's program, however, John Nash used fixed–point theory to prove conditions under which the bargaining problem and noncooperative games can generate a unique equilibrium solution. Noncooperative game theory has been adopted as a fundamental aspect of experimental economics, behavioral economics, information economics, industrial organization, and political economy. It has also given rise to the subject of mechanism design (sometimes called reverse game theory), which has private and public-policy applications as to ways of improving economic efficiency through incentives for information sharing.

In 1994, Nash, John Harsanyi, and Reinhard Selten received the Nobel Memorial Prize in Economic Sciences their work on non–cooperative games. Harsanyi and Selten were awarded for their work on repeated games. Later work extended their results to computational methods of modeling.

Agent-based Computational Economics

Agent-based computational economics (ACE) as a named field is relatively recent, dating from about the 1990s as to published work. It studies economic processes, including whole economies, as dynamic systems of interacting agents over time. As such, it falls in the paradigm of complex adaptive systems. In corresponding agent-based models, agents are not real people but "computational objects modeled as interacting according to rules" ... "whose micro-level interactions create emergent patterns" in space and time. The rules are formulated to predict behavior and social interactions based on incentives and information. The theoretical assumption of mathematical *optimization* by agents markets is replaced by the less restrictive postulate of agents with *bounded* rationality *adapting* to market forces.

ACE models apply numerical methods of analysis to computer-based simulations of complex dynamic problems for which more conventional methods, such as theorem formulation, may not find ready use. Starting from specified initial conditions, the computational economic system is modeled as evolving over time as its constituent agents repeatedly interact with each other. In these respects, ACE has been characterized as a bottom-up culture-dish approach to the study of the economy. In contrast to other standard modeling methods, ACE events are driven solely by initial conditions, whether or not equilibria exist or are computationally tractable. ACE modeling, however, includes agent adaptation, autonomy, and learning. It has a similarity to, and overlap with, game theory as an agent-based method for modeling social interactions. Other dimensions of the approach include such standard economic subjects as competition and collaboration, market structure and industrial organization, transaction costs, welfare economics and mechanism design, information and uncertainty, and macroeconomics.

The method is said to benefit from continuing improvements in modeling techniques of computer science and increased computer capabilities. Issues include those common to experimental economics in general and by comparison and to development of a common framework for empirical validation and resolving open questions in agent-based modeling. The ultimate scientific objective of the method has been described as "test[ing] theoretical findings against real-world data in ways that permit empirically supported theories to cumulate over time, with each researcher's work building appropriately on the work that has gone before."

Mathematicization of Economics

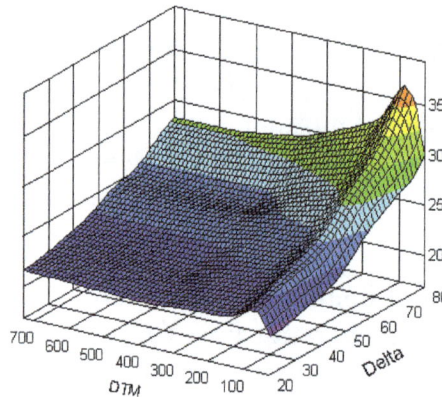

The surface of the Volatility smile is a 3-D surface whereby the current market implied volatility (Z-axis) for all options on the underlier is plotted against strike price and time to maturity (X & Y-axes).

Over the course of the 20th century, articles in "core journals" in economics have been almost exclusively written by economists in academia. As a result, much of the material transmitted in those journals relates to economic theory, and "economic theory itself has been continuously more abstract and mathematical." A subjective assessment of mathematical techniques employed in these core journals showed a decrease in articles that use neither geometric representations nor mathematical notation from 95% in 1892 to 5.3% in 1990. A 2007 survey of ten of the top economic journals finds that only 5.8% of the articles published in 2003 and 2004 both lacked statistical analysis of data and lacked displayed mathematical expressions that were indexed with numbers at the margin of the page.

Econometrics

Between the world wars, advances in mathematical statistics and a cadre of mathematically trained economists led to econometrics, which was the name proposed for the discipline of advancing economics by using mathematics and statistics. Within economics, "econometrics" has often been used for statistical methods in economics, rather than mathematical economics. Statistical econometrics features the application of linear regression and time series analysis to economic data.

Ragnar Frisch coined the word "econometrics" and helped to found both the Econometric Society in 1930 and the journal *Econometrica* in 1933. A student of Frisch's, Trygve Haavelmo published

The Probability Approach in Econometrics in 1944, where he asserted that precise statistical analysis could be used as a tool to validate mathematical theories about economic actors with data from complex sources. This linking of statistical analysis of systems to economic theory was also promulgated by the Cowles Commission (now the Cowles Foundation) throughout the 1930s and 1940s.

Earlier Work in Econometrics

The roots of modern econometrics can be traced to the American economist Henry L. Moore. Moore studied agricultural productivity and attempted to fit changing values of productivity for plots of corn and other crops to a curve using different values of elasticity. Moore made several errors in his work, some from his choice of models and some from limitations in his use of mathematics. The accuracy of Moore's models also was limited by the poor data for national accounts in the United States at the time. While his first models of production were static, in 1925 he published a dynamic "moving equilibrium" model designed to explain business cycles—this periodic variation from overcorrection in supply and demand curves is now known as the cobweb model. A more formal derivation of this model was made later by Nicholas Kaldor, who is largely credited for its exposition.

Application

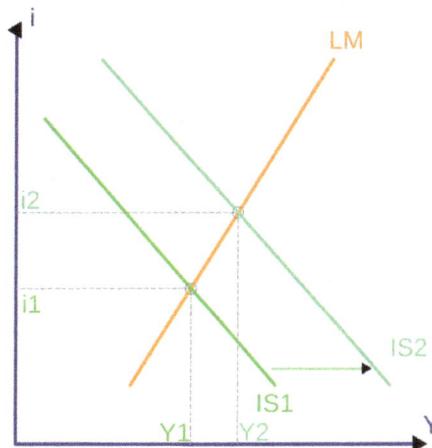

The IS/LM model is a Keynesian macroeconomic model designed to make predictions about the intersection of "real" economic activity (e.g. spending, income, savings rates) and decisions made in the financial markets (Money supply and Liquidity preference). The model is no longer widely taught at the graduate level but is common in undergraduate macroeconomics courses.

Much of classical economics can be presented in simple geometric terms or elementary mathematical notation. Mathematical economics, however, conventionally makes use of calculus and matrix algebra in economic analysis in order to make powerful claims that would be more difficult without such mathematical tools. These tools are prerequisites for formal study, not only in mathematical economics but in contemporary economic theory in general. Economic problems often involve so many variables that mathematics is the only practical way of attacking and solving them. Alfred Marshall argued that every economic problem which can be quantified, analytically expressed and solved, should be treated by means of mathematical work.

Economics has become increasingly dependent upon mathematical methods and the mathematical tools it employs have become more sophisticated. As a result, mathematics has become considerably more important to professionals in economics and finance. Graduate programs in both economics and finance require strong undergraduate preparation in mathematics for admission and, for this reason, attract an increasingly high number of mathematicians. Applied mathematicians apply mathematical principles to practical problems, such as economic analysis and other economics-related issues, and many economic problems are often defined as integrated into the scope of applied mathematics.

This integration results from the formulation of economic problems as stylized models with clear assumptions and falsifiable predictions. This modeling may be informal or prosaic, as it was in Adam Smith's *The Wealth of Nations*, or it may be formal, rigorous and mathematical.

Broadly speaking, formal economic models may be classified as stochastic or deterministic and as discrete or continuous. At a practical level, quantitative modeling is applied to many areas of economics and several methodologies have evolved more or less independently of each other.

- Stochastic models are formulated using stochastic processes. They model economically observable values over time. Most of econometrics is based on statistics to formulate and test hypotheses about these processes or estimate parameters for them. Between the World Wars, Herman Wold developed a representation of stationary stochastic processes in terms of autoregressive models and a determinist trend. Wold and Jan Tinbergen applied time-series analysis to economic data. Contemporary research on time series statistics consider additional formulations of stationary processes, such as autoregressive moving average models. More general models include autoregressive conditional heteroskedasticity (ARCH) models and generalized ARCH (GARCH) models.

- Non-stochastic mathematical models may be purely qualitative (for example, models involved in some aspect of social choice theory) or quantitative (involving rationalization of financial variables, for example with hyperbolic coordinates, and/or specific forms of functional relationships between variables). In some cases economic predictions of a model merely assert the direction of movement of economic variables, and so the functional relationships are used only in a qualitative sense: for example, if the price of an item increases, then the demand for that item will decrease. For such models, economists often use two-dimensional graphs instead of functions.

- Qualitative models are occasionally used. One example is qualitative scenario planning in which possible future events are played out. Another example is non-numerical decision tree analysis. Qualitative models often suffer from lack of precision.

Classification

According to the Mathematics Subject Classification (MSC), mathematical economics falls into the Applied mathematics/other classification of category 91:

Game theory, economics, social and behavioral sciences

with MSC2010 classifications for 'Game theory' at codes 91Axx and for 'Mathematical economics' at codes 91Bxx.

The *Handbook of Mathematical Economics* series (Elsevier), currently 4 volumes, distinguishes between *mathematical methods in economics*, v. 1, Part I, and *areas of economics* in other volumes where mathematics is employed.

Another source with a similar distinction is *The New Palgrave: A Dictionary of Economics* (1987, 4 vols., 1,300 subject entries). In it, a "Subject Index" includes mathematical entries under 2 headings (vol. IV, pp. 982–3):

> Mathematical Economics (24 listed, such as "acyclicity", "aggregation problem", "comparative statics", "lexicographic orderings", "linear models", "orderings", and "qualitative economics")

> Mathematical Methods (42 listed, such as "calculus of variations", "catastrophe theory", "combinatorics," "computation of general equilibrium", "convexity", "convex programming", and "stochastic optimal control").

A widely used system in economics that includes mathematical methods on the subject is the JEL classification codes. It originated in the *Journal of Economic Literature* for classifying new books and articles. The relevant categories are listed below (simplified below to omit "Miscellaneous" and "Other" JEL codes), as reproduced from JEL classification codes#Mathematical and quantitative methods JEL: C Subcategories. *The New Palgrave Dictionary of Economics* (2008, 2nd ed.) also uses the JEL codes to classify its entries. The corresponding footnotes below have links to abstracts of *The New Palgrave Online* for each JEL category (10 or fewer per page, similar to Google searches).

> JEL: C02 - Mathematical Methods (following JEL: C00 - General and JEL: C01 - Econometrics)

> JEL: C6 - Mathematical Methods; Programming Models; Mathematical and Simulation Modeling

> JEL: C60 - General

> JEL: C61 - Optimization techniques; Programming models; Dynamic analysis

> JEL: C62 - Existence and stability conditions of equilibrium

> JEL: C63 - Computational techniques; Simulation modeling

> JEL: C67 - Input–output models

> JEL: C68 - Computable General Equilibrium models

> JEL: C7 - Game theory and Bargaining theory

> JEL: C70 - General

> JEL: C71 - Cooperative games

> JEL: C72 - Noncooperative games

> JEL: C73 - Stochastic and Dynamic games; Evolutionary games; Repeated Games

JEL: C78 - Bargaining theory; Matching theory

Criticisms and Defences

Adequacy of Mathematics for Qualitative and Complicated Economics

Friedrich Hayek contended that the use of formal techniques projects a scientific exactness that does not appropriately account for informational limitations faced by real economic agents.

In an interview, the economic historian Robert Heilbroner stated:

I guess the scientific approach began to penetrate and soon dominate the profession in the past twenty to thirty years. This came about in part because of the "invention" of mathematical analysis of various kinds and, indeed, considerable improvements in it. This is the age in which we have not only more data but more sophisticated use of data. So there is a strong feeling that this is a data-laden science and a data-laden undertaking, which, by virtue of the sheer numerics, the sheer equations, and the sheer look of a journal page, bears a certain resemblance to science . . . That one central activity looks scientific. I understand that. I think that is genuine. It approaches being a universal law. But resembling a science is different from being a science.

Heilbroner stated that "some/much of economics is not naturally quantitative and therefore does not lend itself to mathematical exposition."

Testing Predictions of Mathematical Economics

Philosopher Karl Popper discussed the scientific standing of economics in the 1940s and 1950s. He argued that mathematical economics suffered from being tautological. In other words, insofar that economics became a mathematical theory, mathematical economics ceased to rely on empirical refutation but rather relied on mathematical proofs and disproof. According to Popper, falsifiable assumptions can be tested by experiment and observation while unfalsifiable assumptions can be explored mathematically for their consequences and for their consistency with other assumptions.

Sharing Popper's concerns about assumptions in economics generally, and not just mathematical economics, Milton Friedman declared that "all assumptions are unrealistic". Friedman proposed judging economic models by their predictive performance rather than by the match between their assumptions and reality.

Mathematical Economics as a form of Pure Mathematics

Considering mathematical economics, J.M. Keynes wrote in *The General Theory*:

It is a great fault of symbolic pseudo-mathematical methods of formalising a system of economic analysis ... that they expressly assume strict independence between the factors involved and lose their cogency and authority if this hypothesis is disallowed; whereas, in ordinary discourse, where we are not blindly manipulating and know all the time what we are doing and what the words mean, we can keep 'at the back of our heads' the necessary reserves and qualifications and the adjustments which we shall have to make later on, in a way in which we cannot keep complicated partial differentials 'at the back' of several pages of algebra which assume they all vanish. Too large a proportion of recent 'mathematical' economics are merely concoctions, as imprecise as the initial assumptions they rest on, which allow the author to lose sight of the complexities and interdependencies of the real world in a maze of pretentious and unhelpful symbols.

Defense of Mathematical Economics

In response to these criticisms, Paul Samuelson argued that mathematics is a language, repeating a thesis of Josiah Willard Gibbs. In economics, the language of mathematics is sometimes necessary for representing substantive problems. Moreover, mathematical economics has led to conceptual advances in economics. In particular, Samuelson gave the example of microeconomics, writing that "few people are ingenious enough to grasp [its] more complex parts... *without* resorting to the language of mathematics, while most ordinary individuals can do so fairly easily *with* the aid of mathematics."

Some economists state that mathematical economics deserves support just like other forms of mathematics, particularly its neighbors in mathematical optimization and mathematical statistics and increasingly in theoretical computer science. Mathematical economics and other mathematical sciences have a history in which theoretical advances have regularly contributed to the reform of the more applied branches of economics. In particular, following the program of John von Neumann, game theory now provides the foundations for describing much of applied economics, from statistical decision theory (as "games against nature") and econometrics to general equilibrium theory and industrial organization. In the last decade, with the rise of the internet, mathematical economicists and optimization experts and computer scientists have worked on problems of pricing for on-line services --- their contributions using mathematics from cooperative game theory, nondifferentiable optimization, and combinatorial games.

Robert M. Solow concluded that mathematical economics was the core "infrastructure" of contemporary economics:

Economics is no longer a fit conversation piece for ladies and gentlemen. It has become a technical subject. Like any technical subject it attracts some people who are more interested in the technique than the subject. That is too bad, but it may be inevitable. In any case, do not kid yourself: the technical core of economics is indispensable infrastructure for the political economy. That is why, if you consult [a reference in contemporary economics] looking for enlightenment about the world today, you will be led to technical economics, or history, or nothing at all.

Mathematical Economists

Prominent mathematical economists include, but are not limited to, the following (by century of birth).

19th Century

• Enrico Barone	• Francis Ysidro Edgeworth	• William Stanley Jevons
• Antoine Augustin Cournot	• Irving Fisher	• Léon Walras

20th Century

• Charalambos D. Aliprantis	• Nicholas Georgescu-Roegen	• Andreu Mas-Colell	• Leonard J. Savage
• R. G. D. Allen	• Roger Guesnerie	• Eric Maskin	• Herbert Scarf
• Maurice Allais	• Frank Hahn	• Nimrod Megiddo	• Reinhard Selten
• Kenneth J. Arrow	• John C. Harsanyi	• Jean-François Mertens	• Amartya Sen
• Robert J. Aumann	• John R. Hicks	• James Mirrlees	• Lloyd S. Shapley
• Yves Balasko	• Werner Hildenbrand	• Roger Myerson	• Stephen Smale
• David Blackwell	• Harold Hotelling	• John Forbes Nash, Jr.	• Robert Solow
• Lawrence E. Blume	• Leonid Hurwicz	• John von Neumann	• Hugo F. Sonnenschein
• Graciela Chichilnisky	• Leonid Kantorovich	• Edward C. Prescott	• Albert W. Tucker
• George B. Dantzig	• Tjalling Koopmans	• Roy Radner	• Hirofumi Uzawa
• Gérard Debreu	• David M. Kreps	• Frank Ramsey	• Robert B. Wilson
• Jacques H. Drèze	• Harold W. Kuhn	• Donald John Roberts	• Hermann Wold
• David Gale	• Edmond Malinvaud	• Paul Samuelson	• Nicholas C. Yannelis
		• Thomas Sargent	

References

- Chiang, Alpha C.; Kevin Wainwright (2005). Fundamental Methods of Mathematical Economics. McGraw-Hill Irwin. pp. 3–4. ISBN 0-07-010910-9.

- Schumpeter, J.A. (1954). Elizabeth B. Schumpeter, ed. History of Economic Analysis. New York: Oxford University Press. pp. 209–212. ISBN 978-0-04-330086-2.

- Nicola, PierCarlo (2000). Mainstream Mathmerical Economics in the 20th Century. Springer. p. 4. ISBN 978-3-540-67084-1. Retrieved 2008-08-21.

- Hotelling, Harold (1990). "Stability in Competition". In Darnell, Adrian C. The Collected Economics Articles of Harold Hotelling. Springer. pp. 51, 52. ISBN 3-540-97011-8. OCLC 20217006. Retrieved 2008-08-21

- Gibbons, Robert (1992). Game Theory for Applied Economists. Princeton, New Jersey: Princeton University Press. pp. 14, 15. ISBN 0-691-00395-5.

- Rima, Ingrid H. (1977). "Neoclassicism and Dissent 1890-1930". In Weintraub, Sidney. Modern Economic Thought. University of Pennsylvania Press. pp. 10, 11. ISBN 0-8122-7712-0.

- Heilbroner, Robert L. (1953 [1999]). The Worldly Philosophers (Seventh ed.). New York: Simon and Schuster. pp. 172–175, 313. ISBN 978-0-684-86214-9.

- Gillies, D. B. (1969). "Solutions to general non-zero-sum games". In Tucker, A. W.; Luce, R. D. Contributions to the Theory of Games. Annals of Mathematics 40. Princeton, New Jersey: Princeton University Press. pp. 47–85. ISBN 978-0-691-07937-0.

- Hotelling, Harold (1990). "Note on Edgeworth's Taxation Phenomenon and Professor Garver's Additional Condition on Demand Functions". In Darnell, Adrian C. The Collected Economics Articles of Harold Hotelling. Springer. pp. 94–122. ISBN 3-540-97011-8. OCLC 20217006. Retrieved 2008-08-26

- Nicholson, Walter; Snyder, Christopher (2007). "General Equilibrium and Welfare". Intermediate Microeconomics and Its Applications (10th ed.). Thompson. pp. 364, 365. ISBN 0-324-31968-1.

- Jolink, Albert (2006). "What Went Wrong with Walras?". In Backhaus, Juergen G.; Maks, J.A. Hans. From Walras to Pareto. The European Heritage in Economics and the Social Sciences IV. Springer. doi:10.1007/978-0-387-33757-9_6. ISBN 978-0-387-33756-2.

- Screpanti, Ernesto; Zamagni, Stefano (1993). An Outline of the History of Economic Thought. New York: Oxford University Press. pp. 288–290. ISBN 0-19-828370-9. OCLC 57281275.

- Pontryagin, L. S.; Boltyanski, V. G., Gamkrelidze, R. V., Mischenko, E. F. (1962). The Mathematical Theory of Optimal Processes. New York: Wiley. ISBN 9782881240775.

- Aliprantis, Charalambos D.; Brown, Donald J.; Burkinshaw, Owen (1990). Existence and optimality of competitive equilibria. Berlin: Springer–Verlag. pp. xii+284. ISBN 3-540-52866-0.

- Lester G. Telser and Robert L. Graves Functional Analysis in Mathematical Economics: Optimization Over Infinite Horizons 1972. University of Chicago Press, 1972, ISBN 978-0-226-79190-6.

- Mas-Colell, Andreu (1985). The Theory of general economic equilibrium: A differentiable approach. Econometric Society monographs. Cambridge UP. ISBN 0-521-26514-2. MR 1113262.

- Brockhaus, Oliver; Farkas, Michael; Ferraris, Andrew; Long, Douglas; Overhaus, Marcus (2000). Equity Derivatives and Market Risk Models. Risk Books. pp. 13–17. ISBN 978-1-899332-87-8. Retrieved 2008-08-17.

- Boland, L. A. (2007). "Seven Decades of Economic Methodology". In I. C. Jarvie; K. Milford; D.W. Miller. Karl Popper:A Centenary Assessment. London: Ashgate Publishing. p. 219. ISBN 978-0-7546-5375-2. Retrieved 2008-06-10.

- Keynes, John Maynard (1936). The General Theory of Employment, Interest and Money. Cambridge: Macmillan. p. 297. ISBN 0-333-10729-2.

- Epstein, Roy J. (1987). A History of Econometrics. Contributions to Economic Analysis. North-Holland. pp. 13–19. ISBN 978-0-444-70267-8.

- "Leonid Vitaliyevich Kantorovich — Prize Lecture ("Mathematics in economics: Achievements, difficulties, perspectives")". Nobelprize.org. Retrieved 12 Dec 2010.

Permissions

Index

www.ingramcontent.com/pod-product-compliance
Lightning Source LLC
Chambersburg PA
CBHW061317190326
41458CB00011B/3830